CHAOS IN GRAVITATIONAL N-BODY SYSTEMS

CHAOS IN GRAVITATIONAL
N-BODY SYSTEMS

Proceedings of a Workshop held at La Plata (Argentina),
July 31 – August 3, 1995

Edited by

J. C. MUZZIO

Observatorio Astronómico, Universidad Nacional de La Plata,
La Plata, Argentina

S. FERRAZ-MELLO

Instituto Astronômico e Geofísico, Universidade de São Paulo,
São Paulo, Brazil

and

J. HENRARD

Département de Mathématique, FNDP, Namur, Belgium

Partly reprinted from *Celestial Mechanics and Dynamical Astronomy,*
Volume 64, No. 1, 1996

KLUWER ACADEMIC PUBLISHERS
DORDRECHT / BOSTON / LONDON

Library of Congress Cataloging-in-Publication Data

Chaos in gravitational N-body systems : a workshop held at La Plata,
Argentina, July 31-August 3, 1995 / edited by J.C. Muzzio, S. Ferraz
-Mello and J. Henrard.
      p.   cm.
   ISBN 0-7923-4148-1 (acid free paper)
   1. Many-body problem--Congresses.   2. Celestial mechanics-
-Congresses.   3. Chaotic behavior in systems--Congresses.
I. Muzzio, Juan C.   II. Ferraz-Mello, Sylvio.   III. Henrard, J.
QB362.M3C47 1996
521--dc20   96-28459

ISBN 0-7923-4148-1

Published by Kluwer Academic Publishers,
P.O. Box 17, 3300 AA Dordrecht, The Netherlands.

Kluwer Academic Publishers incorporates
the publishing programmes of
D. Reidel, Martinus Nijhoff, Dr W. Junk and MTP Press.

Sold and distributed in the U.S.A. and Canada
by Kluwer Academic Publishers,
101 Philip Drive, Norwell, MA 02061, U.S.A.

In all other countries, sold and distributed
by Kluwer Academic Publishers Group,
P.O. Box 322, 3300 AH Dordrecht, The Netherlands.

Printed on acid-free paper

Printed in the Netherlands

CONTENTS

PREFACE

The Workshop on Chaos in Gravitational N - Body Systems was held in La Plata, Argentina, from July 31 through August 3, 1995. The School of Astronomy and Geophysics of La Plata National University, best known as La Plata Observatory, was the host institution. The Observatory (cover photo) was founded in 1883, and it has nowadays about 120 faculty members and 70 non-faculty members devoted to teaching and research in different areas of astronomy and geophysics.

It was very nice to see how many people, from young students to well recognized authorities in the field, came to participate in the meeting. This audience success was due to the increasing understanding of the necessity to gather together people from Celestial Mechanics and Stellar Dynamics to explore the problems that exist at the frontier of these two disciplines and their common interest in chaotic phenomena and integrability (the famous Argentine beef was, certainly, also an attraction!).

All the papers of the present volume were refereed. Most were accepted after some revision, while some needed no change at all (compliments to their authors!) and, sadly, a few could not be included. About half a dozen authors did not submit their contributions for publication, mainly because they were already in print elsewhere. Therefore, the special issue of Celestial Mechanics and Dynamical Astronomy includes all the invited lectures of the workshop, while the proceedings volume includes those same lectures plus the bulk of, but not all, the contributions to the meeting.

The Scientific Organizing Committee was formed by S. Ferraz-Mello, C. Froeschlé, J. Henrard, J. Laskar, M. Lecar, J. Makino, G. Meylan and J.C. Muzzio (Chairman), while A. Brunini, P.M. Cincotta, C.M. Giordano, J.A. Núñez (Chairman), M.M. Vergne, H.R. Viturro and F.C. Wachlin made up the Local Organizing Committee. We are very grateful to all of them, as well as to the La Plata students L. Chajet, J. Douglas, E. Gularte, L. Mammana, O.I. Miloni, A. Torres and R. Vallverdú, faculty member S.A. Cora, and non-faculty members S.D. Abal de Rocha, M.C. Fanjul de Correbo, M.C. Visintín and G. Sierra who were all of great help. The meeting was sponsored by the Organization of American States and by the School of Astronomy and Geophysics of La Plata University, and it was supported by the generous contribution of La Plata National University, the International Centre for Theoretical Physics, the Photometry and Galactic Structure Program of the National Research Council of Argentina (CONICET), and the Latin - American Center of Physics. Our warmest thanks to all of them.

J.C. Muzzio   S. Ferraz-Mello   J. Henrard

SPECTRA OF STRETCHING NUMBERS AND HELICITY ANGLES IN DYNAMICAL SYSTEMS

G. CONTOPOULOS and N. VOGLIS
Department of Astronomy, University of Athens
Panepistimiopolis, GR 157 84 Athens, Greece

Abstract. We define a "stretching number" (or "Lyapunov characteristic number for one period") (or "stretching number") $a = \ln \mid \frac{\xi_{t+1}}{\xi_t} \mid$ as the logarithm of the ratio of deviations from a given orbit at times t and $t + 1$. Similarly we define a "helicity angle" as the angle between the deviation ξ_t and a fixed direction. The distributions of the stretching numbers and helicity angles (spectra) are invariant with respect to initial conditions in a connected chaotic domain. We study such spectra in conservative and dissipative mappings of 2 degrees of freedom and in conservative mappings of 3-degrees of freedom. In 2-D conservative systems we found that the lines of constant stretching number have a fractal form.

Key words: Lyapunov characteristic numbers – stretching numbers – helicity angles – conservative and dissipative mappings

1. Stretching Numbers

The maximal Lyapunov characteristic number (LCN)

$$LCN = \lim_{t \to \infty} \frac{1}{t} \ln \mid \frac{\xi_t}{\xi_0} \mid \qquad (1)$$

(where ξ_0, ξ_t are the infinitesimal deviations from a given orbit at times 0 and t respectively) provides a good characterization of chaos. The LCN is zero in ordered regions while it is positive in chaotic regions. However in practice the limiting value of LCN is reached after a long time that may be unrealistic. E.g. in galactic dynamics in order to obtain good results we need in general t of the order of 10^5 - 10^6 periods, while the age of the Universe is of the order of 10^2 periods.

In recent years there has been much interest in finite time Lyapunov numbers (e.g. Fujisaka 1983, Grassberger and Procaccia 1984, Benzi *et al.* 1985, Udry and Phenniger 1988, Sepulveda et al 1989, Ababarnel *et al.* 1992). Most of the information about a dynamical system is provided when t is small, rather than large. In mappings the smallest t is 1 period. The 1-period Lyapunov characteristic number was introduced by Froeschlé *et al.* (1993) and by Lohinger *et al.*. (1993), and was studied in detail by Voglis and Contopoulos (1994) both for conservative and dissipative mappings.

The quantity

$$a_i = \ln \mid \frac{\xi_{i+1}}{\xi_i} \mid \qquad (2)$$

Celestial Mechanics and Dynamical Astronomy **64**: 1–20, 1996.
© 1996 *Kluwer Academic Publishers. Printed in the Netherlands.*

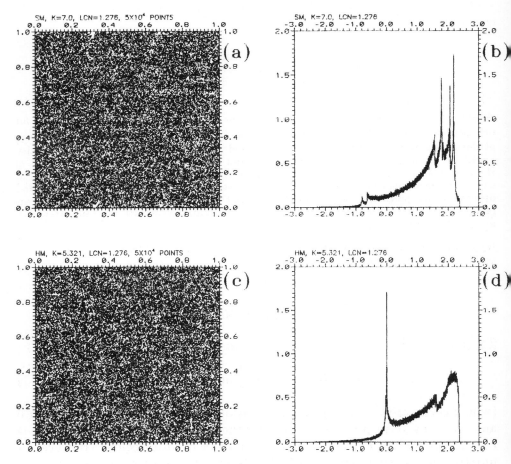

Figure 1. (a) The distribution of 5×10^4 consequents in the standard map for $K = 7$, and initial conditions $(x = 0.1, y = 0.5, \xi_x = 1, \xi_y = 0)$. (c) The same distribution in the Hénon conservative map for $K = 5.321(b = 1)$ and the same initial conditions. The corresponding spectra are shown in (b) and (d) respectively. The Lyapunov characteristic numbers in both cases are equal LCN=1.276, but the spectra are completely different.

is called a stretching number. The distribution of the stretching numbers (spectrum) gives the fractional number $\frac{dN}{N}$ of values of a in an interval $(a, a + da)$, divided by da:

$$S(a) = \frac{\Delta N}{N \, da} \tag{3}$$

as a function of a.

The spectrum gives more information about a system than the Lyapunov characteristic number, or the distribution of the consequents on a Poincaré surface of section. In Fig.1 we compare the spectra of two different mappings, that have the same LCN. The first spectrum (Fig. 1b) corresponds to the standard map

$$x_{i+1} = x_i + y_{i+1} \qquad y_{i+1} = y_i + \frac{K}{2\pi} \sin 2\pi x_i \qquad (mod1), \qquad (4)$$

while the second spectrum (Fig. 1d) corresponds to the conservative Hénon map

$$x_{i+1} = 1 - K'x_i^2 - y_i \qquad y_i = bx_i \qquad (mod1), \qquad (5)$$

with $b = 1$.

The corresponding distributions of the consequents on the plane (x, y) are given in Figs. 1a and 1c. The values $K = 7$ and $K' = 5.321$ were chosen in such a way that the Lyapunov numbers are equal LCN=1.276. Furthermore both systems look completely chaotic (in fact there are some very small islands that are not seen in Figs. 1a and 1c). But although the two usual criteria of chaos (LCN and distribution of consequents) are the same, the spectra are quite different. This means that there are basic differences in the structure of these two systems.

The LCN is the average value of the stretching numbers a_i.

The usefulness of the spectra is based on the fact that the spectra are invariant (a) with respect to the initial conditions along the same orbit, (b) with respect to the direction of the initial deviation ξ_0 , and, more important, (c) with respect to initial conditions in a connected chaotic domain (in the case of chaotic orbits).

In the case of ordered orbit the last invariance is replaced by (c') invariance with respect to initial conditions along the same invariant curve (or torus). But the spectra of various invariant curves are different. Furthermore, as we see in Fig. 1, the spectra of different systems are also different.

2. Helicity Angles

The stretching numbers give much information about a dynamical system, but an equally important information is provided by the "helicity angles" namely the angles ϕ_i of the vectors ξ_i with a given direction. These angles are defined in the interval $(-180^\circ, 180^\circ)$.

The name "helicity angle" was chosen because in more degrees of freedom the vectors ξ_i are not in the same plane.

The spectrum of the helicity angles is defined in the same way as the spectrum of the stretching numbers. Namely $S(\phi)$ is the fractional number $\frac{dN}{N}$ of the helicity angles in the interval $(\phi, \phi + d\phi)$, divided by $d\phi$.

In Fig.2b,d,f we give three spectra, corresponding respectively to a chaotic orbit (2a), a set of two closed invariant curves in islands (2c) and an invariant curve extending from $x = 0$ to $x = 1$ (2e).

All these spectra are invariant with respect to initial conditions along each orbit. In each case (2b,d,f) the spectrum of the first 10^6 iterations (continuous line) coincides with the spectrum of the next 10^6 iterations (dots). In the chaotic case the spectrum is also invariant with respect to the initial direction of the deviation ξ and with respect to the initial conditions in the same connected chaotic domain. The first two spectra (2b,d) are periodic in ϕ with period $180°$. This means that it is equally probable that the vector ξ has a direction ϕ and $\phi + 180$. In the chaotic case this is due to the fact that two initially opposite directions of ξ give always opposite ξ, therefore, if the two spectra are identical, they have to be periodic in ϕ with period $180°$. In the case of islands the vector ξ tends to be tangent to the invariant curve, and the symmetry of the islands with respect to the origin guarantees the equal probabilities of ϕ and $\phi + 180°$.

In the case 2f, however, there is no such symmetry around the origin, but the tangent along the invariant curve has only relatively small variations with respect to its mean value. Thus the spectrum is localized. The spectrum is invariant with respect to all initial deviations ξ above the invariant curve. But if the initial deviation is below this curve we find the above spectrum displaced by $180°$.

The average value of $< \phi >$ is constant in the chaotic domain, but is varies smoothly if we change the initial conditions inside a region of islands. This is seen clearly in Fig. 3. Namely we calculate the average value $< \phi >_n$ at many points along a line passing through a chaotic domain and a set of islands, for not very large number of iterations n. E.g. in Fig. 3 we take $n = 10^4$ and find the values of $< \phi >_n$ at successive points along a straight line parallel to the axis x. The various values of $< \phi >_n$ in the chaotic domain have some dispersion around the same value $< \phi >$ (calculated for a much larger number of periods, say 10^6). The constancy of $< \phi >$ is due to the invariance of the spectrum with respect to the initial conditions in the same chaotic domain.

When the value of x goes from the chaotic domain to an island of stability the value of $< \phi >_n$ changes abruptly, and as long as x is inside the island it changes smoothly. In Fig. 3 we notice that the dispersion of the values of $< \phi >_n$ is much smaller in the region of the island (almost insignificant). The secondary minima and maxima that we see on both sides of the main minimum are symmetric, indicating that they are not dispersion features but secondary islands inside the main island. In Fig. 3 we see also a different very narrow but deep minimum which is due to a small island that can be seen in the phase portait (x, y).

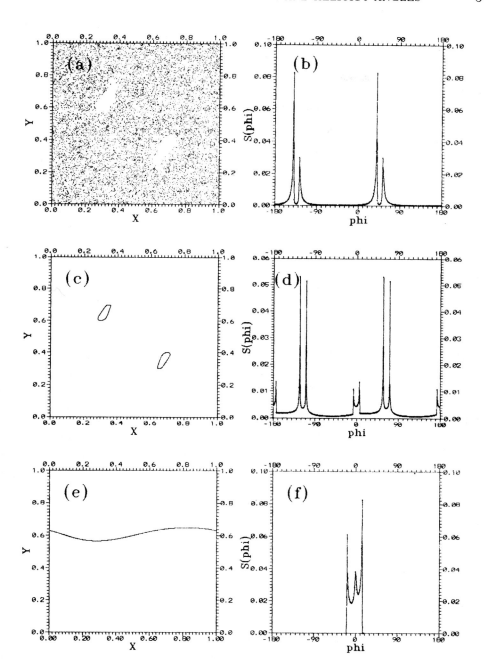

Figure 2. (a) The distribution of 10^6 consequents in the standard map for $K = 5$ and initial conditions $(x = 0.1, y = 0, \xi_x = 1, \xi_y = 0)$. (c) The same for $K = 5$ and initial conditions $(x = 0.65, y = 0.34, \xi_x = 1, \xi_y = 0)$. (e) The same for $K = 0.5$ and initial conditions $(x = 0.1, y = 0.6, \xi_x = 1, \xi_y = 0)$. The corresponding specta of the helicity angles are shown in (b),(d) and (f) respectively.

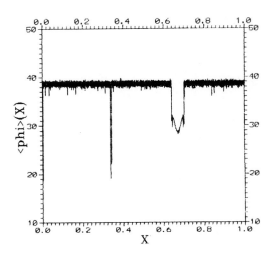

Figure 3. The average values of the helicity angle $< \phi >_n$ for $n = 10^4$, in the standard map for $K = 5$, at successive values of x with a step 0.001 from $x = 0$ to $x = 1$ and $y = 0.34$.

If we take n smaller, e.g. $n = 10^3$, or $n = 10^2$, the dispersion of the values of $< \phi >_n$ is larger, but we can still clearly distinguish the chaotic domain, where the average $< \phi >$ is constant, from the ordered domain, where $< \phi >$ varies smoothly. We find that the dispersion of $< \phi >_n$ is larger in the chaotic domain than in the ordered domain.

Thus we may use this method to identify the chaotic and ordered domains, in a system with divided phase space, without excessive numerical calculations. In this way the calculation of Lyapunov exponents, which is much more time consuming, is avoided. This method of distinguishing ordered and chaotic domains is somewhat similar to the method of Laskar et al.(1992, 1993), who find the frequencies at successive points and their irregular or smooth variation. Our method is simpler because it does not need a Fourier analysis (or a variant thereof) to find the frequences. The helicity angle is sufficient to separate the chaotic from the ordered domains.

3.   Lines of Constant Stretching Number

It has been argued that the invariance of the chaotic spectra is obvious because a chaotic orbit goes close to every point of a connected chaotic domain. This argument is based on the assumption that at each point of phase space one can define a largest eigenvector, to which the deviation ξ_i tends for large i.

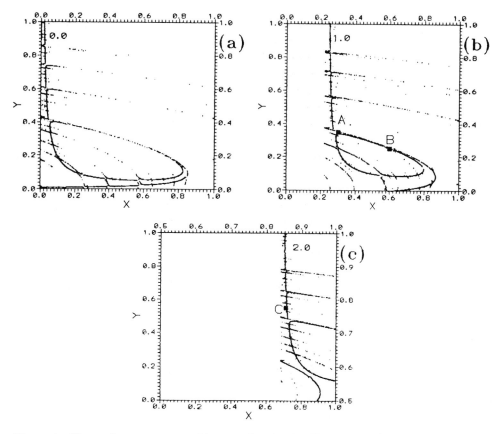

Figure 4. Lines of constant stretching number in the Hénon map (for the orbit of Fig. 1c). (a) $a_0 = 0$, (b) $a_0 = 1$, (c) $a_0 = 2$. The boxes have dimensions (0.01×0.01) and their left lower corners are at $A(0.30, 0.34)$, $B(0.60, 0.24)$ and $C(0.71, 0.55)$.

However the eigenvectors at many points of phase space are complex. Furthermore we will present evidence that in the neighbourhood of many (perhaps all) points of phase space the stretching number takes an infinity of widely different values.

To this purpose we have calculated the lines of constant stretching number a in the conservative Hénon map (5) for $K' = 5.321$ and $b = 1$. In Figs 4a,b,c we mark the lines corresponding to $a_0 = 0$, $a_0 = 1$, and $a_0 = 2$. Namely we mark the consequents of a particular orbit calculated for 10^8 periods that are very close to a_0, namely for a in the interval $(a_0 - 0.0005, a_0 + 0.0005)$.

We see that each figure 4a,b,c, contains a basic, almost vertical, line, that is broken at many points, and turns to the right in the lower part of each figure, forming more complicated structures there.

There are also many nearly horizontal parallel lines that have a small negative inclination. In the lower parts of the figures these lines have a larger inclination and they are curved downwards.These nearly horizontal lines are generated near the points where the vertical line is broken.

They extend mainly to the right, all the way to the boundary of the figure $(x = 1)$. Some lines extend to the left, but reach only a small distance from the main, almost vertical, line (Figs. 4b,c). Only in Fig. 4a $(a = 0)$ the lines to the left reach the boundary $x = 0$. Finally these lines in the lower part of the figure (small y) reach the boundary $y = 0$ (Figs 4a,b, while in Fig. 4c the situation is not clear; only one line reaches certainly the boundary $y = 0$).

The almost horizontal lines are defined by a number of scattered points. Some lines contain many points, but others contain only a few points. However we are certain of the existence of these lines, because when we extend the calculations to longer periods more points appear along each line.

It is remarkable that (a) these lines are invariant in the sense that different initial conditions define the same lines, and (b) that more lines appear when the calculations extend to longer intervals of time. It seems probable that the number of these lines is infinite and the "vertical" lines are broken at infinite points. This, however, has to be checked with more detailed calculations. At any rate it is obvious that the densities of points along different lines are different.

What is certain from our calculations is that the nearly horizontal lines from small a pass through the gaps of the vertical lines for larger a. E.g. the regions A and B, of size (0.01×0.01), in Fig. 4b contain points both from $a \simeq 1$ and $a \simeq 0$. However the region C (Fig.4c) seems to contain only points from the vertical line $a = 2$ and similar vertical lines from nearby values of a.

The spectra of the values of a of the regions A,B and C are given in Fig. 5, together with the chaotic invariant spectrum for $N = 10^8$ periods. The spectra A,B,C are multiplied by 10^4 to become conspicuous.

We notice, first, that the number 10^8 is large enough so that the invariant spectrum is clean, i.e. it has extremely small noise. Then we remark that the local spectra A and B are terminated abruptly on the right, but extend considerably to the left. This is consistent with the fact that the regions A and B contain many almost horizontal lines coming from their left (smaller values of a), but only a few lines coming from the right (larger values of a).

As regards the spectrum C we see that it is very peaked at the value $a = 2$ with very small thickness (in fact it contains an infinity of values of a close to $a = 2$). The spectrum C has very weak extensions to the left and to

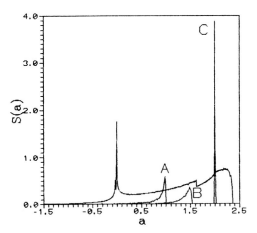

Figure 5. Spectra of the values of a from points of the orbit of Fig. 1c (calculated for 10^8 periods) belonging to the boxes A,B,C of Figs. 4b,c. The full spectrum is also given.

the right, which indicate that, even in this case, there may be some parallel lines coming from regions of different a (mainly smaller a) that pass through the region C of Fig. 4c.

It is evident that many points in phase space have in their neighbourhood an infinity of very different values of a. This is evident in the regions A and B, but it may be true also in the region C and perhaps near every point of the space (x, y).

In any case the fact that the lines of Figs 4a,b,c are broken at so many points indicates that these lines are, in fact, fractals. It is obvious that the loci of constant a are not smooth curves that cover in a continuous way the plane (x, y). Thus the fact that an orbit is ergodic in a chaotic domain does not imply that it goes through practically the same values of a.

This means that the usual ergodic theorem (Sinai 1989) cannot guarantee the invariance of the spectra. There must be an extended form of ergodic theorem, valid for fractal sets, that should "explain" the invariance of the spectra, that we found empirically. Much further work is needed in this direction.

4. Dissipative Systems

The stretching numbers and the helicity angles can be defined also in dissipative systems. E.g. if the parameter b in the Hénon map (5) is absolutely smaller than 1, the system is dissipative and in general the successive conse-

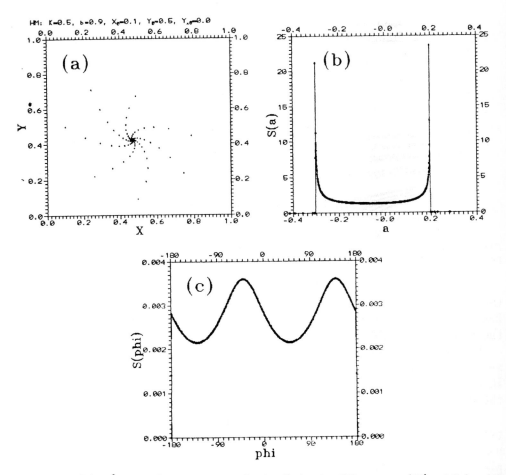

Figure 6. (a) 10^6 successive consequents in the dissipative Hénon map ($K' = 0.5, b = 0.9$ and initial conditions $x = 0.1, y = 0.5, \xi_x = 1, \xi_y = 0$). The points approach a point attractor. The corresponding spectra of stretching numbers and helicity angles are shown in (b) and (c).

quents tend to a point (point attractor), a curve (line attractor), or a strange attractor.

In all these cases we can find invariant spectra for the stretching numbers and the helicity angles. In Fig. 6 we see the case of a point attractor. After about 100 scattered points, 10^6 calculated points are in the central dot of Fig. 6a, which seems to have no structure. However the corresponding spectra for the stretching numbers (Fig. 6b) and for the helicity angles (Fig. 6c) are well defined, and they are invariant, i.e. the spectra for the first 10^6 iterations (solid lines) and the next 10^6 iterations (dots) coincide (the

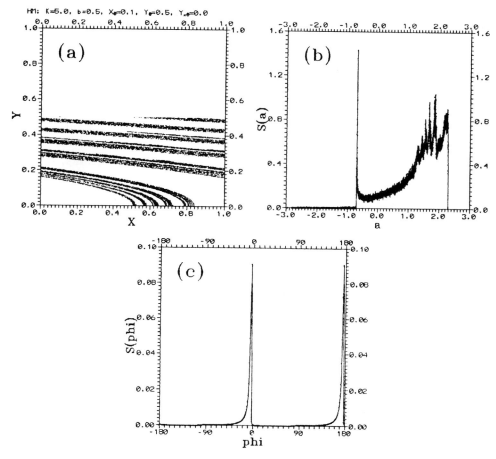

Figure 7. (a),(b),(c) As in Fig.6 for a strange attractor ($K' = 5$, $b = 0.5$, $x = 0.1$, $y = 0.5$, $\xi_x = 1$, $\xi_y = 0$).

dots are distinguished in Fig.6b, but not in Fig.6c, because they fall exactly on the solid line). The spectrum of the helicity angles (6c) has a period of $180°$, as in the conservative cases 2b,d. We notice that the average value of a (Lyapunov characteristic number) is negative.

In Fig. 7 we see similar results for a strange attractor (Fig. 7a). The corresponding spectra of stretching numbers (Fig. 7b) and of helicity angles (Fig. 7c) are again invariant, as in Figs.6b,c. In this case the spectrum of stretching numbers for 10^6 periods has some noise, while the spectrum of helicity angles has practically no noise. The dots representing the spectrum of the next 10^6 periods are seen both in Fig.7b and 7c. The spectrum 7c has again a periodicity of $180°$. Now the Lyapunov characteristic number is positive.

5. Resonance Overlap Diffusion vs. Arnold Diffusion

In systems of more than two degrees of freedom the KAM surfaces do not separate the phase space and diffusion from one chaotic region to another is possible. This is usually called Arnold diffusion (Arnold 1964, Arnold and Avez 1967, Lichtenberg and Lieberman 1983) and it is due to the interaction of the various resonances that constitute the "Arnold web".

Arnold diffusion exists even for arbitrarily small perturbations, but its time scale is extremely long. When the perturbations are large the diffusion is much faster. Thus some authors (Chirikov 1979, Chirikov *et al.*. 1979, Laskar 1993) have distinguished between Arnold diffusion and resonance overlap.

The overlapping of resonances as a mechanism that introduces chaos is well known since 1966 (Rosenbluth *et al.*. 1966, Contopoulos 1966, Zaslavskii and Chirikov 1972, Chirikov 1979). Large degree of chaos is due to the intersection of the homoclinic tangles (chaotic zones) around different unstable periodic orbits. This introduces heteroclinic intersections between the asymptotic curves of unstable periodic orbits of different types, that exist in regions of phase space far from each other.

In 2 degrees of freedom such intersections do not appear for small perturbations, because the various homoclinic tangles are separated by closed KAM curves. But in systems of 3 or more degrees of freedom such intersections appear for arbitrarily small perturbations. On the other hand the regions of resonance overlap are extremely small for small perturbations, but become significant for larger perturbations.

We will present evidence here that the transition from the very slow Arnold diffusion to the much faster resonance overlap diffusion is quite sudden.

We study a system of 2 coupled standard maps

$$x_1' = x_1 + y_1', \qquad y_1' = y_1 + \frac{K}{2\pi} \sin 2\pi x_1 - \frac{\beta}{\pi} \sin 2\pi(x_2 - x_1)$$

$$(mod 1) \qquad (6)$$

$$x_2' = x_2 + y_2', \qquad y_2' = y_2 + \frac{K}{2\pi} \sin 2\pi x_2 - \frac{\beta}{\pi} \sin 2\pi(x_1 - x_2)$$

This 4-dimensional map corresponds to a Poincaré surface of section of a system of 3 degrees of freedom. If the coupling parameter β is zero the system (6) represents two independent 2-D maps. But if $\beta \neq 0$ each 2-D map is influenced by the other.

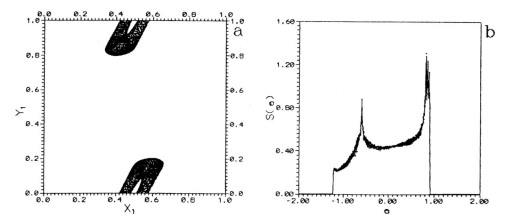

Figure 8. (a) The distribution of the consequents of the map (6) with $K = 3$, $\beta = 0.3$ for an orbit in the "ordered" region (initial conditions: $x_1 = 0.55$, $y_1 = 0.1$, $x_2 = 0.62$, $y_2 = 0.2$, $\xi_{x1} = \xi_{y1} = \xi_{y2} = 0$, $\xi_{x2} = 1$) on the plane (x_1, y_1) for 10^6 periods. (b) The corresponding spectrum of stretching numbers.

In Fig. 8a we see the distribution of 10^6 consequents of the map (6) in the case $\beta = 0.3$ on the plane (x_1, y_1). The initial conditions are in the ordered regions of the two maps, and the corresponding invariant curves on the $(x_1, y_1), (x_2, y_2)$ planes are shown as solid lines in Fig.8a. The distribution of the consequents on the (x_1, y_1) plane shows some dispersion, but probably this is not due to chaos, but to projection effects. Our opinion is based on the fact that the region covered by points has a very well defined boundary, and there is no diffusion to the region outside this boundary.

The corresponding "ordered" spectrum of stretching numbers is shown in Fig. 8b. This spectrum is different from the spectra of the uncoupled case $\beta = 0$ (both ordered and chaotic). If the initial conditions are in the chaotic domains of the (x_1, y_1) and (x_2, y_2) maps the distribution of the consequents on the (x_1, y_1) plane is shown in Fig. 9a and the corresponding spectrum in Fig. 9b. This spectrum is quite different from the ordered spectrum 8b, but it is invariant in the sense that different initial conditions give the same spectrum if (x_1, y_1) is in the chaotic domain. Even if the initial conditions (x_1, y_1) are in the ordered domain the spectrum is the same as 9b. Thus the chaotic spectrum of stretching numbers is truly invariant.

When $\beta = 0.30513$ (Fig. 10a) the orbit remains for about 10^5 periods in the same area as in Fig. 8a, but then it diffuses and fills the whole available space. The spectrum for 0.5×10^5 periods (Fig. 10b, line a) is a transient one. After that time the spectrum changes (line b) and tends to a final spectrum (line c) that is invariant. This spectrum (c) does not change appreciably if the

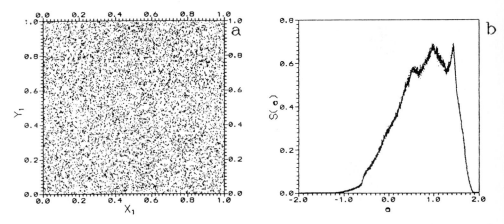

Figure 9. (a),(b) As in Fig.8 for an orbit in the chaotic region, with $K = 3, \beta = 0.3$ and initial conditions $x_1 = 0.1$, $y_1 = 0.5$, $x_2 = 0.2$, $y_2 = 0.6$, $\xi_{x1} = \xi_{y1} = \xi_{y2} = 0$, $\xi_{x2} = 1$.

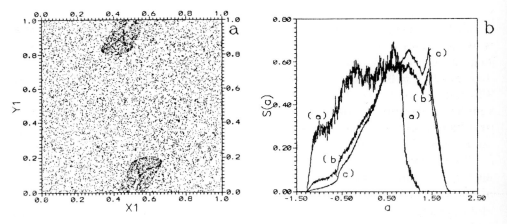

Figure 10. (a) The distribution of 10^6 consequents of the map (6) with $K = 3, \beta = 0.30513$ for the same initial conditions as in Fig. 8a. (b) The corresponding spectrum of stretching numbers for 0.5×10^5 periods (line a), 10^6 periods (line b), and 10^7 periods (line c).

time increases further. Thus it is a really invariant spectrum. Furthermore this invariant spectrum only slowly changes for increasing β.

The escape time (the time needed to escape from the boundaries of the confined region) is a function of β. It increases as β decreases, according to the empirical law (Fig. 11)

$$\log_{10} T = 4.94 - 4160(\beta - \beta_c) \qquad (7)$$

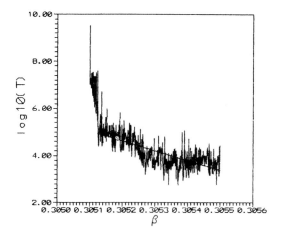

Figure 11. The transition time as a function of the coupling parameter β.

which is valid for $\beta > \beta_c$, where the critical value of β_c is $\beta_c = 0.305124$. Thus the increase of T is exponential. E.g. for $\beta = 0.306$, T becomes equal to $T = 20$ and for larger β the diffusion is extremely fast.

On the other hand when β decreases below the value 0.305124 the escape time increases abruptly, extremely faster than given by the line (7) (Fig. 11).

E.g. for $\beta = 0.305124$, the escape time is about $T = 10^5$, for $\beta = 0.30512$, $T = 10^6$, for $\beta = 0.30511$, $T = 10^7$, and for $\beta = 0.30510016$, $T = 3 \times 10^9$. For smaller β we found that T is larger than 10^{10}, and its calculation becomes practically impossible.

We identify the diffusion leading to the relatively slow increase of T (although exponential) for $\beta > 0.305124$ as resonance overlap diffusion, while the much slower diffusion leading to the abrupt increase of T for $\beta < 0.305124$ as Arnold diffusion.

From Fig. 11 we conclude that the transition from the two regimes is quite sudden, although it is continuous. The continuity of the two regimes is natural because Arnold diffusion is also due to resonance overlap. In fact the chaotic regions along different intersecting lines of the Arnold web overlap at their intersections. However it seems that when the thickness of these chaotic regions (which is an increasing function of β) goes beyond a certain limit the diffusion increases substantially and the overlap is much more important. Thus the value of $dT/d\beta$ changes discontinuously and we can clearly distinguish between the two regimes of Arnold diffusion and resonance overlap diffusion.

As the escape time is extremely long in the Arnold regime (much longer that the time scale of any physical process), we can consider the "ordered"

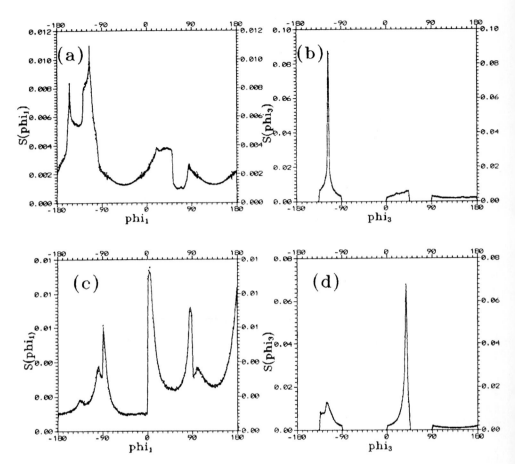

Figure 12. Spectra of the helicity angles ϕ_1 and ϕ_3 in an ordered case (data of Fig.8) (a),(b), and in a chaotic case (data of Fig.9) (c) (d). Solid lines: the first 10^6 periods. Dots: the next 10^6 periods.

spectrum (Fig.8b for $\beta = 0.3$) as invariant. The chaotic spectrum 9b is also invariant and rather similar to the spectrum (c) of Fig. 10b.

The same distinction between ordered and chaotic spectra of stretching numbers in 4-D maps for relatively small coupling β, appears also in the spectra of the helicity angles. In 4-D maps the vector ξ is not planar, therefore it defines 3 helicity angles. In the present paper we define the angles ϕ_1, ϕ_2, ϕ_3 of the axis x_2 with the projections of the vector ξ on the planes $(x_1, x_2), (y_1, x_2)$ and (y_2, x_2) respectively. All these angles are defined in the interval $(-180°, 180°)$. In Fig. 12a,b we give the spectra of the helicity angles ϕ_1, ϕ_3 in an ordered region (a) and in a chaotic region (b) for K=3 and

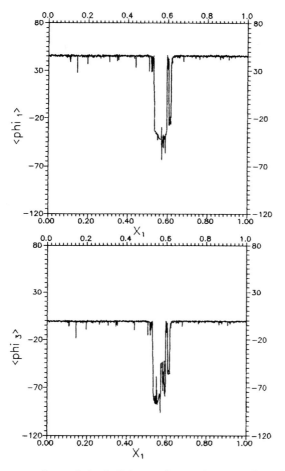

Figure 13. The average values of the helicity angles $< \phi_1 >_n$ and $< \phi_3 >_n$ for $n = 1\text{C}$ in the map (6) for $K = 5$ and $\beta = 0.3$ at successive values of x_1 with a step 0.001 fro1 $x_1 = 0$ to $x_1 = 1$ and $y_1 = 0.1, x_2 = 0.62, y_2 = 0.2, \xi_{x1} = \xi_{y1} = \xi_{y2} = 0, \xi_{x2} = 1$.

$\beta = 0.3$ and the same initial conditions as the spectra of stretching numbe 8b and 9b.

These spectra are invariant in the same way as the corresponding spectr of the stretching numbers. Namely the first spectrum (12a) is in fact transient and it is expected to evolve to the spectrum (12b) after an extremely long time. However this long time is beyond any possible calculation. Thus for all practical purposes we can consider the spectrum (12a) as invariant. In fact if we calculate an orbit for 10^6 periods we find two spectra (for ϕ_1 and ϕ_3 respectively, solid lines) and the spectra for the next 10^6 periods (dots) are on the same lines.

The fact that the helicity angles give different spectra in the ordered and in the chaotic domains, allows us to separate the ordered and chaotic domains using the same method as in 2-D systems. In Fig. 13 we give the average helicity angles $< \phi_1 >_n$ and $< \phi_3 >_n$, i.e. the average angles of the x_2 axis with the projections of the vector ξ on the planes (x_1, x_2) and (y_2, x_2) for $n = 10^5$. The initial conditions are taken along the straight line $y_1 = 0.2, x_2 = 0.1, y_2 = 0.62$ and values of x_1 increasing by a step $\Delta x = 0.001$ from $x_1 = 0$ to $x_1 = 1$. The quantities $< \phi_1 >_n$ and $< \phi_3 >_n$ (and also $< \phi_2 >_n$) have some dispersion around the constant values $< \phi_1 >$ and $< \phi_3 >$ (and $< \phi_2 >$) in the chaotic domain, but they have a smooth variation in ordered regions. The values of $< \phi_1 >_n$, $< \phi_2 >_n$ and $< \phi_3 >_n$ change abruptly at the boundaries of the main ordered region of Fig. 13. Inside this ordered region they seem to have some irregular variations, but if we magnify this part of Fig. 13 we see that the variation is smooth. Furthermore in the chaotic regions of Fig.13 we see some long vertical lines that correspond to small islands of stability. If the number of iterations n becomes smaller the noise in the values of ϕ_1, ϕ_2, ϕ_3 becomes larger, but the transition from the chaotic to the ordered domain is again marked by an abrupt change of ϕ_1, ϕ_2, ϕ_3. Thus we can distinguish between chaotic and ordered domains of 3-D systems in the same way as in 2-D systems.

6. Conclusions

The invariant spectra of stretching numbers and helicity angles give much more information about a dynamical system than the Lyapunov character- istic number (LCN) or the distribution of the consequents on a Poincaré surface of section. We give an example of two different chaotic systems that have the same LCN and the same coarse grained distribution of consequents, but very different spectra.

The helicity angles provide a way to distinguish between ordered and chaotic domains in a divided phase space. This method is extremely fast because it does not require the calculation of frequencies in order to dis- tinguish between order and chaos. The lines of constant stretching numbers along any given orbit characterize the chaotic phase space. They are frac- tals forming an infinity of lines. Thus near many points of the phase space (perhaps all) there are other points with very different stretching numbers.

Invariant spectra of stretching numbers and helicity angles appear also in dissipative systems. We give two examples, one with a point attractor and another one with a strange attractor.

In the case of 3 degrees of freedom, or 4-dimensional maps on a Poincaré surface of section, we found again invariant spectra of stretching numbers and helicity angles. We studied two coupled standard maps when each map has both ordered and chaotic domains. In all cases there is a diffusion that leads to a large chaotic domain. We found that the time needed for a transition to this large chaotic domain is extremely long for a small coupling β, but it is relatively fast for a coupling β above a critical value β_c. We identify the case of small $\beta(< \beta_c)$ as Arnold diffusion, and the case of $\beta > \beta_c$ as resonance overlap diffusion. In the second case the transition time decreases exponentially with $(\beta - \beta_c)$ but in the first case the transition time increases much faster for decreasing β.

When the initial conditions of both maps are in ordered regions the spectra of stretching numbers and helicity angles are transient, while if one at least initial condition is in the chaotic domain the spectra are invariant. However in the case of Arnold diffusion the transient spectra can be considered as invariant for extremely long times. Thus even in cases of 3 degrees of freedom we distinguish between chaotic and ordered domains. The helicity angles provide a clear distinction between chaos and order, as in the case of 2 degrees of freedom.

Acknowledgements

We thank Mr. C. Efthymiopoulos for much help in calculating Figs. 8,10,11. This research was supported in part by the EEC Human Capital and Mobility Program EPB4050 PL 930312.

References

Ababarnel, H.D.I., Brown, R. and Kennel, M.B.: 1991, *Int. J. Mod. Phys.* **B5**, 134.
Arnold, V.I.: 1964, *Dokl. Akad. Nauk SSSR* **156**, 9.
Arnold, V.I. and Avez, A.: 1967, *Problèmes Ergodiques de la Mécanique Classique*, Gauthier-Villars, Paris.
Benzi, R., Paladin, G., Parisi, G. and Vulpiani, A.: 1985. *J. Phys. A.* **18**, 2157.
Chirikov, B.V.: 1979, *Phys. Rep.* **52**, 263.
Chirikov, B.V., Ford, J. and Vivaldi, F.: 1979, in Month, M. and Herreira, J.C. (Eds), "Nonlinear Dynamics and the Beam-Beam Interaction", Amer. Inst. Phys., N. York.
Contopoulos, G.: 1966, in Nahon, F. and Hénon, M. (eds) "Les Nouvelles Méthodes de la Dynamique Stellaire", CNRS, Paris ≡ *Bull. Astron. Ser.* **3**, 2, 223.
Froeschlé, C., Froeschlé, Ch. and Lohinger, E.: 1993, *Celest. Mech. Dyn. Astron.* **56**, 307.
Fujisaka, H.: 1983, *Prog. Theor. Phys.* **70**, 1264.
Grassberger, P. and Procaccia, I.: 1984, *Physica* **D13**, 34.
Laskar, J.: 1993, *Physica* **D67**, 257.
Laskar, J., Froeschlé, C. and Celletti, A.: 1992, *Physica* **D56**, 253.
Lichtenberg, A.J. and Lieberman, M.A.: 1983, *Regular and Stochastic Motion*, Springer Verlag, N.York.
Lohinger, E., Froeschlé, C. and Dvorak, R.: 1993, *Cel. Mech. Dyn. Astron.* **56**, 315
Rosenbluth, M.N., Sagdeev, R.A., Taylor, J.B. and Zaslavskii, M.: 1966, *Nucl. Fusion* **6**, 297.

Sepulveda, M.A., Badii, R. and Pollak, E.: 1989, *Phys.Rev.Lett.* **63**, 1226.
Sinai, Ya.G.: 1989, Dynamical Systems II, Springer, Berlin, Heidelberg, N. York.
Udry, S. and Pfenniger, D.: 1988, *Astron. Astrophys.* **198**, 135.
Voglis, N. and Contopoulos, G.: 1994, *J. Phys.* **A27**, 4899.
Zaslavskii, G. M. and Chirikov, B.V.: 1972, *Sov. Phys. Uspekhi* **14**, 549.

ON THE MEASURE OF THE STRUCTURE AROUND THE LAST KAM TORUS BEFORE AND AFTER ITS BREAK-UP

C. FROESCHLÉ

Observatoire de Nice B.P.229, 06304 Nice cedex 4

E. LEGA

Observatoire de Nice B.P.229, 06304 Nice cedex 4
and
CNRS LATAPSES, 250 Rue A.Einstein, 06560 Valbonne

Abstract. Using mappings as model problem we study the structure around the last invariant KAM torus for different values of the perturbing parameter. We used the standard map for the analysis of the hierarchical structure existing around invariant KAM tori applying two complementary methods: the Laskar's frequency map analysis and the sup-map analysis. We recover and extend the theoretical prediction recently given by Morbidelli and Giorgilli about the existence of a neighborhood almost completely full of slave tori around a chief torus. We then make tests about the diffusion in order to measure the barrier to diffusion which still remains after the break-up of the last KAM torus.

Key words: KAM tori – diffusion – symplectic maps – standard map

1. Introduction

Symplectic mappings have been widely used to study the dynamics associated to Hamiltonian systems. Since the pioneering work of Hénon (1969), mappings have been used even without an immediate purpose for celestial mechanics, to study fundamental issues, like the stability connected with the existence of invariant tori or the diffusion processes. Recent works have been made on the last few years in this spirit, see for example: Henrard and Morbidelli (1993), Lohinger and Froeschlé (1993), Froeschlé *et al.* (1993), Voglis and Contopoulos (1994), Petit and Froeschlé (1994), Froeschlé and Lega (1995). In the domain of celestial mechanics, mappings have been devised to study the dynamics of comets and asteroids. The last papers using this approach are those of Hadjidemetriou (1991,1992,1993a,1993b) while for earlier work, in the same spirit, the reader can refer to the review of Froeschlé (1992).

Using the standard map as a model problem this paper deals with the measure of the structure around the last KAM torus before its break-up, i.e. the study of the distribution of invariant KAM tori around the golden one, and with the study of the diffusion process for values of the perturbing parameter such that no more invariant tori exist. The first part of the study relies on a theoretical breakthrough recently made by Morbidelli and Giorgilli (1995) proving the super-exponential stability of invariant tori. As

Celestial Mechanics and Dynamical Astronomy **64**: 21–31, 1996.
© 1996 *Kluwer Academic Publishers. Printed in the Netherlands.*

usual in the theory of dynamical systems, the results are rigorously proved assuming that the perturbation is small enough. The numerical experiments show, however, that invariant tori persist up to much larger perturbation magnitudes. Therefore, it is interesting to check numerically if the super-exponential stability and the other properties outlined in Morbidelli and Giorgilli's theorem persist up to the value of the perturbation for which the torus actually breaks up. Moreover, one would like to have a numerical indication about the size of the super-exponentially stable region existing around a torus. Is the super-exponential stability just an asymptotic result, or does it concern a macroscopic region of physical interest?

Our numerical computations show in a striking way that the description of the dynamics given, for small perturbations, by Morbidelli and Giorgilli's result is true in reality as long as the invariant torus persists. Moreover, the size of the structure described by Morbidelli and Giorgilli around the invariant torus shrinks to 0 like $\exp(-\epsilon_c/(\epsilon_c-\epsilon))$ when the size of the perturbation ϵ tends to the threshold value ϵ_c corresponding to the torus break-up. This implies that, when the perturbation magnitude is a little bit smaller than the break–up threshold, the size of such structure is macroscopic.

In section 2 we recall the result by Morbidelli and Giorgilli and the main ideas of their approach. In section 3 we discuss our numerical experiments, based on three different methods, and their significance.

2. Super-exponential stability of invariant tori

In their investigation of the dynamics in the vicinity of an invariant KAM torus, Morbidelli and Giorgilli started from the so called Kolmogorov normal form (Kolmogorov,1954).

According to Kolmogorov's construction, one can introduce suitable action angle variables P, Q, such that, in the neighborhood of the invariant torus $P = 0$, the Hamiltonian writes:

$$H(P,Q) = \omega \cdot P + O(P^2)f(Q) \ .$$

The Kolmogorov normal form shows in an non-equivocally way that in the vicinity of the invariant torus the significant perturbation parameter is the distance $|P|$ from the torus itself.

Therefore, in the ball $|P| < \rho$ one can introduce new action angle variables J_ρ, ψ_ρ such as to reduce the local perturbation to its optimal size, which, assuming analytic Hamiltonians, is exponentially small with $1/\rho$, i.e.

$$H(J_\rho, \psi_\rho) = \omega \cdot J_\rho + H_0(J_\rho) + \epsilon_\rho H_1(J_\rho, \psi_\rho)$$

with H_0 quadratic in J_ρ and $\epsilon_\rho \sim \exp(-1/\rho)$.

At this point, it is enough to remark that, provided ρ is small enough, ϵ_ρ is smaller than the threshold for the applicability of Arnold's version of KAM theorem (Arnold, 1963) in the ball $|J_\rho| < \rho$. This allows to prove that in the vicinity of the central torus at $P = 0$ there exist an infinity of invariant tori, the volume of the complement decreasing to zero exponentially with $1/\rho$.

On the other hand, provided $H_0(J_\rho)$ is convex in $J_\rho = 0$, if ρ is small enough, the local perturbation parameter ϵ_ρ is also smaller than the threshold for the applicability of Nekhoroshev's theorem. This allows to prove that the diffusion of the actions J_ρ must be bounded by ϵ_ρ^b for all times up to $\exp(1/\epsilon_\rho)$, which, by substitution gives the super-exponential estimate $\exp[\exp(1/\rho)]$. The hypothesis of local convexity is a very natural one. It means indeed that on a given energy surface, the torus with given frequency ratios is locally unique.

The picture provided by Morbidelli and Giorgilli's result is therefore the following. The tori given by Kolmogorov's theory are *master* tori, surrounded by a structure of *slave* tori, which accumulate in an exponential way to the central master torus. These slave tori are all n–dimensional Diophantine ones, but they are characterized by a very small Diophantine constant γ (we recall that a frequency ω is said to be Diophantine if it satisfies the relation $|k \cdot \omega| > \gamma/|k|^{n+1}$ for all integer vectors k and some positive γ); for this reason, they could not be found directly by Kolmogorov's construction. Moreover, diffusion among this structure of slave tori is super-exponentially slow, so that chaotic orbits can enter in, or escape from, only in a time proportional to $\exp[\exp(1/\rho)]$.

The interest of this result is double. On the one hand, this makes open the set of invariant tori from all practical point of view; this is important for what concerns the compatibility of KAM theorem with the errors in initial conditions of numerical experiments. On the other hand, a direct consequence of the local super-exponential stability is that invariant tori can form, even in three or more degrees of freedom, a kind of impenetrable structure which orbits cannot penetrate for an exceedingly long time, very large even with respect to the usual Nekhoroshev's estimates.

3. Numerical measures

3.1. THE STRUCTURE OF THE NEIGHBORHOOD OF THE GOLDEN TORUS

In order to test the structure of invariant tori around a chief torus we have taken as a model problem the standard map. We recall the set of equations for this mapping:

$$\begin{cases} y_{i+1} = y_i + \epsilon \sin(x_i) & y \in \Re \\ x_{i+1} = y_{i+1} + x_i & mod(2\pi) \end{cases} \tag{1}$$

We have computed for a set of initial conditions regularly spaced on the line $x = 0$ both the rotation number (Lega & Froeschlé 1995) associated to each initial condition and the sup of the action y. As already pointed out by Laskar *et al.* (1992), the existence of KAM tori corresponds to monotonic variations of the rotation number as a function of y. Conversely, islands correspond to constant frequencies and chaotic regions correspond to either noisy or simply non monotonic variations of frequencies. All this features appear clearly in Figure 1 a ,b ,c. The perturbing parameter is $\epsilon = 0.9715$. The same monotonic variation appear when considering, instead of the rotation number as a function of y, the sup of the action. Actually we will use the fact that:

1. The vertical line $x = 0$ is intersecting the invariant KAM curves only once.
2. If two initial conditions are such that $y_1 < y_2$ and y_2 lead to a KAM curve then $\sup\{y_1\} < \sup\{y_2\}$ where $\{y_i\}$ stands for the set of iterates of the initial condition y_i.

This two properties have been used by Celletti (1990) to study resonances in the spin-orbit problem. Therefore any inversion of monotony, which remains unchanged when the number of iterations increases drastically (a good approximation of ∞), indicates the non existence of KAM invariant curves. Figs. 1 a' ,b' ,c' show the variation of the $\sup\{y_i\}$ for the same initial conditions taken in Figs. 1 a ,b ,c.

As already stated a monotonic variation of the sup corresponds to a set of invariant KAM tori, a jump followed by a plateau indicates the crossing of a chaotic zone and the crossing of nested islands gives rise to v-shaped structures. However, since a drastic increase of the number of iterations induces the disappearance of some of the v-shaped spots, such spots do not correspond to the crossing of an island. In this case the v-shaped structure is the signature of the presence of a nearby rational frequency ω. Invariant tori are still there but an ergodic covering of this tori is very slow since the corresponding rotation numbers are closed to ω and the detection of the sup is very sensitive to the covering of the invariant curve. We get also the indication that the island corresponding to the frequency ω is very tiny, which is not at all in contradiction with the KAM and Morbidelli-Giorgilli theorems. A deeper analysis of this phenomena is under study using the theory of Diophantine approximations.

On each plot the origin correspond to the golden torus, i.e. the torus whose rotation number is equal to the golden number $\nu_o = \frac{1}{2}(3 - \sqrt{5})$. We observe that the majority of orbits of Figs.1 a, a' correspond to chaotic

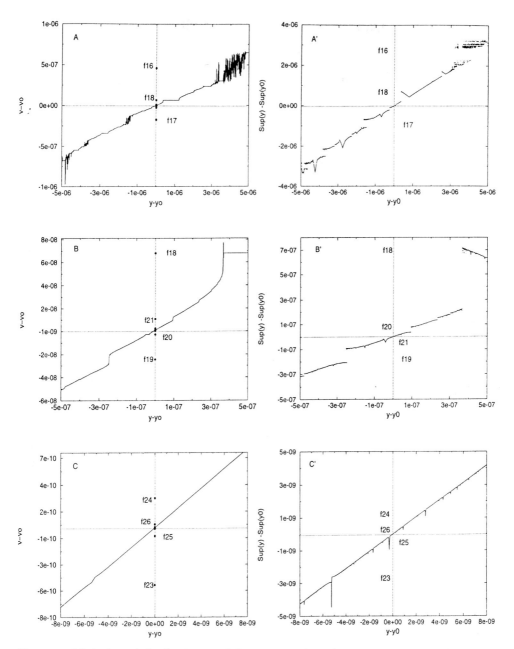

Figure 1. Variation of the fundamental frequency ν (a,b,c) and of the sup of the action (a',b',c') for the standard mapping, with $\epsilon = 0.9715$, in the vicinity of the golden rotation number ν_o which corresponds to the origin of the axes. The Fibonacci terms are indicated on each figure by the set of points f_i.

regions and islands. The situation changes drastically in Figs.1 b, b': the noisy variation of ν corresponding to strong chaotic regions, as well as the fuzzy plateau on the sup, has disappeared; islands and crossing of hyperbolic points are still there, but their relative measure in the action variables is now definitely smaller than the relative measure of tori. This phenomenon is strongly enhanced in the last magnification: up to a step size of $\Delta y = 1.6\,10^{-11}$ (Fig. 1 c) we only see one hyperbolic point and a large continuous region of tori. It is clear that the magnifications represented in Figs. 1 b, c show a completely different regime and in Fig. 1 c only the chief torus and its "slaves" appear with a density which seems to be in full agreement with the prevision of the Morbidelli Giorgilli theorem.

The same conclusions hold when looking at the variation of the sup, in particular the change of regime from Fig.1 a' to Fig.1 c'. However we stress the fact that tiny islands still appear in Fig.1 c' showing the sensitivity of what we can call the sup-map (instead of the frequency-map). Of course the sup-map does not carry all the information contained in the frequency-map. This map allows in particular to detect characteristic sequences of rational frequencies as well as the location of the golden torus. We think that the two maps should be used as complementary tools.

We have indicated on each plot the values of frequencies corresponding to the Fibonacci sequence, i.e. the set of the successive terms obtained when developing the golden number through the continued fraction process. In order to test the exponential decrease of the volume occupied by the complement of the set of tori (V_c), as a function of the distance (ρ) to the chief torus, we have measured the size of the Fibonacci islands. Such islands are the largest ones and therefore they fill the major part of V_c.

Figure 2 a shows the variation of the size of the Fibonacci islands as a function of the distance ρ in a log-log diagram. The distance ρ is the absolute value: $|y_c - y_o|$ where y_c is the action of the center of the island and y_o is the action of the golden torus. In the Morbidelli-Giorgilli regime we have also taken into account the hyperbolic points corresponding to the Fibonacci chain of islands. We have made an estimate of the dimension of the corresponding islands through the jump in frequency which occurs when crossing the hyperbolic point.

On Fig.2 a the change of regime is drastic: at the 19th term of the Fibonacci sequence we enter in the Morbidelli-Giorgilli regime where the size of the perturbation decreases exponentially with the distance from the golden torus. Let us emphasize that the measure has been done for a value of the perturbation parameter: $\epsilon = 0.9715$ very close to the critical one: $\epsilon_c = 0.971635$. When approaching ϵ_c the exponential part bends to the linear regime. Actually when $\epsilon = \epsilon_c$ we detect, up to the precision of our computation, only the linear regime as shown on Fig.2 a for which the points

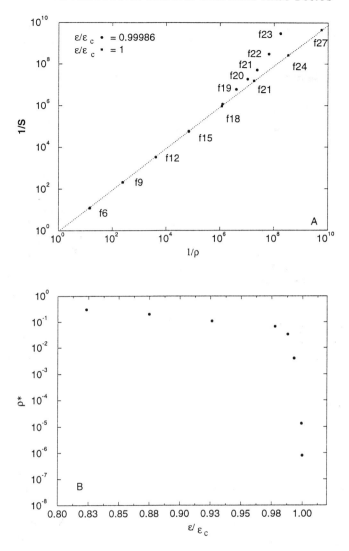

Figure 2. Variation of the size of the Fibonacci islands as a function of the distance to the golden torus. b) Variation of the threshold distance ρ^* as a function of the perturbing parameter ϵ/ϵ_c.

are equally spaced in a log-scale. This seems to corresponds to the already known (Greene 1981, Contopoulos 1993) scale invariance at ϵ_c.

Using the same technique we have estimated the distance $\bar{\rho}^*$, at which we enter in the Morbidelli-Giorgilli regime, for a set of different values of the perturbing parameter. Fig.2 b shows our result on the variation of $\bar{\rho}^*$ as a function of ϵ/ϵ_c. After a linear decrease of $\bar{\rho}^*$, up to $\epsilon = 0.95$, we observe a

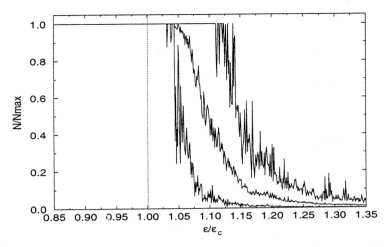

Figure 3. Variation of the number of iterations $N/Nmax$ necessary for a set of 36 points in the neighborhood of the origin to reach $|y| = 10$. Left curve: minimum value of $N/Nmax$, central curve: mean value, right curve: maximum value.

sharp drop of $\bar{\rho}^*$ up to $\bar{\rho}^* = 8\,10^{-7}$ for $\epsilon = 0.9715$. It seems therefore that all the slave tori disappear at once, like $\exp(-\epsilon_c/(\epsilon_c - \epsilon))$ when approaching the critical value ϵ_c.

3.2. DIFFUSION REGIME AFTER THE DISAPPEARANCE OF THE GOLDEN TORUS

In principle when ϵ becomes greater then ϵ_c all KAM tori have disappeared and therefore no barrier prevent the diffusion of the action from the chaotic zone associated with the hyperbolic point $(0,0)$ all over the phase space. For the same set of initial conditions taken in the neighborhood of the origin we have made $N_{max} = 10^6$ iterations increasing the perturbing parameter from $\epsilon/\epsilon_c = 0.85$ to $\epsilon/\epsilon_c = 1.35$. Fig.3 shows the variation of the mean number of iterations (N/N_{max}) necessary for the set of points to reach $|y| = 10$. The minimum as well as the maximum number of iterations for the set of points are also plotted. We observe a sharp decrease of N/N_{max} for values of $\epsilon_d/\epsilon_c > 1.03$ (ϵ_d stands for the perturbing parameter at which diffusion through all the phase space occurs).

This behaviour is very similar to what is observed for higher dimensional mapping when going from Nekhoroshev regime to a resonance overlapping-one. Of course ϵ_d will move to ϵ_c when increasing the number N_{max}. However let us remark that even for 10^6 iterations the difference between ϵ_d and ϵ_c is quite large.

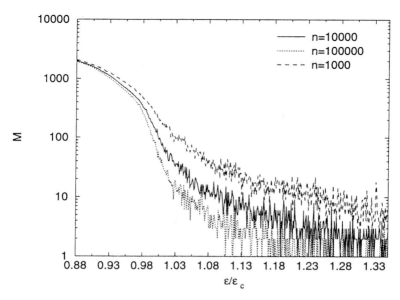

Figure 4. Variation of the number of steps M necessary for the pseudo-orbit to reach the value of the action $y = 10$. The jump Δy is equal to 10^{-3} and the number n of iterations changes from one curve to the other by a factor 10 as indicated.

In order to get information, not only on the range of ϵ, but even on the size of the phase space where such a barrier is effective, we computed as a function of ϵ a pseudo-orbit defined as follow:

- Starting at an initial condition P_0 we consider the sup$\{P_0\}$ of the action after n iterations.

- Then the new initial condition is sup$\{P_0\} + \Delta y$ and the process is iterated until the action reaches the value $y = 10$ (as taken for the previous case).

The number of steps M necessary to reach $y = 10$ gives a measure of the difficulty to get through the barrier which remains after the golden torus breaks-up. Fig.4 and 5 show respectively the variation of M for different values of n and $\Delta y = 10^{-3}$ and the variation of M for $n = 10^4$ and different values of Δy.

Fig.4 shows the decrease of M when ϵ/ϵ_c increases with a region of steep decline for ϵ/ϵ_c going from 0.95 to 1.05. This feature repeat for any value of n. The curves are shifted to the left becoming closer as n goes from 10^3 to 10^6. This comfort our feeling about the existence of a kind of limiting curve of the same shape when $n \to \infty$. We can therefore analyze the mapping with this jump-frog method concentrating now the attention on different values of Δy.

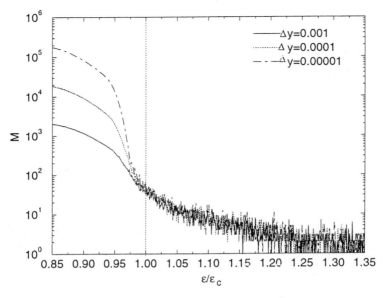

Figure 5. Same as Fig.4. The umber n of iterations is $n = 10^4$ while Δy changes from one curve to the other by a factor 10 as indicated.

Fig.5 shows for three different values of Δy ($\Delta y = 10^{-3}, 10^{-4}, 10^{-5}$), the variation of M with ϵ/ϵ_c. Again we have steep decline curves with noisy queues, whose separation goes from 10 to zero when increasing ϵ/ϵ_c. For $\epsilon/\epsilon_c = 0.85$ it is clear that the majority of steps concerns the passage through invariant curves since $M_{\Delta y/10} \simeq 10\, M_{\Delta y}$. The three curves almost superpose for $\epsilon/\epsilon_c = 0.98$. This is less than the critical value but corresponds exactly to the sharp drop observed on Fig.2 b showing the sudden break-up of all slave tori when approaching the critical perturbing parameter. This kind of analysis, sensitive to the dynamics of the break-up of tori is therefore complementary to the previous one which gave us results on the barrier to diffusion after the break-up of all the invariant tori.

4. Conclusion

The breakthrough provided by the Morbidelli-Giorgilli theorem seems, at first glance, to be out of the range of the numerical experiments. This is still our opinion concerning the super-exponential character of the diffusion. Using both the frequency map and the sup-map analysis we have first confirmed the existence of a chief torus surrounded by slaves. Then we have shown that such a structure exists even for values of the perturbing parameter close to the one for which no KAM tori survive. Figure 2 b shows the variation of the threshold distance $\bar{\rho}^*$ as a function of the perturbing parameter. The behavior observed for $\bar{\rho}^*(\epsilon)$ might deserve further theoretical studies.

We have also studied the passage through the critical value using pseudo-orbits generated by the jump-frog method. It appears clearly that the diffusion does not occur easily after the disappearance of the last KAM torus but there exist in the neighborhood of ϵ_c a sizable interval in action corresponding to the existence of cantori-like structures.

References

Arnold, V.I. (1963): "Proof of a Theorem by A.N. Kolmogorov on the invariance of quasi-periodic motions under small perturbations of the Hamiltonian", *Russ. math. Surv.*, **18**, 9.

Celletti, A. (1990): "Analysis of resonances in the spin-orbit problem in Celestial Mechanics: the synchronous resonance (Part I)", *J. of Applied mathematics and Physics (ZAMP)*, **41**, 174.

Froeschlé, C. (1992), Mapping in Astrodynamics, in *Chaos, Resonance and Collective dynamical phenomena in the solar system*, S.Ferraz-Mello (ed.) Kluwer Academic Publishers, 375-390.

Froeschlé, C., Froeschlé, Ch. and Lohinger,E. (1993): "Generalized Lyapunov characteristic indicators and corresponding Kolmogorov like entropy of the standard mapping", *Celestial Mech.* , **56**, 307-314.

Froeschlé, C. and Lega, E. (1995): "Polynomial approximation of Poincaré maps for Hamiltonian systems", *Earth, Moon and Planets*, in press.

Hadjidemetriou, J.D. (1991): *"Mapping models for Hamiltonian Systems with applications to resonant asteroid motion"*, in Predictability, Stability and Chaos in N-body Dynamical Systems, A.E. Roy (ed.), 157-175, Kluwer Publ.

Hadjidemetriou, J.D. (1992): "The elliptic restricted problem at the 3:1 resonance", *Celest. Mech.*, **53**, 153-181.

Hadjidemetriou, J.D. (1993a): "Asteroid motion near the 3:1 resonance", *Celest. Mech.*, **56**, 563-599.

Hadjidemetriou, J.D. (1993b): "Resonant motion in the restricted three body problem", *Celest. Mech.*, **56**, 201-219.

Henrard,J. and Morbidelli,A. (1993): "Slow crossing of a stochastic layer", *Physica D*, **68**, 187-200.

Hénon,M. (1969): "Numerical study of quadratic area-preserving mapping", *Quarterly of Applied Mathematics*, **27**, 292-306.

Kolmogorov, A.N (1954): "On the conservation of conditionally periodic motions under small perturbation of the Hamiltonian", *Dokl. Akad. Nauk. SSSR*, **98**, 469.

Laskar, J., Froeschlé, C. and Celletti, A. (1992): "The measure of chaos by the numerical analysis of the fundamental frequencies", *Physica D*, **56**, 253-269.

Lega, E. and Froeschlé, C. (1995): "Numerical measures of the structure around an invariant KAM torus using the frequency map analysis", *submitted to Physica D*.

Lohinger,E. and Froeschlé C. (1993): "Fourier analysis of local Lyapunov characteristic exponents for satellite-type motions", *Celestial Mech.*, **57**, 369-372.

Morbidelli, A. and Giorgilli, A. (1995): "Superexponential Stability of KAM tori", *J. Stat. Phys.*, **78**, 1607-1616.

Petit, J.M. and Froeschlé,C. (1994): "Polynomial approximations of Poincaré maps for Hamiltonian systems. II", *Astron. Astrophys.*, **282**, 291-303.

Voglis, N. and Contopoulos,G.J (1994): "Invariant spectra of orbits in dynamical systems", *J. Phys. A: Math. Gen.*, **27**, 4899-4909.

SOME COMMENTS ON NUMERICAL METHODS FOR CHAOS PROBLEMS

R. H. MILLER
Astronomy Center, University of Chicago,
5640 Ellis Ave., Chicago 60637, USA

Abstract. Hamiltonian systems with chaotic regions are particularly slippery to treat numerically. Numerical treatments can introduce nonphysical features. Simple examples illustrate some of the pitfalls. Integer, or discrete, arithmetic is a favorite "workaround." While it does not cure chaos, it clarifies the interaction of computational methods with the underlying mathematical structure.

Be forewarned: I won't give any prescription that is guaranteed to give a good and reliable method to handle chaotic problems numerically. Instead, I'll stress a few of the concerns and describe one or two pitfalls.

Key words: Chaotic motion – numerical methods – discrete arithmetic

1. Early Concerns

The original discovery that trajectories in the full $6n$–dimensional phase space (Γ–space) of the gravitational n–body problem separate exponentially with the time was reported 30 years ago (Miller 1964). The separation e–folds amazingly fast—in times short compared to a typical orbital period. Some of the history and motivation were recalled at the Geneva conference (Miller 1994).

Today, we would describe the problem as chaotic, but that term was not popular back then. We would call the e–folding rate of trajectory separation a Lyapunov number today, although the fact that the phase space is unbounded means that, technically, it is not quite correct to use that term. The process is sometimes referred to as an "initial value instability," indicating that the results of a computation depend sensitively on initial conditions.

As an experimenter, I was concerned whether the results were reliable, even though I argued that the effect was physical. It was a reassuring when it was confirmed by others. So far as I am aware, Myles Standish's PhD thesis (1968) was the earliest confirmation. Lots of others have confirmed it since. Later, I found that Krylov (1979) had described the physical effect much more elegantly. Standish showed that softened interactions between the particles decrease the e–folding rate, confirming that the strong forces in near encounters make a major contribution to trajectory separation. Softening means replacing the $1/r$ interaction between pairs of particles by $(1/\sqrt{r^2 + b^2})$; b is called a softening length.

Celestial Mechanics and Dynamical Astronomy **64**: 33–42, 1996.
© 1996 *Kluwer Academic Publishers. Printed in the Netherlands.*

Progress in chaos came from finding the simplest problems that exhibit the effect, and then pursuing developments toward ever more complicated systems. My concern was with the dynamics of star clusters. While clusters show the exponential trajectory separation symptomatic of chaos, star clusters are much too complex to provide a good ground to study the phenomenon of chaos. I missed the chance to make some great discoveries by concentrating on stellar dynamical problems.

But the question what the phenomenon of chaos implied for numerical studies of star clusters and other stellar dynamical systems was opened, and it remains a serious question even today. Exponential trajectory separation was found while seeking a computationally efficient way to study the failure of microscopic reversibility in numerically computed systems. The Liouville theorem is probably lost in these computed systems as well.

On the other hand, this is a *microscopic* irreversibility, while the computed systems *look* perfectly reasonable. Exponential trajectory separation need not keep computed systems from representing reasonable macroscopic wholes.

Some properties are well computed in spite of trajectory separation. Energy, centroid position, centroid momentum, and angular momentum (the usual ten first integrals of the system) are pretty reliably computed. The number of particles in the system is reliably computed.

On the other hand, some macroscopic properties are very poorly computed. Lecar (1968) provided the most dramatic examples. He compared results from eleven different workers, each of whom integrated the same 25-body system, starting from identical initial conditions.

This raises a troubling question: some quantities are computed reliably, some not. How do you know about the quantity you want to study? Nobody would undertake an n-body integration to verify that energy is conserved. In spite of pleas to the numerical analysts for help with this problem (Miller 1974), the only recipe that has been suggested to date is to run a number of similar integrations, taking statistics on the scatter of results (Miller 1967). This approach has been particularly emphasized by Haywood Smith (1979), who argued, "In general it appears to be much more efficient to calculate a large number of examples at relatively small n than one or two at large n; results of the same accuracy can be obtained with much less cost of computer time."

But the real concern was that the trajectory of a computed system might explore the phase space in a different way from a mathematical trajectory. If so, time averages over computed systems need not represent proper ensemble averages over the corresponding physical systems. The picture we have today that the phase space is carved into large numbers of metrically transitive regions, some integrable and some not, had not yet been worked out at the time, so the question is better defined (but still unanswered) today.

An example comes with the binaries frequently formed in n−body calculations, but seldom found in open star clusters (open clusters may contain a few hundred stars; the situation seems to be different with globular clusters that may have 10^4 to a few million stars). Might the calculations be giving us incorrect estimates of rates of binary formation because calculated trajectories explore the phase space differently from a true physical trajectory? Heggie (1975) argued that there are "hard" and "soft" binaries. The "hard" binaries are tightly bound and become ever more tightly bound by energy exchanges with passing stars, while "soft" binaries are loosely bound and soon broken up. In this picture, a "hard" binary could, in principle, represent an escape trajectory and change the topology of the phase space. Do we map that property correctly with practical calculations? These simple questions remain unanswered.

The ten first integrals define (at least for short times) a hypersurface of $(6n - 10)$ dimensions in the phase space. Differences between a computed system and the mathematically correct solution may be approximated by studying differences between two computed systems, as is usually done in estimating rates of trajectory separations. Suppose each of these systems has the same values of all ten first integrals. The difference vector between the two systems can explore the $(6n - 10)$−dimensional hypersurface reasonably freely, but it is more tightly constrained normal to the surface. Numerical checks show that this is, in fact, the case (Miller 1971a). The difference vector tends to lie parallel to the trajectory. It swings to and fro, alternately lying forward along the trajectory and then backward. Two systems whose initial separation lies along the trajectory have a difference-vector that retains its direction along the trajectory in the mathematical formulation of the problem. It does not in the numerical calculation.

One might imagine trying to improve the calculation by constraining the first integrals exactly. This has been done, but it doesn't help with the exponentiation. Trajectories separate exponentially, at nearly the same rate, even if both systems are constrained to the hypersurface (Miller 1971b) (but see Nacozy 1971 for a different opinion). Further, the possibility of using the first integrals as a check on the integration has been lost. The arguments of Greenspan (1973, 1974) and of Marciniak (1985) that energy-conserving difference schemes will save the day for the gravitational n−body problem are flawed because of this property of the chaotic solutions.

The possibility of confining the system to the "integral hypersurface" on which the first integrals have pre-assigned values suggests a variant on n−body integrations. An evolutionary calculation might be mimicked in a Monte Carlo sense by making random small changes in the phase coordinates, adjusting the system at each sampling to return it to the integral hypersurface. While the notion is attractive from the point of view of investigating properties of the phase space, this does not seem to produce a useful replacement for $n-$body integrations. The amount of calculation required to force the system onto the integral hypersurface is as great, or possibly even greater, than that required to do a careful integration in the first place.

2. Attempts to Cope with Chaos

We attempted to cope with chaos in the gravitational $n-$body problem by reformulating the problem to ensure exact reversibility and a Liouville theorem. The new formulation was described as a "game" which is played on a piece of graph paper (Miller and Prendergast 1968). A point moves over the graph paper according to the following rules.
(1) It moves between locations whose coordinates, (x, u) are integers.
(2) It may move from one allowed location to another, but it must always alternate a move in which only x changes with one in which only u changes. Let the value of x immediately following the nth step be $x^{(n)}$, and let the value of u following its next change after $x^{(n)}$ was changed be $u^{(n+1/2)}$.
(3) The value of $x^{(n+1)}$ is given by

$$x^{(n+1)} = x^{(n)} + Tu^{(n+1/2)}. \tag{1}$$

(4) A table gives the rule for changing u according to the present value of x. Let the value read out of this table following the nth step be $Tf^{(n)}$ (an integer). Then the value of $u^{(n+1/2)}$ is given by

$$u^{(n+1/2)} = u^{(n-1/2)} + Tf^{(n)}. \tag{2}$$

The pair of moves makes up a complete step. The constant, T, appearing in these equations is dimensionally a time. It is normally taken to be unity and not written explicitly. An example, approximating a harmonic oscillator with $f = -x$, was worked out in detail in the 1968 paper. Its trajectory closes in six complete steps, and it would retrace those same phase locations forever thereafter.

It is clear that the system is reversible with appropriate care about the times to which x and u refer. Prendergast derived the scheme from a variational principle, which provides a simple and direct means to build time-reversible difference schemes. The scheme represented by Eq's (1) and (2) is known today as the time centered leapfrog. It is well known to be symplectic, which also follows automatically from the suitable variational principle.

Suppose many points on the lattice are occupied, the occupants of each moving about according to these rules without regard for occupants at the other points. Occupants of each point on a given row at $u = $ const are moved the same distance to the right by the value of u. Similarly, occupants at each point in a given column at $x = $ const are moved up or down by f, each particle moving the same distance. Even if the table of f's changes from step to step, it is clear that occupants of no two points on the lattice ever map onto one point after a complete step, and it is equally clear that the contents of one point never split and wind up occupying two points. Occupants move from point to point as the game proceeds, but the *same* occupants remain together and the number of occupants is strictly constant. Constant numbers of occupants is the content of the Liouville theorem, here applied to a lattice. Statement of the Liouville theorem in terms of Lebesgue measure includes this case. We have strict measure preservation.

The content of the collision-free Boltzmann equation is contained in a shearing of the phase space in one direction followed by a shearing in the other direction. The number of occupants of any given phase cell does not change as the image of that cell moves about the phase plane. Extension to two or three spatial dimensions (to four or six phase dimensions) is immediate. In computer terms, the Boltzmann equation is integrated by a set of shifts in the computer registers. Our first formulation, as described in the 1968 paper, actually made use of shifts. Two additional features were introduced into the gravitational n–body problem in that paper: representation by a different data structure, and representation as a kind of cellular automaton. It is sometimes useful to think of the problem in these terms.

Systems with zero or one occupant per phase cell are often described as "particle pushing." Workers in plasma physics have allowed several occupants; the process is then described as "distribution pushing." Nearly full spaces could represent a Fermi-Dirac gas, or some of the particle physicist's para-statistics systems. Extensions of the idea to fluid dynamics led to a method that has been called the "beam scheme."

The resulting calculations have some interesting properties which arise principally from the point of view concerning the relations between the underlying physics and the computer model. The physics is expressed in certain invariance properties which are exactly incorporated into the model. An exactly calculable model results, a model that is pleasing to work with because of the feeling of precision that it affords. We have gone directly from the physics to the computer, without the intermediary of some differential equations.

A batch of particles moves about in response to the forces they feel. The physics of this problem is the batch of particles and the forces generated by their interactions. It does not live in the differential equations, which at best approximate what is going on. Then the differential equations are further approximated in the computer.

Not all the physical invariance properties can be built into a calculation; a matter of taste is involved in the selection of those deemed most important. We have stressed a Liouville theorem and reversibility over things like energy conservation. Marciniak (1985) argues for using difference schemes which conserve energy exactly. But there is more physics in the gravitational $n-$body problem than energy conservation.

Stellar systems occupy the phase space very sparsely, however, and this suggests reverting to representations closer to the traditional formulations of $n-$body problems. This can be done while retaining the pleasing properties of the calculation. Calculations are also nicely balanced in two senses: the accuracy with which forces are represented matches the lattice spacing, and the number of computer cycles used in calculating forces is about the same as that used in pushing particles.

We used a rather coarse carving of the phase space into a lattice, but the same principles apply with arbitrarily fine carvings, provided that permitted phase locations are constrained to be integer multiples of some basic length unit. It must be recalled that all numbers that can be represented in a computer are rational—they are a tiny subset of the rationals. So any computer representation is always on a lattice like that considered here—the difference is that it need not be a uniformly spaced lattice. Measure preservation is lost with floating-point arithmetic. Whenever a boundary is crossed that leads to a change in the exponent in a computer representation, it is possible either to merge occupants of several phase points onto one phase point, or a split is suggested. We thought a bit about what measure preservation implied numerically (Miller 1971), but Peter Lax (1971) provided a more precise discussion.

The real advantage of this formulation turned out to be that it allowed computations with reasonable numbers of particles. We used 100 000 to something over a million around 1968. We used it to represent point particles moving on a plane, to study disk galaxies, in some papers around 1969 to 1970. Some of the numerical properties of the difference scheme, and the general formulation of the potential solver in terms of Fourier transform methods (using Ewald summation methods for periodic spaces) were explored in more detail in Miller (1970). Interactions are softened as in §1, of course, to allow potentials to be represented on the lattice. The softening length must be on the order of the lattice spacing, but often is considerably larger.

The sad thing is that, while providing a comforting feeling that at least we tried, all this effort did little to help cope with chaos.

2.1. INTEGER ARITHMETIC

The virtues of using integer arithmetic for certain computations, including chaotic systems, have been rediscovered by several workers. Those that I know of include Rannou (1974), several papers by C. F. F. Karney of Princeton Plasma Physics (the most recent that I know of is Karney 1986), and some papers by Earn and Tremaine (1991, 1992) and by David Earn (1994).

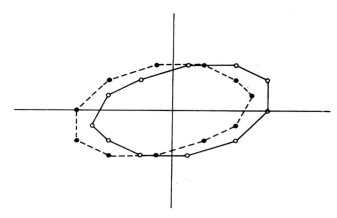

Figure 1. Phase Trajectories of a Harmonic Oscillator using Integer Arithmetic. The system is not ergodic.

One gets a sense of déjà vu on reading these papers. Karney (1986) gives some useful prescriptions, especially on how to decompose a rotation into two successive shears to map integer coordinates into integer coordinates while preserving measure. This is not all that easy with integer lattices laid out according to cartesian coordinates. Greenspan (1973, 1974) also argued for treating computer calculations as discrete, implying integers.

There are some consequences, not all evident, that follow from using integer arithmetic. Most follow for any computed system because of the limited ability to represent numbers. Some are described here.

The continuity properties usually associated with the Liouville theorem no longer apply. A "contiguous" region in the phase space quickly develops a diffuse boundary when numeric values are confined to a lattice. The thickness of the diffuse boundary grows as a calculation proceeds, and it quickly reaches several lattice spacings.

It is sometimes claimed that the use of integer arithmetic eliminates roundoff, or that the results are exact. Neither is true. A point on an integer lattice will not, in general, map to an integer point under some kind of mathematical operation. Roundoff is the process of forcing it back to a value the computer is capable of representing. And the process cannot be exact if the point has been moved. What integer arithmetic does, when imposed on most calculations, is to reduce the relative measure of the fuzzy regions in which it may be uncertain which neighboring representable point the computer will choose. It does this by coarsening the level of numeric representations, so the fuzzy regions look smaller. But the shift to a representable value, which is the essence of roundoff, remains. These comments do not apply if the calculation is designed so only integers appear, as in the "game" of §2.

Several interesting anomalies arise with the use of integer arithmetic. One is illustrated in Fig. 1, which shows the phase space of a harmonic oscillator which is periodic in 11 integration steps. The spring constant is small enough that the force rounds to the same integer value at several neighboring lattice points. This allows disjoint orbits, each periodic with the same period and each with the same "energy." Two of these orbits are shown. A time average over one yields a different centroid position from a time average over the other, while an ensemble average gives an average of the two. The system is not ergodic. The error arising from the failure to be ergodic is on the order of the graininess of the integer representation.

3. Difference Schemes can Make a Difference

An unfortunate choice of difference schemes can make a qualitative difference and yield misleading results. Fuller details are given for the following example in Miller (1991). Some years ago, a "simple predictor-corrector" was suggested for numerical studies in galaxy dynamics. It runs:

$$
\begin{aligned}
x_p &= x_n + T u_n, \\
u_{n+1} &= u_n + \frac{T}{2}(f(x_n) + f(x_p)), \\
x_{n+1} &= x_n + \frac{T}{2}(u_n + u_{n+1}) = x_p + \frac{T^2}{4}(f(x_n) + f(x_p)).
\end{aligned}
\tag{3}
$$

Values at the old timestep are denoted by n, those at the new by $n+1$, and at the predicted position by p, and T is used for the integration timestep. Perceived advantages for this scheme are that velocities and positions are given at the same time, unlike the leapfrog (but the leapfrog can be rearranged to do this as well).

This method troubled me for several years, but when symplectic methods became popular, I checked it to see if it were symplectic. It is, so long as the forces are no more complicated than a harmonic oscillator. Something more complicated is needed to test it. The simplest physical system more complicated than a harmonic oscillator is a physical pendulum, a pendulum that may rotate clear around its pivot point.

The force for the physical pendulum may be written, $T^2 f = K \sin \theta$ with $K > 0$, so $\theta = 0$ when the pendulum points upward. Change the notation a bit more, using I_n for $T u^{(n-1/2)}$ and so on, and the leapfrog (Eq's (1) and (2)) takes the form,

$$
\begin{aligned}
I_{n+1} &= I_n + K \sin \theta_n, \\
\theta_{n+1} &= \theta_n + I_{n+1}.
\end{aligned}
\tag{4}
$$

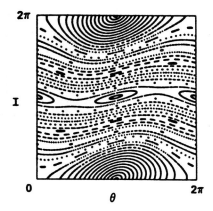 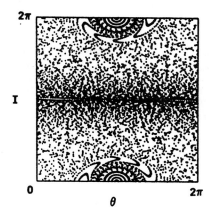

Figure 2. (Left Panel) Standard Map for $K = 0.5$. (Right Panel) Map produced by the "predictor-corrector" of Eq. (5) for the same value of K. Reprinted from *Journal of Computational Physics* by permission of the Publisher.

This system will be recognized as the standard, or Chirikov, map that has appeared several times already at this workshop.

The predictor-corrector difference equations for the physical pendulum read:

$$\theta_p = \theta_n + I_n,$$
$$I_{n+1} = I_n + \frac{K}{2}(\sin\theta_n + \sin\theta_p), \tag{5}$$
$$\theta_{n+1} = \theta_n + \frac{1}{2}(I_n + I_{n+1}).$$

Trouble lurks already in these equations. If $I = \pi$ then $(\sin\theta_n + \sin\theta_p) = 0$ identically so I is constant and $\theta_{n+1} = \theta_n + \pi$ for any value of θ_n whatsoever. Every point on the line, $I = \pi$, is a period-2 fixed point whatever the value of K. The line of fixed points cuts the phase plane into two parts. There are only two period-2 fixed points in the standard map.

This line of fixed points is an attracting set, as becomes evident from a map of the phase plane (Figure 2). This property can be confirmed algebraically (Miller 1991). Attracting sets are forbidden for bounded Hamiltonian systems like the physical pendulum. This unphysical feature was introduced by the difference scheme.

The moral is that one must be quite careful in treating these problems. Checks whether difference schemes are symplectic can be quite challenging, and it is even more challenging to design tests to see how a scheme acts when tested with potentials that push it beyond stages in which it is symplectic.

References

Earn, David J. D.: 1994, 'Symplectic integration without roundoff error', in V. G. Gurzadyan and D. Pfenniger, (eds.), *Proceedings of the Workshop on Ergodic Concepts in Stellar Dynamics*, Vol. 430 of Lecture Notes in Physics, Springer, Berlin, Heidelberg, pages 123–130.

Earn, David J. D. and Tremaine, Scott: 1991, 'Exact solutions for Hamiltonian systems using integer arithmetic', in B. Sundelius, (ed.), *Dynamics of Disk Galaxies*, S-412 96 Göteborg, Sweden, Department of Astronomy, Göteborgs University, pages 137–141.

Earn, David J. D. and Tremaine, Scott: 1992, 'Exact numerical studies of Hamiltonian maps. Iterating without roundoff error', *Physica D* **56** 1–22.

Greenspan, D.: 1973, *Discrete Models*, Addison-Wesley, New York.

Greenspan, D.: 1974, *Discrete Numerical Methods in Physics and Engineering*, Academic Press, New York.

Heggie, D. C.: 1975, 'Binary evolution in stellar dynamics', *MNRAS* **173** 729–787.

Karney, Charles F. F.: 1986, 'Numerical techniques for the study of long-time correlations', *Particle Accelerators* **19** 213–221.

Krylov, N. S.: 1979, *Works on the Foundations of Statistical Physics*, Princeton University Press, Princeton, N.J.

Lax, P. D.: 1971, 'Approximation of measure preserving transformations', *Comm.Pure&Appl.Math.* **24** 133–135.

Lecar, M.: 1968, 'A comparison of eleven numerical integrations of the same gravitational 25–body problem', *Bull.Astron.(Paris), Ser. 3*, **3** 91–104.

Marciniak, A.: 1985, *Numerical Solutions of the n−Body Problem*, D. Reidel, Boston.

Miller, R. H.: 1964, 'Irreversibility in small stellar dynamical systems', *ApJ* **140** 250–256.

Miller, R. H.: 1967, 'An experimental method for testing numerical stability in initial-value problems', *J. Comp. Phys.* **2** 1–7.

Miller, R. H.: 1970, 'Gravitational n-body calculation in a discrete phase space', *J. Comp. Phys.* **6** 449–472.

Miller, R. H.: 1971, 'Experimental studies of the numerical stability of the gravitational n-body problem', *J. Comp. Phys.* **8** 449–463.

Miller, R. H.: 1971, 'Partial iterative refinements', *J. Comp. Phys.* **8** 464–471.

Miller, R. H.: 1974, 'Numerical difficulties with the gravitational n-body problem', in D. G. Bettis, (ed.), *Proceedings of the Conference on the Numerical Solution of Ordinary Differential Equations*, Vol. 362 of Lecture Notes in Mathematics, Springer, Berlin, pages 260–275.

Miller, R. H.: 1991, 'A horror story about integration methods', *J. Comp. Phys.* **93** 469–476.

Miller, R. H.: 1994, 'Core motions and global chaotic oscillations', in V. G. Gurzadyan and D. Pfenniger, (eds.), *Proceedings of the Workshop on Ergodic Concepts in Stellar Dynamics*, Vol. 430 of Lecture Notes in Physics, Springer, Berlin, Heidelberg, pages 137–150.

Miller, R. H.: 1971, 'On the computation of measure-preserving transformations', Quarterly Report No. 31, Institute for Computer Research, University of Chicago, November 1, 1971, (unpublished).

Miller, R. H., and Prendergast, K. H.: 1970, 'Stellar dynamics in a discrete phase space,' *ApJ* **151** 699–709.

Nacozy, Paul E.: 1971, 'The use of integrals in numerical integrations of the n−body problem', *Ap&SS* **14** 40–51.

Rannou, F.: 1974, 'Numerical study of discrete plane area-preserving mappings', *A&A* **31** 289–301.

Smith, Haywood: 1979, 'The dependence of statistical results from n−body calculations on n', *A&A* **76** 192–199.

Standish, E. M.: 1968, *Numerical Studies of the Gravitational Problem of n Bodies*, PhD thesis, Yale University.

INFORMATION ENTROPY
An Indicator Of Chaos

J. A. NÚÑEZ, P. M. CINCOTTA and F. C. WACHLIN
PROFOEG - Facultad de Ciencias Astronómicas y Geofísicas, Paseo del Bosque (1900)
La Plata, Argentina.

Abstract. In this paper we present an indicator of chaos that takes advantage of the Information Entropy concept. We develop the mathematical formulation and test the results of its application with those obtained by other methods for the 2D Hénon–Heiles system and the 3D Contopoulos, Galgani and Giorgilli potential.

Key words: Information entropy – Chaos

1. Introduction

Although "chaos" is now undergoing an exponential development in many subjects, some references to it may be found in the remote past. Let us say, for instance, that already in 1873 had Maxwell (Maxwell 1991) addressed concepts like "instabilities" and "sensitivity to initial conditions". Moreover, Poincaré is often referred to as the "father" of chaos. However, people began to think seriously in problems like stability, predictability and chaos only a few decades ago. In fact, it was in the early sixties when Moser (1962, 1963) demonstrated the KAM theorem. Since then, a lot of effort has been devoted to develop the mathematical tools for coping with chaos.

In particular, the history of chaos within Stellar Dynamics is rather short, although Hénon had already introduced it in his famous paper (Hénon and Heiles 1964). Probably, the "large time-scale" for the growth of chaos in Stellar Dynamics is due to the wide-spread wrong idea that chaotic motion was irrelevant in stellar systems. Among others, Merritt and Fridman (1995) -MF95 hereafter- showed that in strongly triaxial stellar systems the phase space structure is dominated by chaos. Also Cincotta *et al.* (1995a,b) showed, in a somewhat "naive" way, that chaotic motion plays an important role in the dynamics of real stellar systems.

One of the matters that still remains open is that of finding a "good indicator of chaos". The ones most frequently used are the well-known Lyapunov exponents but, in some cases there exist limitations for their computation, as claimed out by many authors. In particular, when we are interested in the orbital structure of a stellar system, we need some other technique for computing the degree of stochasticity of a large number of orbits (see for example, MF95, Cincotta *et al.* 1995a). Another popular method is the Poincaré

Surface of Section, though useful only for 2D systems; a kind of generalization of this technique to 3D systems seems to be a poor indicator of chaos
(again, see MF95).

Laskar (1990, see also Laskar, Froeschlé and Celletti, 1992) has developed
a new technique called "frequency analysis" which should help to understand
the global structure of the orbits in a variety of stellar systems. But, up to
now, this method has been mainly applied to the solar system (Dumas and
Laskar, 1993) and it is barely beginning to be used when dealing with stellar
systems (Papaphilippou and Laskar, 1995).

In this paper we present what we call the Information Entropy (IE hereafter) as an indicator of chaos. We take from Fraser and Swinney (1986)
some aspects of the mathematical background but we use it in a different
context. The paper is divided in two: in the first part (Section 2) we develop
the mathematical formalism of the IE and, in the second one (Section 3) we
give two examples of its application. In particular, we compare the results
given by the IE for the Hénon-Heiles potential with those obtained by per-
forming the Poincaré Surface of Section. We also compute the IE for a 3D
hamiltonian system, namely the one given by Contopoulos (1978).

2. The Method

Let us consider a manifold M provided with a measure μ. Let A be the set
$\{a_i, i = 1, ..., n\} \subset M$. Following Fraser and Swinney (1986), we define the
entropy of A as:

$$S(A) = - \sum_{i=1}^{n} \mu(a_i) \ln \mu(a_i). \tag{1}$$

Consider another set $B = \{b_i, i = 1, ..., n\} \subset M$. The conditional entropy of
A relative to b_j is given by

$$S(A/b_j) = - \sum_{i=1}^{n} \mu(a_i/b_j) \ln \mu(a_i/b_j), \tag{2}$$

where

$$\mu(a_i/b_j) = \frac{\mu(a_i, b_j)}{\mu(b_j)} \tag{3}$$

is the conditional probability of a_i for a given b_j, and $\mu(a_i, b_j)$ the joint
probability of a_i and b_j. Thus, the conditional entropy of A relative to B is
defined as the average value of $S(A/b_j)$:

$$S(A/B) = \sum_{j=1}^{n} \mu(b_j) S(A/b_j). \tag{4}$$

From (2), (3) and (4) it follows

$$S(A/B) = - \sum_{i,j} \mu(a_i, b_j) \ln \mu(a_i, b_j) + \sum_{j} \mu(b_j) \ln \mu(b_j), \tag{5}$$

where the fact that $\mu(b_j) = \sum_i \mu(a_i, b_j)$ has been used. The first term in (5) is the joint entropy of both sets $S(A, B)$ and the second one is $-S(B)$. Then, defining the *information entropy* I as the symmetric conditional entropy $I(A, B) \equiv \frac{1}{2}[S(A/B) + S(B/A)]$, we obtain

$$I(A, B) = S(A, B) - \frac{1}{2}[S(A) + S(B)]. \tag{6}$$

It can be proved that

$$I(A, B) \geq 0 \qquad \forall A, B \tag{7}$$

and

$$I(A, B) = 0 \qquad if \ A = B. \tag{8}$$

The last result is a very important one, since it shows that I vanishes when both sets are equal. Thus, statistically speaking, we can state the mathematical results given by (7) and (8) as follows: $I(A, B) \approx 0$ *if A and B are strongly correlated*, while $I(A, B) > 0$ *whenever A and B are not correlated.*

Let us now discuss the physical meaning of these results. For that purpose, let us consider a hamiltonian system with N degrees of freedom and two nearby orbits whose initial conditions are $\mathbf{x_0}$ and $\mathbf{y_0} = \mathbf{x_0} + \delta$, where $\mathbf{x_0} = (\mathbf{q_0}, \mathbf{p_0})$ and $||\delta|| \ll 1$. Solving the equations of motion for both initial conditions, we obtain the two sets: $\mathcal{X} = \{\mathbf{x}(t), t \in U\}$ and $\mathcal{Y} = \{\mathbf{y}(t), t \in U\}$, $U \subset \mathbf{R}$. Thus, *if $\mathbf{x}(t)$ is regular, then $\mathcal{X}$ and $\mathcal{Y}$ will be strongly correlated for all t and $I(\mathcal{X}, \mathcal{Y}) \approx 0$.* On the other hand, *if $\mathbf{x}(t)$ is chaotic, then, for some value t_u, the sets $\mathcal{X}_d = \{\mathbf{x}(t), t > t_u\}$ and $\mathcal{Y}_d = \{\mathbf{y}(t), t > t_u\}$ will be uncorrelated and $I(\mathcal{X}_d, \mathcal{Y}_d) > 0$.*

In other words, for short times we expect the same behavior of the indicator I for both regular and chaotic orbits. On the other hand, for large values of t, we expect a positive value of I for chaotic orbits while $I \approx 0$ for regular ones. Then, we have found an indicator of chaos which should show the nature of the orbits earlier than either the Lyapunov exponents or the KS entropy.

In order to compute the measure μ, let us consider two alternatives:

a) Given the two sets $\mathcal{X}^i = \{x^i(t), t \in U\} \subset \mathcal{X}$ and $\mathcal{Y}^i = \{y^i(t), t \in U\} \subset \mathcal{Y}$, let $x^i(t)$ and $y^i(t)$ be the i-th component of $\mathbf{x}(t)$ and $\mathbf{y}(t)$, respectively. If we introduce a partition in the set $\mathcal{X}^i \times \mathcal{Y}^i$ of length ϵ and, if $n(\alpha, \beta)$ is the number of points within the cell (α, β) and $N = \sum_{\alpha, \beta} n(\alpha, \beta)$, we define the measure of the element (α, β) as

$$\mu_{\alpha\beta} = \frac{n(\alpha, \beta)}{N} \tag{9}$$

and the individual probabilities as $\mu_\alpha = \sum_\beta \mu_{\alpha\beta}$. Then, we obtain the quantity I^i for each component of $\mathbf{x}(t)$ and $\mathbf{y}(t)$. This way of computing the measure is widely used. Nevertheless, it is well known that such a mechanism of counting points within cells is not an efficient one (for more efficient alternatives see, for example, Cohen and Procacia 1985). We will not discuss this choice for the measure any further but consider an alternative definition which uses all the information involved in the phase space trajectory instead of just regarding one component of the phase space vector.

b) The length of the curves $\mathbf{x}(t)$ and $\mathbf{y}(t)$ should have information about the "degree" of correlation of both trajectories. In other words, we should expect the length of both orbits to be nearly the same for a regular orbit. Instead, due to the exponential divergence, the rate of length (some sort of speed) will be rather different for a chaotic orbit. Therefore, when the length of the partition goes to zero ($\epsilon \to 0$), we can define a probability density distribution as the normalized length per unit time: $\dot{\ell}_x, \dot{\ell}_y, \dot{\ell}_{xy}$, where

$$\dot{\ell}_x = \frac{|\dot{\mathbf{x}}|}{L_x}, \qquad \dot{\ell}_y = \frac{|\dot{\mathbf{y}}|}{L_y}, \qquad \dot{\ell}_{xy} = \frac{\sqrt{\dot{\mathbf{x}}^2 + \dot{\mathbf{y}}^2}}{L_{xy}}. \tag{10}$$

L_x and L_y are the total length of the curves $\mathbf{x}(t)$ and $\mathbf{y}(t)$, respectively, and L_{xy} is the total length of the parametric curve $[\mathbf{x}(t), \mathbf{y}(t)]$. Then we compute the entropies (up to time $t = T$) in the following way:

$$S(\mathcal{X}) = -\int_0^T \dot{\ell}_x \ln \dot{\ell}_x dt, \quad S(\mathcal{Y}) = -\int_0^T \dot{\ell}_y \ln \dot{\ell}_y dt, \tag{11a}$$

$$S(\mathcal{X}, \mathcal{Y}) = -\int_0^T \dot{\ell}_{xy} \ln \dot{\ell}_{xy} dt \tag{11b}$$

and, finally, the IE as

$$I(\mathcal{X}, \mathcal{Y}) = S(\mathcal{X}, \mathcal{Y}) - \frac{1}{2}[S(\mathcal{X}) + S(\mathcal{Y})]. \tag{12}$$

In order to test whether we obtain the expected results with the entropy thus defined, let us consider the asymptotic behavior of I:

i) *Regular orbits (and chaotic orbits for short times)*
It may be shown that

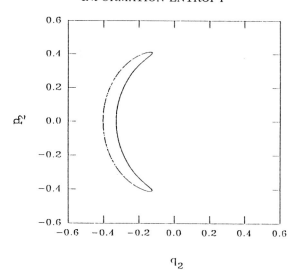

Figure 1. Poincaré surface of section of a regular orbit in the Hénon–Heiles system. Initial conditions are $q_1 = 0$, $q_2 = -0.33$, $p_2 = 0.02$ and energy= 0.125.

$$\dot{\ell}_y \approx \dot{\ell}_x + O(||\delta||) \qquad \dot{\ell}_{xy} \approx \dot{\ell}_x + O(||\delta||) \qquad (13)$$

and, as expected (due to the correlation between the two sets),

$$I(\mathcal{X}, \mathcal{Y}) \approx O(||\delta||) \qquad (14)$$

where $||\delta|| \ll 1$.

ii) Chaotic orbits (for long times)
It is not difficult to show that, in this case, $\dot{\ell}_y$ and $\dot{\ell}_{xy}$ are independent of $\dot{\ell}_x$ and

$$I(\mathcal{X}, \mathcal{Y}) \approx f(\lambda) \qquad (15)$$

where λ is the time scale of the exponential divergence of both orbits. Equations (14) and (15) show that the IE defined by (11a), (11b) and (12) gives us the expected results. But to use I as an indicator of chaos, we must be sure that it satisfies certain properties, v. g., the indicator should take very different values when applied to regular and chaotic orbits. These issues will be addressed in the next section.

3. Applications of the IE

In order to study the behavior of the indicator we chose a well known dynamical system, that is, the Hénon–Heiles system

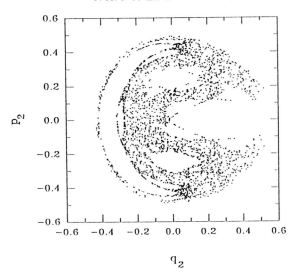

Figure 2. Same as Fig. 1 for a chaotic orbit. Initial conditions are $q_1 = 0$, $q_2 = -0.31$, $p_2 = 0.02$ and energy= 0.125.

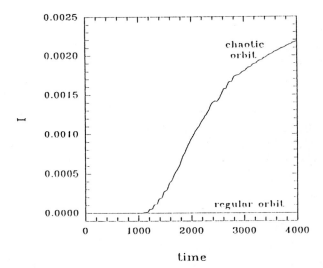

Figure 3. Evolution of the information I as a function of time for a regular and chaotic orbit.

$$H = \frac{1}{2}\sum_{i=1}^{3}(q_i^2 + p_i^2) + q_1^2 q_2 - \frac{q_2^3}{3}. \tag{16}$$

We set the energy equal to the value corresponding to what Hénon and Heiles called the "surprising case", that is $E = 0.125$. This energy level has

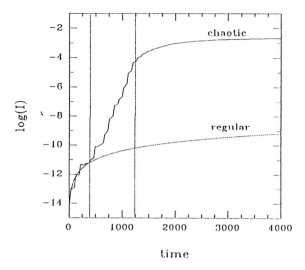

Figure 4. Evolution of the logarithm of the information $\log(I)$ as a function of time for a regular and chaotic orbit.

the advantage over other choices, that the orbital structure becomes rather complex involving regular and chaotic zones, which is what we need to test the IE as an indicator of chaos. Therefore, let us choose two orbits from the system, one that is regular and one that is chaotic. Figs. 1 and 2 show the Poincaré Surfaces of Section obtained by plotting the successive points of the trajectory which lie in the (q_2, p_2) plane and satisfy the conditions $q_1 = 0$, $p_1 > 0$, for the regular and chaotic orbits, respectively. Fig. 1 was obtained from the initial condition $q_1 = 0$, $q_2 = -0.33$, $p_2 = 0.02$ and the integration until a time of $t = 74250$ (which is not enough to close perfectly the curve). Fig. 2 was obtained by the integration until $t = 59110$ of the initial condition $q_1 = 0$, $q_2 = -0.31$, $p_2 = 0.02$. Both figures contain 3000 points.

Let us now apply the IE to these orbits. Fig. 3 shows the evolution of the value of I as a function of time for both cases. The behavior of the indicator is quite different: I remains at a value of zero for the regular orbit but begins to increase soon, at $t = 1000$, in the chaotic case. Thus we confirm the assumptions made in the former section. It also seems to be possible (at least as suggested by the figure) to define a "time of uncorrelation" (t_u) as the time required by an orbit to show its chaotic nature. We were tempted to take $t_u \approx 1000$ but, in a second thought, we reckoned the need of having a better detailed picture of the evolution of I and plotted the same graph in a logarithmic scale (see Fig. 4). Fig. 4 shows several features that were hidden by the lineal representation: a first interval of time, until $t \approx 400$, where the value of I evolves similarly for the regular and chaotic orbits; from $t \approx 400$

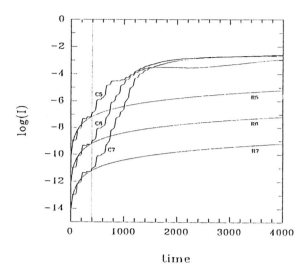

Figure 5. Dependence of the information on the parameter $||\delta||$. Curves are labeled by following the convention: C, chaotic orbit; R, regular orbit; 5, 6 and 7 correspond to $||\delta||$ values of 10^{-5}, 10^{-6} and 10^{-7}, respectively.

until $t \approx 1250$ both curves begin to differ, the regular orbit IE increases its value very slowly while the chaotic orbit IE increases steeply in a linear way; after $t \approx 1250$ both indicators approximate "asymptotic" values that differ by several orders of magnitude. It is still possible to define an uncorrelation time but, now, the manifestation of chaos shows to be earlier ($t_u \approx 400$).

It is important to recognize that the indicator depends on one parameter, δ, that is, the separation between the two nearby orbits which we integrate to analyze the correlation. The Hénon–Heiles system at the energy considered takes a range of about unity in coordinates and conjugate momenta and therefore, since δ should be a small displacement, we try three different choices for the value of its norm: $||\delta|| = 10^{-8}$, 10^{-7} and 10^{-6}. Fig. 5 shows the change of $\log(I)$ with time for all cases. There is a quite different behavior of the indicator when we change $||\delta||$ for chaotic orbits and when we do it for regular orbits: chaotic orbits tend to a *same value* while regular orbits do not* (as predicted by equation 14). On the other side, the separation between the asymptotic value of I for chaotic motion and regular motion is larger the smaller the value of $||\delta||$. Finally, we see in Fig. 5 that the time of uncorrelation does not depend on the value of $||\delta||$ (see the vertical dotted line on the figure), which is a very important feature if we would like to

---

* This point can be well understood thinking of the sensitivity on initial conditions in the chaotic regime: it does not matter how close the initial conditions are taken, the sets of points of both orbits will be equally uncorrelated. Regular orbits naturally will be more correlated the closer we set the initial condition.

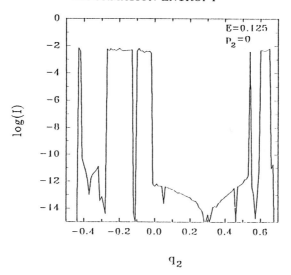

Figure 6. Final value of the information I as a function of the initial condition in q_2 for the Hénon–Heiles potential. We set the energy to 0.125 and, initially, $q_1 = 0$, and $p_2 = 0$. All integrations were followed until 1000 time units.

propose a relation between t_u and the maximum exponent of Lyapunov (see later). Thus we can adopt the asymptotic value of I as an indicator of the regime of the orbit, taking a small value of $||\delta||$ in order to ensure a large separation between the levels.

To check the method of the IE with a whole set of initial conditions, not only with two initial conditions as before, we apply it once more to the Hénon–Heiles potential, taking $||\delta|| = 10^{-7}$, $q_1 = 0$, $p_2 = 0$, $E = 0.125$ and $-0.5 \leq q_2 \leq 0.7$, with $p_1 > 0$. Fig. 6 shows the $\log(I)$ as a function of q_2. The two regimes can be clearly distinguished in the figure.

Having tested the indicator, it would be of interest to apply it also to a three degrees of freedom system. We took for this purpose the dynamical system studied by Contopoulos, Galgani and Giorgilli (1978) whose Hamiltonian is

$$H = \frac{1}{2} \sum_{i=1}^{3} \omega_i (q_i^2 + p_i^2) + q_1^2 q_2 + q_1^2 q_3, \tag{17}$$

where $\omega_1 = 1$, $\omega_2 = 1.4142$ and $\omega_3 = 1.7321$. Contopoulos *et al.* (1978) analyzed 9 orbits at the same energy surface, $E = 0.09$, and they found some of them to be chaotic and some to be regular, but also they could distinguish different degrees of chaoticity. Figs. 10 and 11 of their paper show the calculated Lyapunov exponents for the 9 orbits: orbits 1 and 2 seem to be regular and orbits 3 to 9, chaotic, but after some detailed study,

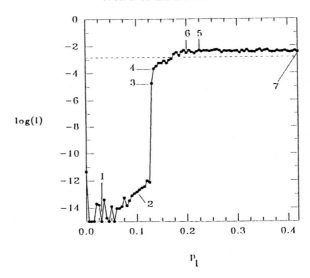

Figure 7. Final value of the information *I* as a function of the initial condition in p_1 for the Contopoulos *et al.* system. All trajectories started at the condition $q_1 = q_2 = q_3 = p_3 = 0$ and energy= 0.09, and the integrations were followed by 20000 time units.

the authors concluded that orbits 3 and 4 corresponded to a set of chaotic regime different than that of orbits 5 to 9. We reproduced the experiment but applying our indicator to a set of 85 initial conditions (integrated until $t = 20000$ time units) that involves 7 of the 9 conditions studied by Contopoulos *et al.* (1978). We fixed the energy at $E = 0.09$, took $q_1 = q_2 = q_3 = p_3 = 0$ and $0 \leq p_1 \leq 0.425$. The results of this implementation of the IE are summarized in Fig. 7 where we have marked the particular orbits studied earlier with their corresponding number. Initial conditions 1 and 2 are clearly regular orbits since they reach final (asymptotic) values several orders of magnitude less than the others; orbits 3 to 7 are chaotic and we also obtain a quantitative difference between orbits 3 and 4 from orbits 5 to 7. Thus, we have obtained the same results as Contopoulos *et al.* (1978) but by means of the IE, an indicator that needs about one order of magnitude less time than the Lyapunov exponents to show the regime of the orbits.

4. Conclusions

We have shown that the IE may be used as a nice indicator of chaos. The mathematical formalism is easy and it is completely justified within the framework of the information theory. We could predict analytically its asymptotic behavior that was confirmed by the numerical integrations. It has the advantage over other methods (like Poincaré Surfaces of Section or

Lyapunov exponents) that it is very easy to compute, independently on the number of degrees of freedom of the system, and that very long-term integrations are not necessary to obtain its asymptotic values. This point has been confirmed by the results of the application of the IE to every initial condition studied by Contopoulos *et al.* (1978), the IE needing one order of magnitude less time than the Lyapunov exponents to detect the actual nature of the trajectories. In the case of 3D systems, it is sensitive enough to detect different degrees of chaoticity (as Contopoulos *et al.* 1978 pointed out). Moreover, we would like to remark that the "time of uncorrelation" t_u is independent of the initial separation of the two orbits $||\delta||$ and it seems to be related to the time scale of exponential divergence of both orbits λ. This fact seems to suggest that t_u might be related to the Lyapunovs' time. What still remains to be done is to give a rigorous formalism to the computation of the entropy and to check the behavior of the IE when applied to actual dynamical systems. These issues will be addressed in a future paper.

Acknowledgements

We are very grateful to Claudia Giordano for her valuable comments and suggestions. This work was partly supported by Fundación Antorchas, Consejo Nacional de Investigaciones Científicas y Técnicas de la República Argentina, and the Facultad de Ciencias Astronómicas y Geofísicas de la Universidad Nacional de La Plata.

References

Cincotta, P., Núñez, J. and Muzzio, J.: 1995a, 'Stochastic motion in a central field with a weak non–rotating bar perturbation' in A. E. Roy and B. A. Steves, ed(s)., *FROM NEWTON TO CHAOS*, Plenum:New York, 537-543.
Cincotta, P., Muzzio, J. and Núñez, J.: 1995b, *Chaos, Solitons and Fractals* 6, 89-93.
Cohen, A. and Procaccia, I.: 1985, *Phys. Rev. A* 31, 1872.
Contopoulos, G., Galgani, L. and Giorgilli, A.: 1978, *Phys. Rev. A* 18, 1183-1189.
Dumas, H. S. and Laskar, J.: 1993, *Phys. Rev. Lett* 70, 2975-2979.
Fraser, A. and Swinney, H.: 1986, *Phys. Rev. A* 33, 1134.
Hénon, M. and Heiles, C.: 1964, *AJ* 69, 73-79.
Laskar, J.: 1990, *Icarus* 88, 266-291.
Laskar, J., Froeschlé, C. and Celletti, A.: 1992, *Physica D* 56, 253-269.
Maxwell, J.: 1991, *Matter and Motion* , Dover: New York.
Merritt, D. and Fridman, T.: 1995, *ApJ* , to be published.
Moser, J.: 1962, *Nachr. Akad. Wiss. Gottingen, Math. Phys.* K1, 1.
Moser, J.: 1963, *Nonlinear Problems* , Univ. of Wisconsin.
Papaphilippou, Y. and Laskar, J.: 1995, *A&A* , to be published.

CHAOS AND ELLIPTICAL GALAXIES

D. MERRITT

Department of Physics and Astronomy, Rutgers University

Abstract. Recent results on chaos in triaxial galaxy models are reviewed. Central mass concentrations like those observed in early-type galaxies – either stellar cusps, or massive black holes – render most of the box orbits in a triaxial potential stochastic. Typical Liapunov times are 3-5 crossing times, and ensembles of stochastic orbits undergo mixing on timescales that are roughly an order of magnitude longer. The replacement of the regular orbits by stochastic orbits reduces the freedom to construct self-consistent equilibria, and strong triaxiality can be ruled out for galaxies with sufficiently high central mass concentrations.

Key words: Chaotic motion – elliptical galaxies – triaxial models

1. Introduction

Stellar dynamics was one of the first fields where chaos was understood to be important, but the relevance of chaos to galaxies has never been completely clear. Three arguments have commonly been cited to show that chaos is unlikely to be important in determining the structure of real galaxies. First, it is pointed out that many reasonable potentials contain only modest numbers of stochastic orbits. This is always true for the potentials of rotationally symmetric models, and there is even a class of non-axisymmetric models for which the motion is completely regular, including the famous "Perfect Ellipsoid" (Kuzmin 1973; de Zeeuw & Lynden-Bell 1985). Second, it is noted that stochastic orbits often behave much like regular orbits over astronomically interesting timescales. Therefore (it is argued) one need not make a sharp distinction between regular and stochastic orbits when constructing an equilibrium model. Third, following the successful construction by Schwarzschild (1979, 1982) and others of self-consistent triaxial equilibria, it has generally been assumed that the regular orbits – which are confined to narrow parts of phase space and thus have definite "shapes" – are the only useful building blocks for real galaxies.

Recent theoretical work, combined with an improved understanding of the central structure of early-type galaxies, have exposed weaknesses in each of these arguments. The phase space of a triaxial model that looks similar to real elliptical galaxies is typically strongly chaotic; the regular box orbits (the "backbone" of elliptical galaxies) usually do not exist. Furthermore the stochastic orbits in such models often behave chaotically on timescales that are short compared to the age of the universe; typical Liapunov times are just 3-5 orbital periods, and characteristic mixing timescales are roughly ten

Celestial Mechanics and Dynamical Astronomy **64**: 55–67, 1996.
© *1996 Kluwer Academic Publishers. Printed in the Netherlands.*

times longer. A few attempts have now been made to build self-consistent
triaxial models with realistic density profiles, and one finds that a large fraction of the mass must typically be placed on the stochastic orbits. In models
with strong central mass concentrations, the generally spherical shape of
the stochastic orbits, combined with their dominance of phase space, can
preclude a self-consistent equilibrium.

Here I review this work and present some speculations on its relevance
for the structure and evolution of elliptical galaxies.

2. Imperfect Ellipsoids

Kuzmin (1973) showed that there is a unique, ellipsoidally-stratified mass
model for which the corresponding potential has three global integrals of the
motion, quadratic in the velocities. The density of Kuzmin's model is given
by

$$\rho = \rho(m) = \frac{\rho_0}{(1 + m^2)^2}, \qquad m^2 = \frac{x^2}{a^2} + \frac{y^2}{b^2} + \frac{z^2}{c^2}, \qquad (1)$$

with $a \geq b \geq c$ the axis lengths that define the ellipsoidal figure. The oblate
case, $a = b$, was discovered by Kuzmin already in 1956 in his classic study of
separable models of the Galaxy; the fully triaxial case was rediscovered by de
Zeeuw & Lynden-Bell in 1985, who christened Kuzmin's model the "Perfect
Ellipsoid." Kuzmin (1973) and de Zeeuw (1985) showed that all the orbits
in the Perfect Ellipsoid potential fall into one of only four families. Three
of these families are "tube" orbits that respect an integral analogous to the
angular momentum about the long (x) or short (z) axis; tube orbits avoid the
center, as do almost all of the orbits in an arbitrary axisymmetric potential.
The fourth family of orbits, the "boxes," are unique to triaxial potentials.
Box orbits have filled centers and touch the equipotential surface at eight
points, one in each octant. Self-consistent triaxial models (Schwarzschild
1979; Statler 1987) tend to weight the box orbits heavily, since these orbits
have mass distributions that mimic that of the underlying mass model.

Kuzmin's density law was arrived at via mathematical manipulations and
it is hardly surprising that the Perfect Ellipsoid bears little resemblance to
the distribution of light or mass in real elliptical galaxies. The discrepancy
is particularly great near the center, where Kuzmin's law predicts a large,
constant-density core. The luminosity densities in real elliptical galaxies are
always observed to rise monotonically at small radii (Ferrarese *et al.* 1994;
Møller *et al.* 1995; Lauer *et al.* 1995). The steepest cusps are seen in low-
luminosity galaxies like M32, which has $\rho \propto r^{-1.63}$ near the center (Gebhardt
1995). More luminous ellipticals like M87 have cusps with power-law indices
flatter than -1 (Merritt & Fridman 1995). The cusps in these luminous
galaxies are sufficiently shallow that the power-law nature of the profile is

obscured when the cusp is observed through the outer layers of the galaxy; as a result, such galaxies are sometimes still described as having "cores" (e.g. Kormendy *et al.* 1995) even though their central densities diverge as power laws.

Although the Perfect Ellipsoid does not describe real galaxies very well, its complete integrability makes it useful as a starting point for constructing more realistic models. Consider the mass model

$$\rho(m) = \frac{\rho_0}{(m_0^2 + m^2)(1 + m^2)}, \tag{2}$$

which might be called the "Imperfect Ellipsoid." For $m_0 = 1$, Eq. (2) reduces to Kuzmin's law, while for $m_0 = 0$ the Imperfect Ellipsoid has a $\rho \propto r^{-2}$ central density cusp like those observed in some elliptical galaxies. Thus varying m_0 from 1 to 0 takes one from a fully integrable but nonphysical model, to a strongly nonintegrable but realistic model.

Every box orbit in the Perfect Ellipsoid will eventually pass arbitrarily close to the center, and a small enough value of m_0 might be expected to destroy the integrability of at least some of these orbits. The dissolution of the box orbits can be tracked by computing their Liapunov characteristic numbers, defined in the usual way as finite-time averages of the rate of exponential divergence of intially close trajectories. Of the three positive Liapunov numbers, one is always close to zero; the other two Liapunov numbers σ_1 and σ_2 should tend to zero only if the orbit is regular. Fig. 1 shows the distributions of σ_1 and σ_2 for sets of 192 orbits that were dropped with zero velocity from an equipotential surface in the potential corresponding to Eq. (2), with $c/a = 0.5$, $b/a = 0.79$ and various values of m_0. The orbits have energies equal to that of the long-axis orbit that just touches the ellipsoidal shell dividing the model into two equal-mass parts. The Liapunov numbers were computed over 10^2, 10^3 and 10^4 dynamical times T_d, defined as the full period of the long-axis orbit.

A large fraction of the orbits with box-like initial conditions are stochastic, and this fraction is not tremendously dependent on m_0. Even for $m_0 = 0.1$ — which is much too large to describe real elliptical galaxies — more than half of the equipotential surface generates stochastic orbits, and this fraction increases to $\sim 150/192$ when m_0 has the physically more realistic values of 10^{-2} or 10^{-3}.

Fig. 1 shows that the distribution of Liapunov numbers for the stochastic orbits always evolves toward a single narrow peak – presumably because every stochastic orbit at a given energy moves in the same stochastic sea. One can therefore identify, after sufficiently long integrations, the unique numbers $\sigma_1(E)$ and $\sigma_2(E)$ that characterize stochastic motion at any given energy in these models. Expressed in units of the inverse dynamical time T_d, Fig. 1 shows that $\sigma_1 \approx 0.2 - 0.3$ and $\sigma_2 \approx 0.04 - 0.1$. Thus, divergence

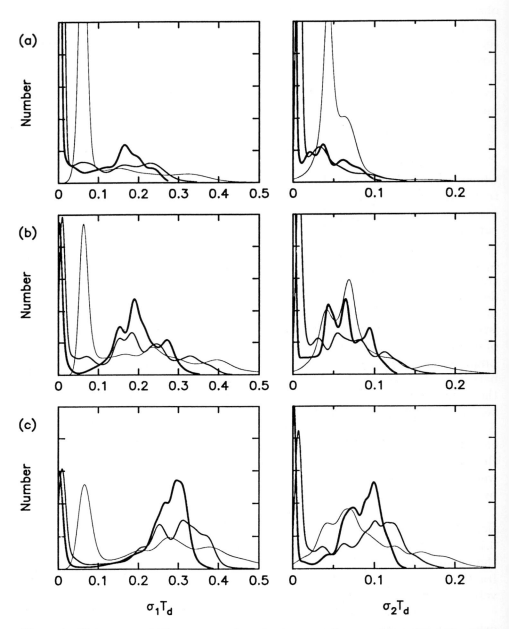

Figure 1. Histograms of Liapunov numbers for isoenergetic ensembles of box-like orbits in the triaxial potential corresponding to Eq. 2, with $m_0 = 10^{-1}$ (a), $m_0 = 10^{-2}$ (b) and $m_0 = 10^{-3}$ (c). $c/a = 0.5$ and $b/a = 0.79$. Orbits were integrated for 10^2 (thin lines), 10^3 (medium lines) and 10^4 (solid lines) dynamical times.

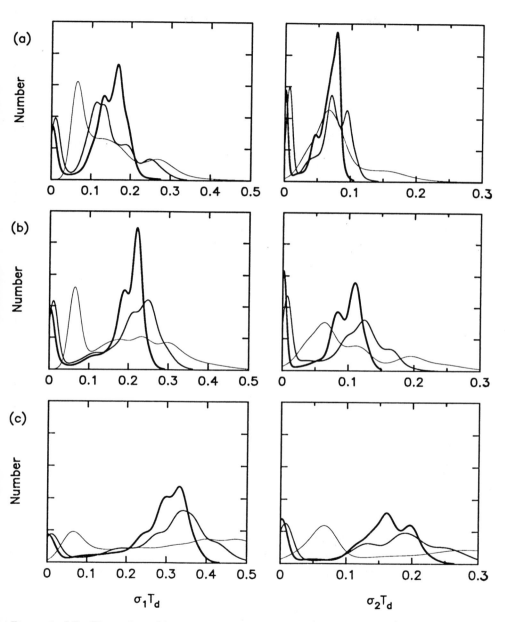

Figure 2. Like Fig. 1, for orbits in the potential of Eq. 2 with $m_0 = 10^{-1}$. A central point mass containing 0.1% (a), 0.3% (b) and 1% (c) of the galaxy mass has been included.

between nearby stochastic orbits takes place on timescales that are only $3-5$ times longer than the dynamical time.

3. Central Singularities

These results demonstrate that chaos can be important in triaxial models without singular densities or divergent forces. However there is increasingly strong evidence that early-type galaxies *do* contain central singularities, in the form of massive dark objects, possibly black holes (Ford *et al.* 1994; Miyoshi *et al.* 1995). Fig. 2 shows how the distribution of Liapunov numbers in the Imperfect Ellipsoid with $m_0 = 0.1$ is changed by the addition of a central point mass containing 10^{-2}, 3×10^{-3} or 10^{-3} of the total galaxy mass. (The ratio of black hole mass to luminous galaxy mass is thought to be about 5×10^{-3} for M87 and 2.5×10^{-3} for M32.) Even the smallest of these "black holes" induces the majority of the box-like orbits to behave chaotically, with $\sigma_1 T_d \approx 0.15$ or greater.

Singular forces also occur in galaxies where the stellar density increases more rapidly than $1/r$ toward the center, as is the case in low-luminosity ellipticals (Kormendy *et al.* 1995). Schwarzschild (1993) and Merritt & Fridman (1996) have investigated the orbital structure of triaxial potentials with $\rho \propto r^{-1}$ and r^{-2} density cusps and find that the box-like phase space is large- ly chaotic. The long-axis orbit, which generates the box orbits in integrable triaxial potentials, is unstable at most energies when the central density increases more rapidly than $r^{-0.5}$ (Merritt & Fridman 1995). Much of the stochasticity is driven by instabilities out of the principal planes and was missed in earlier studies that restricted the motion to two dimensions.

It is possible that the majority of elliptical galaxies contain either a cen- tral black hole, a steep central cusp, or both. If so, then chaos and triaxiality are strongly linked.

4. Mixing

An important quantity is the time required for an initially non-random dis- tribution of stars in stochastic phase space to "mix" into a time-invariant state. This process is essentially irreversible, and if it occurs on a timescale that is much shorter than the age of a galaxy, we would expect the stochas- tic parts of phase space to have reached a nearly steady state by now. On the other hand, if the mixing time exceeds a galaxy lifetime, then triaxial galaxies might still be slowly evolving as the stochastic orbits continue to mix.

Astronomers often talk loosely about "phase mixing" of regular orbits, but dynamicists discuss mixing only in the context of chaos. In fact one can define a hierarchy of dynamical systems in terms of their degree of irreversibility or randomness (e.g. Zaslavsky 1985). The lowest rung on the ladder is occupied by "ergodic" systems; an ergodic trajectory has the property that the average time spent in any phase-space volume is proportional to that volume. Thus an ergodic coin is one that falls equally often on heads and tails. That ergodicity is a very weak form of randomness is shown by the example of a coin that falls according to a predictable sequence, e.g. heads followed by tails followed by heads, etc. In the same way, even regular orbits, which are highly predictable, are ergodic, since the time-averaged density of a regular orbit is constant on its torus (e.g. Binney & Tremaine 1987). *

A stronger, and physically more appealing, sort of random behavior is "mixing". A mixing system is one in which any small part of phase space will eventually spread itself uniformly over the entire accessible space, i.e. in which the coarse-grained density, evaluated at a *given* time, approaches a constant (e.g. Krylov 1979). Mixing is always associated with chaos, and regular orbits do not mix – any small patch on the torus simply translates, unchanged, around the torus. "Phase mixing" as understood by astronomers always involves sets of orbits on *different* tori, whose trajectories gradually move out of phase due to differences in period. Phase mixing is an intrinsically slow process and in fact has no well-defined timescale, since two regular trajectories that are sufficiently close will never pull apart.

Mixing implies the approach of an initially non-uniform phase space density toward a coarse-grained equilibrium. Fig. 3 shows the configuration-space density corresponding to such an equilibrium. It was computed by evolving a set of 10^4 particles in the stochastic phase space of the model of Eq. (2), with $m_0 = 10^{-3}$, for 200 dynamical times; at this point the coarse-grained density defined by the ensemble had ceased to evolve significantly. The result – which might be called an "invariant density" (e. g. Kandrup & Mahon 1994) – looks remarkably like a regular box orbit: the density is elongated along the long (X) axis of the figure and is low around the short and intermediate axes. This is because stochastic orbits act, from turning point to turning point, very much like box orbits (Gerhard & Binney 1985), until they come close enough to the center to be perturbed onto another box-like trajectory, etc. Thus the invariant density corresponding to the full stochastic phase space at a given energy is similar to that of a superposition of box-like orbits (Merritt & Fridman 1996).

The approach to an invariant, coarse-grained density via mixing can take place on a surprisingly short timescale, a point emphasized by Kandrup and coworkers (Kandrup & Mahon 1994; Mahon *et al.* 1995) who investigated chaotic motion in two-dimensional potentials. They estimated mixing timescales by coarse-graining the configuration space and computing the

* The term "ergodic" is also used to describe motion that visits every point on the energy surface – a definition that is too restrictive to be of much use for stellar dynamics.

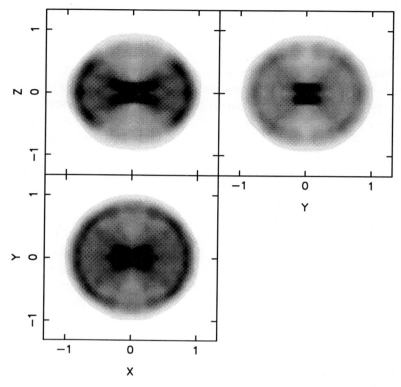

Figure 3. Invariant density of an isoenergetic ensemble of 10^4 stars in the chaotic phase space of a triaxial model with $m_0 = 10^{-3}$, $c/a = 0.5$ and $b/a = 0.79$. The X [Z] axis is the long [short] axis of the triaxial figure. Plotted are the densities near each of the three principal planes.

average deviation between the occupation numbers of an evolving ensemble, and the occupation numbers of the invariant density toward which the ensemble was evolving. The "distance" so defined decreased roughly exponentially with time, with a time constant of ~ 20 dynamical times in a modified Plummer potential. Experiments in strongly stochastic triaxial potentials (Merritt & Valluri 1996) show that the mixing timescale is similar, of order $30 - 50$ dynamical times, or roughly ten Liapunov times — rather less than a galaxy lifetime. These results suggest that the stochastic orbits in the inner regions of triaxial galaxies ought to be "fully mixed." That is, the entire stochastic phase space at a given energy should be viewed as a single "orbit" with a well-defined shape and density distribution, like that of Fig. 3.

It is intriguing that one can define a mixing timescale for stochastic orbits but not for regular ones. It is nonetheless commonly assumed that the distribution of stars along regular orbits reaches an equilibrium state in just a few dynamical times after the formation of a galaxy. The mechanism that is usually invoked to produce this mixing – violent relaxation, i.e. the rapid mixing of phase space that takes place during the collapse and virialization of a galaxy (Lynden-Bell 1967) – is of course a chaotic process. But there is no obvious reason why violent relaxation should mix the regular parts of phase space while leaving the stochastic parts unmixed. Thus the most likely model for a galaxy might be one in which stochastic phase space is fully mixed from the start, in the same way that the regular parts of phase space are usually assumed to be. While this would be a difficult proposition to test, Habib *et al.* (1995) have shown how the addition of even a very small, time-dependent component to the potential can sometimes greatly enhance the mixing rate in stochastic systems. The much larger perturbations that are present during galaxy formation would presumably be even more effective.

5. Trapped Stochastic Orbits

In nearly integrable potentials, the stochastic orbits are strongly hampered in their motion by the surrounding invariant tori and can mimic regular orbits for many oscillations (e.g. Karney 1983). A certain number of these "trapped" stochastic orbits are expected to be present in any potential containing both regular and stochastic trajectories. Goodman & Schwarzschild (1981) investigated the motion in a weakly chaotic, triaxial potential with a large core and found that virtually all the stochastic orbits were trapped for 10^2 dynamical times. They coined the term "semistochasticity" to describe this phenomenon.

Trapped stochastic orbits are less important in the strongly chaotic triaxial models discussed here. Fig. 4 is a map of the regular and stochastic regions in initial condition space for the two triaxial potentials used to construct the histograms in Figs. 1c and 2b. Open circles represent starting points, on the equipotential surface, corresponding to trapped stochastic orbits; these were defined as orbits that have non-zero Liapunov numbers but which look essentially regular in configuration space for 10^3 dynamical times or longer. (A roughly equivalent definition is: Orbits whose Liapunov numbers lie in the tail that extends leftward from the main stochastic peaks of Figs. 1c and 2b.) There are nearly as many of these trapped stochastic orbits as there are fully regular ones (although they comprise a small fraction of all stochastic orbits), which suggests that trapped orbits might play about as important a role as regular orbits in real galaxies. Similar numbers of trapped orbits are seen in triaxial models with weak cusps (Merritt & Fridman 1996).

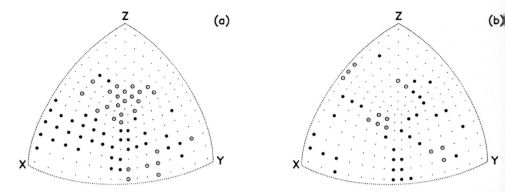

Figure 4. Starting points of box-like orbits on one octant of the equipotential surface. Small dots are stochastic orbits; large dots are regular orbits; circles are trapped stochastic orbits. (a) Model of Fig. 1c ($m_0 = 10^{-3}$); (b) model of Fig. 2b ($m_0 = 10^{-1}$; $M_{BH} = 3 \times 10^{-3}$).

The mixing times quoted above were for ensembles of non-trapped, stochastic orbits. It seems certain that an ensemble of points in a trapped region of stochastic phase space would take a longer time to mix. The slow mixing of trapped stochastic orbits might introduce a long timescale into galaxy evolution.

6. Jeans's Theorem

Jeans's theorem states that the distribution function describing an equilibrium galaxy must depend only on the isolating integrals of motion. Jeans did not consider the case of a galactic potential containing both regular and chaotic orbits, but his theorem is easily generalized by introducing the concept of an "invariant density" defined above. Thus: an equilibrium galaxy is one that is representable as a superposition of time-invariant components. The latter include the uniformly-populated tori of regular phase space, but also the invariant densities that result from a uniform population of stochastic phase space at any energy.

As Fig. 3 illustrates, invariant densities in stochastic phase space can have considerable structure. This suggests that such components might be useful building blocks for real galaxies – though presumably not as useful as regular orbits, which are confined to smaller regions of phase-space and have a wider variety of shapes. Furthermore, if the relatively short mixing times described above are the rule in realistic triaxial potentials, then nature would be forced to incorporate the stochastic orbits via invariant densities, rather than via some more general, non-uniform population of stochastic phase space.

A rather different point of view was presented in a provocative paper by Binney (1982), who proposed that Jeans's theorem does not apply to systems containing stochastic orbits. Binney's idea was that mixing implies ergodicity, and only regular orbits can rigorously be shown to be ergodic. But regular orbits, while ergodic, do not mix; and it is mixing, not ergodicity, that is relevant for the approach to equilibrium. Thus one is almost tempted to turn Binney's argument on its head: stochastic orbits are "good" in the sense that they have a built-in mechanism that promotes mixing; regular orbits are "bad" in the sense that their motion is quasi-periodic and hence non-mixing (at least in a fixed potential). In any case, neither ergodicity nor mixing appear to be essential properties of orbits in order for Jeans's theorem to be valid.

7. Self-Consistent Models

Before the publication of Schwarzschild's (1979, 1982) self-consistent models, triaxiality was generally considered to be an unlikely phenomenon because orbits in non-axisymmetric potentials were assumed to lack non-classical integrals of motion, i.e. to be chaotic. Schwarzschild showed that the motion in a triaxial potential with a large core is essentially fully regular, and that the four main families of regular orbits provided sufficient freedom to construct self-consistent equilibria. The work summarized above suggests roughly the opposite conclusion, namely that the motion in realistic, coreless triaxial potentials is largely chaotic. Does this mean that triaxial galaxies can not exist?

Schwarzschild (1993) himself first investigated this question by constructing self-consistent, scale-free ($\rho \propto r^{-2}$) triaxial models with six different choices of axis ratios. Schwarzschild's scale-free models were designed to represent the outer parts of galactic halos, and so he integrated individual orbits for only 55 dynamical times, roughly the age of the universe at ~ 10 kpc from the center of a galaxy. Many of his self-consistent solutions required the inclusion of stochastic orbits; however, since he allowed different stochastic orbits of the same energy to have different occupation numbers, the stochastic phase space was not uniformly populated and hence his models were not precisely stationary. Schwarzschild in fact showed that this non-uniform population of stochastic phase space implied that his models would evolve in shape in another 50 dynamical times.

Merritt & Fridman (1996) constructed non-scale-free triaxial models with Dehnen's density law,

$$\rho(m) = \rho_0 m^{-\gamma}(1+m)^{-(4-\gamma)} \qquad (3)$$

both for $\gamma = 1$ ("weak cusp") and $\gamma = 2$ ("strong cusp"). Both models had $c/a = 0.5$ and $b/a = 0.79$. For neither model could self-consistency be achieved using only the regular orbits. Solutions constructed in the same way as Schwarzschild's (1993) – i.e. with arbitrary occupation numbers assigned to "different" stochastic orbits of the same energy – existed for both mass models, but real galaxies constructed in this way would evolve quickly near their centers due to mixing of the stochastic orbits. Solutions that were "fully mixed" near the center could be found for the weak-cusp model but not for the strong-cusp model. Thus, fully stationary phase-space populations do not always exist for triaxial mass models with strong central concentrations – real galaxies with the same density profiles would have to be weakly triaxial, axisymmetric, or slowly evolving.

Although no one has yet attempted the construction of self-consistent triaxial models containing central black holes, Figures 2 and 4 suggest that chaos will play at least as strong a role in such models as it does in models with a strong cusp.

Given that stationary triaxial models are somewhat harder to construct than earlier believed, it is interesting to ask whether the *observational* evidence for triaxiality is compelling. The answer is probably "no", at least for the majority of elliptical galaxies (e.g. Gerhard 1994). The case is strongest for high-luminosity ellipticals, which have shallow cusps and low densities, hence long dynamical times – factors which would tend to reduce the mixing rates of stochastic orbits and thus to favor triaxiality. Low-luminosity ellipticals have steeper cusps and shorter dynamical times; interestingly, the kinematics of these galaxies have long been known to be crudely consistent with axial symmetry (Davies *et al.* 1983).

8. Acknowledgements

M. Valluri aided in the preparation of the figures. I thank her, H. Kandrup and G. Quinlan for useful discussions.

9. Discussion

G. Contopoulos: You have considered only nonrotating potentials. But if you include rotation, the classification of orbits is different. E.g. there are no box orbits, but close to the center there are elliptic orbits with loops. In our models of barred galaxies most orbits near the center are regular, while most chaotic orbits are near corotation.

D. Merritt: It is important to investigate whether slow figure rotation can strongly affect the degree of chaos in triaxial models. The fact that the axial orbits become loops when the figure rotates does not mean that the stochasticity will vanish, since an orbit need not pass exactly through the center in order to be stochastic.

References

Binney, J. J.: 1982, *Mon. Not. R. Astron. Soc.*,**201**, 15
Binney, J. and Tremaine, S.: 1987, *Galactic Dynamics*,Princeton University Press, Princeton
Davies, R. L., Efstathiou, G., Fall, S. M., Illingworth, G. and Schechter, P. L.: 1983, *Astrophys. J.*,**266**, 41
de Zeeuw, P. T.: 1985, *Mon. Not. R. Astron. Soc.*,**216**, 273
de Zeeuw, P. T. and Lynden-Bell, D.: 1985, *Mon. Not. R. Astron. Soc.*,**215**, 713
Ferrarese, L., van den Bosch, F. C., Ford, H. C., Jaffe, W. and O'Connell, R. W.: 1994, *Astron. J.*,**108**, 1598
Gebhardt, K.:1995, private communication.
Ford, H. C., Harms, R. F., Tsvetanov, Z. I., Hartig, G. F., Dressel, L. L., Kriss, G. A., Bohlin, R. C., Davidsen, A. F., Margon, b. and Kochhar, A. K.: 1994, *Astrophys. J.*,**435**, L27
Gerhard, O.: 1996, in *Galactic Dynamics and N-Body Simulations*,G. Contopoulos & A. Spyrou eds, 6th European EADN Summer School
Gerhard, O. and Binney, J. J.: 1985, *Mon. Not. R. Astron. Soc.*,**216**, 467
Goodman, J. and Schwarzschild, M.: 1981, *Astrophys. J.*,**245**, 1087
Goodman, J. and Schwarzschild, M.: 1981, *Astrophys. J.*,**245**, 1087
Habib, S., Kandrup, H. E. and Mahon, M. E.:1995, *Phys. Rev. E*, in press.
Kandrup, H. E. and Mahon, M. E.: 1994, *Phys. Rev. E*,**49**, 3735
Karney, C. F. F.: 1983, *Physica*,**8D**, 360
Kormendy, J., Dressler, A., Byun, Y.-I., Faber, S. M., Grillmair, C., Lauer, T. R., Richstone, D. and Tremaine, S.: 1995, *ESO/OHP Workshop on Dwarf Galaxies*, G. Meylan & P. Prugniel eds, ESO: Garching
Krylov, N. S.: 1979, *Works on the Foundations of Statistical Physics*,Princeton University Princeton: Princeton
Kuzmin, G. G: 1956, *Astr. Zh*,**33**, 27
Kuzmin, G. G: 1973, *Dynamics of Galaxies and Clusters*,T. B. Omarov ed., Akad. Nauk. Kaz. SSR: Alma Ata, 71
Lauer, T., Ajhar, E. A., Byun, Y.-L., Dressler, A., Faber, S. M., Grillmair, C., Kormendy, J., Richstone, D. & Tremaine, S.:1995, preprint
Lynden-Bell, D.: 1967, *Mon. Not. R. Astron. Soc.*,**136**, 101
Mahon, M. E., Abernathy, R. A., Bradley, B. O. and Kandrup, H. E.: 1995, *Mon. Not. R. Astron. Soc.*,**275**, 443
Merritt, D. and Fridman, T.: 1995, *A. S. P. Conf. Ser. Vol. 86, Fresh Views of Elliptical Galaxies*,A. Buzzoni, A. Renzini & A. Serrano eds, Astron. Soc. of the Pacific: Provo, 13
Merritt, D. and Fridman, T.: 1996, *Astrophys. J.*,**456**, in press
Merritt, D. and Valluri, M.:1996, in preparation
Miyoshi, M., Moran, J. Herrnstein, J., Greenhill, L., Nakai, N., Diamond, P. and Inoue, M.: 1995, *Nature*,**373**, 127
Schwarzschild, M.: 1979, *Astrophys. J*,**232**, 236
Schwarzschild, M.: 1982, *Astrophys. J*,**263**, 599
Schwarzschild, M.: 1993, *Astrophys. J*,**409**, 563
Statler, T.: 1987, *Astrophys. J.*, **321**,113
Zaslavsky, G. M.: 1985, *Chaos in Dynamic Systems*, Harwood: London, 28

SINKING, TIDALLY STRIPPED, GALACTIC SATELLITES

JUAN C. MUZZIO

Facultad de Ciencias Astronómicas y Geofísicas de la Universidad Nacional de La Plata, and Programa de Fotometría y Estructura Galáctica del Consejo Nacional de Investigaciones Científicas y Técnicas de la República Argentina

Abstract. Tidally interacting galaxies offer an interesting field for the investigation of chaotic phenomena in stellar systems. When the galaxies are gravitationally bound, and one of them is much larger than the other, the latter can be regarded as a satellite of the former. The study of their dynamics is somewhat simplified in this case, which presents well observed examples in nature (e.g., globular clusters). Galactic satellites suffer orbital decay due to dynamical friction, a process that may be greatly enhanced in the presence of chaotic motions. Besides, the satellite is stripped by the field of tidal forces and, in the long run, it will disintegrate completely. Modern observations are able to show the signature of these processes taking place at present.

Key words: galaxies – globular clusters – galactic dynamics

1. Introduction

The investigation of the dynamics of interacting stellar systems is a very difficult problem where, despite some brave analytical studies, most results are derived from numerical simulations (see, e.g., Muzzio 1994). The case of galactic satellites is of special interest within this class of objects because, on the one hand, some simplifications are possible and, on the other hand, they offer examples of well observed objects, such as the globular clusters. If we remember that those observations are usually adjusted to models, like those of King or Michie, derived from Jeans strong theorem (Binney and Tremaine 1987) whose hypotheses, in turn, demand regular motions, it is easy to understand the importance of knowing the relevance of chaos in galactic satellites. This is no idle question, as can be immediately realized if we remember that not even the energy (or Jacobi) integral is conserved in the case of a satellite in a typical elongated orbit.

Here we will review the evidence of chaotic motions in interacting systems (Section 2), some remaining problems in our understanding of dynamical friction for sinking satellites (Section 3), the tidal stripping process (Section 4) and some recent work on the subject being done at La Plata Observatory (Sections 5 and 6).

2. Satellites and chaos

The presence of chaos in the restricted three body problem was noticed early by Jefferys (1966) and Hénon (1966a,b). The case of equal masses was considered in those investigations but Hill's case (one of the main masses vanishingly small), which is more akin to the problem of galactic satellites, was also found by Hénon (1970) to exhibit chaotic orbits in small regions of the phase space; the chaotic regions are larger in the elliptic, rather than circular, restricted three body problem (Winter 1995). More recently, Rix and White (1989) also found irregular motions in the neighborhood of a pair of elliptical galaxies on a circular orbit.

Stewart (1993) considered two equal galaxies on circular orbits around each other and computed the maximum Liapunov exponents for a sample of orbits of massless particles in their neighborhood. His galaxies were modelled as Plummer (or Schuster) spheres and the exponents turned out to be larger for the steeper potentials. For reasonable values of mass and orbit size, he found Liapunov times as short as one - fifth or one - tenth of the age of the Universe. Particles released at rest deep inside the potential troughs, however, turned out to have Liapunov times larger than that age, i.e., they are in "regular" orbits for practical purposes.

Cincotta et al. (1995) showed that weak bar - like perturbations in a spherical field are capable of generating significant chaos, provided that the stellar orbits are of low angular momentum, and speculated that irregular motions might be important even in such "regular" systems as globular clusters.

Finally, Pfenniger (1986) showed that, in the presence of chaotic motions, relaxation and dynamical friction can be several orders of magnitude larger than in nearly integrable systems. He advised caution when considering the results derived from simple theoretical models, as they are usually based on integrable or near integrable potentials.

3. Why are galactic satellites sinking?

The orbital decay of galactic satellites is caused by dynamical friction, a process investigated by Chandrasekhar (1943) as due to the exchange of momentum between the satellite and the material particles that make up the galaxy within which the satellite is moving. Thus, in this view dynamical friction is a purely local effect.

A different picture was suggested by Kalnajs (1972) who computed the drag force on the satellite caused by the wake of particles deflected by the satellite and accumulated behind it; in other words, his is a global view of the dynamical friction process. Considering the galaxy as an infinite homogeneous medium, Kalnajs obtained the same result as Chandrasekhar but, when he considered a satellite moving in a disk of stars, the drag turned out to be zero! Tremaine (1981) found no drag either on a satellite moving in a spherical galaxy, and showed that the fault was the neglect of resonances which are crucial to dynamical friction in the global analysis. This line was followed by Tremaine and Weinberg (1984) and Weinberg (1986 and 1989).

Numerical experiments yielded some contradictory results. Lin and Tremaine (1983) found good agreement with Chandrasekhar's equation. White (1983) argued that such an equation was not enough for a complete picture and favored a global view. Bontekoe and van Albada (1987) found excellent agreement with Chandrasekhar and supported the local approach. Zaritsky and White (1988) attributed to a subtle technical detail the discrepancy of White with Bontekoe and van Albada, and recognized that Chandrasekhar's local approach was adequate. Nevertheless, Hernquist and Weinberg (1989) and Prugniel and Combes (1992) still maintain that a global view is necessary.

Despite the discrepancies, there are some undisputed facts. First, all authors agree in that essentially the same results are obtained whether the medium where the satellite is moving is self - consistent or not, a finding that poses no problem to the local view, but that may be not so easy to reconcile with global theories. Second, artificially fixing the center of the galaxy, rather than allowing it to move around the common center of mass, drastically accelerates the orbital decay of satellites. It is worth recalling that Hernquist and Weinberg's (1989) definition of self - consistency differs from that of all other authors and, instead of referring to particle - particle interaction, corresponds to motion of the galaxy.

One can not help but wonder how realistic global theories can be. In a spherical system made up of N particles the random distribution causes, for example, a quadrupole of intensity $N^{1/2}/2$. That is equivalent, for a galaxy of 10^{12} stars, to a satellite of 5×10^5 stars, i.e., the size of a globular cluster. Besides, what global response can we expect from 200 (or 20,000) globular clusters? If the advocates of global theories are right, then it would be very difficult indeed to estimate the effects of dynamical friction in real galaxies.

A word of caution is needed due to the widespread use of satellites 1/10 the mass of the galaxy in numerical experiments which were, but no longer are, necessary to make the lengthy integrations in the computers of the past. The reasons were that the decay time increases as the satellite mass decreases and, besides, the satellite has to be much more massive than the particles that make up the galaxy to avoid excessively noisy results. The 1/10 mass ratio offered thus a way to limit both the integration time and the number of particles in the galaxy, thus saving computing time. The problem is that real satellites such as dwarf galaxies or globular clusters are much lighter (about 10^{-6} to 10^{-3} the galactic mass), so that the results of those numerical experiments might tend to perpetuate unrealistic concepts. One must always remember that: 1) Orbital decay is very slow for real satellites and the main effect takes place (by orders of magnitude) at the pericenter (see, e.g., Fig. 7 of Pesce et al. 1992); 2) That does not necessarily mean that orbits are quickly "circularized" as most numerical simulations assume when using unrealistic circular orbits (Cora et al. 1995); 3) Satellites as massive as 1/10 the galaxy's mass substantially alter the structure of the galaxy during their orbital decay (Bontekoe and van Albada 1987 and Cora et al. 1995); 4) Satellite disruption drastically alters the picture based purely on dynamical friction (Prugniel and Combes 1992) and massive, more extended, satellites are the ones most affected.

4.   Why are galactic satellites stripped?

There are several processes that contribute to the disintegration of galactic satellites:

a) Tidal limitation. For a satellite on a circular orbit in a galaxy, common wisdom says that the material outside the Roche lobe gets lost. King's (1962) idea of using this limit to impose a tidal radius to the satellite was a very important and useful one. Nevertheless, tidal radii are less well defined than the Roche lobe approximation suggests. For example, stars on retrograde orbits around the satellite can remain stable at much larger distance from the satellite's center than stars on direct orbits can, due to the effect of the Coriolis force (Innanen 1979). Thus, the satellite is more easily stripped of stars on direct orbits.

Simple theory can give only order of magnitude estimates of the mass that remains within the tidal radius, an important parameter in the investigation of the dynamical evolution of clusters of galaxies. Such estimates were provided by Merritt (1988) who, in order to improve them, also performed numerical experiments that allowed him to derive more accurately the mass that remains bound to the galaxies in a cluster.

b) Tidal torquing. Contrary to common wisdom, it does not depend on the satellite being deformed (McGlynn and Borne 1991). Due to symmetry, the effects of direct and retrograde torques cancel on first approximation but, on second approximation, the motion relative to the torque is faster for the retrograde stars than for the direct ones. Thus, the torque accelerates direct stars and deccelerates retrograde ones yielding, as a result, a net direct acceleration.

c) Encounters. Let us forget satellites for a minute, and consider a galaxy that suffers an encounter with another. In the case of fast encounters we have the impulse approximation of Spitzer (1958) which assumes that the stars do not change their relative position during the encounter and that just get a kick that changes their velocities an amount DV, from their initial velocity V_i, before the encounter, to a final velocity V_f after it. Averaging over the whole galaxy, one gets the change in its internal kinetic energy:

$$< V_f^2 > - < V_i^2 >= 2 < V_i.\Delta V > + < \Delta V^2 > \tag{1}$$

The first part of the right hand term is of first order in the perturbation, but it is usually assumed to average to zero because of symmetry. Nevertheless, when material is lost as a result of the encounter, there is a preferred direction and the first order may be significant (Richstone 1975).

For slow encounters, the orbit reacts as a whole and the adiabatic approximation, also pioneered by Spitzer (1958) can be used. Fig. 1 of Aguilar and White (1986) beautifully shows the the coupled effects of the impulse approximation in the outer parts of a galaxy that suffers an encounter, with the adiabatic process that affects its inner regions.

The net result is that the system gets an injection of energy: some stars may escape, and the rest relax towards a new equilibrium where, according to the virial theorem, the kinetic energy equals the new total energy.

Let us get back now to the satellites and their tidal limitation. In the case of elongated orbits, King's original idea had been to derive the tidal radius in the worst possible situation, i.e., at pericenter. The results of the past few years tend to support King's tidal limitation idea for circular orbits, but tidal stripping due to "encounters" with the central part of the galaxy at pericenter seems to offer a better model for elongated orbits and, as a result, truncation is less severe than previously thought (Allen and Richstone 1988).

In additon to the stars that escape due to the sudden kick, there is also the (second order) tidal heating which increases the energy of the stars in each succesive pericenter passage, leads to a continuous slow loss of stars, and forms the apogalactic wings to which we will refer below.

d) Disk Shocking. Satellites of disk galaxies can get another injection of energy from the varying field when crossing the galactic plane (Ostriker et al. 1972). It affects mainly the stars distant from the center of the satellite.

Weinberg (1994 a, b & c) applied recent mathematical results to show that the usual adiabatic criterium (based on the harmonic oscillator) does not apply in general, and that a slowly varying perturbation may cause significant evolution in stellar systems. He improved the theory of tidal shocking and implemented a Fokker - Planck code with disk shocking that leads to significant differences with the standard theory.

e) Relaxation effects. Besides all the extrinsic effects described above, there is also the intrinsic relaxation effect due to star encounters within the satellite. The relaxation time for systems in equilibrium is much longer than the Hubble time for dwarf galaxies, but comparable to it for globular clusters.

As stars get lost, relaxation contributes to replenish the reservoir of stars whose energy and angular momentum put them on the verge of getting lost, and that will escape later on. It also changes the structure of the satellite, concentrating the innermost parts and expanding the halo.

Some recent numerical simulations show the result of all these different processes in action. Oh et al. (1992) developed a code which uses a Fokker -

Planck approach for the inner regions of a satellite and a restricted three - body approach for its halo; it is not self - consistent, and accurate for small mass losses only. Their simulations used 5,000 particles, and the code was tested with isolated models. Their Figs. 8 and 9 show nicely the expansion due to the relaxation effects.

Oh and Lin (1992) subsequently used the code to investigate the evolution of globular clusters in the presence of a galactic field. In the absence of relaxation, they found good agreement with King's tidal radius for both circular and eccentric orbits. When relaxation effects were included, they found a self - similar evolution of the density profile in the inner regions, and that tidal radii could extend well beyond the theoretical limit. They noticed no phase dependence of the limiting radius. Velocity distributions tended to be isotropic in the core, mainly radial at intermediate radii and isotropic near the tidal radius. The isotropization of the outer parts was also aided by disk shocking. Escapers came preferentially from prograde orbits and prolonged tidal interaction could induce apparent retrograde rotation of the outer parts, which might extend to the interior in models with highly anisotropic velocity distributions.

Grillmair (1992) used a tree - code, so that self - consistency and mass losses were no problem in his study. He simulated the satellites with aniso-tropic Jaffe models and used 64K particles in most cases; his results provide an excellent picture of the evolution of a galactic satellite. Considering ini-tially an isolated satellite, he investigated the effect of truncation, in one case eliminating the stars beyond a certain radius and in another those with energies larger than some value; it turned out that in the former case the cluster recovered its original size much faster than in the latter. When the galactic field was included, the satellites were truncated at the tidal radius, as could be expected, but they exhibited shallow, orbital phase dependent, wings beyond that radius. Despite the initial preference for radial orbits, Grillmair found that tidal torquing, and the loss of particles on radial orbits caused a strong tangential velocity component. Particles on prograde orbits gradually outnumbered those on retrograde ones when going from the cen-ter of the satellite toward its outer parts. Grillmair's Fig. 3.28 dramatically shows how prograde motion increases as the satellite approaches perigalacti-ca, to suddenly decrease due to the particles that are lost thereafter. As the satellite describes one orbit after another around the galaxy, it suffers a gen-eral expansion, modulated by slight compresions near perigalactica followed by expansions at apogalactica. Grillmair also obtained deep photometry of a sample of globular clusters that allowed him to push the star counts well below than previous studies and to reveal the tidal wings in the profiles of several of his clusters.

Piatek and Pryor (1995) investigated whether the tidal effects can sim-ulate an apparently large mass to luminosity ratio for dwarf spheroidals.

They found that tides produce large ordered (rather than random) motions. In particular, they cause apparent rotation, due to unbound stars. Similar results were obtained by Oh et al. (1995), who also found that the survival of dwarf spheroidals for a Hubble time depends on its size being smaller or larger than the tidal radius.

5. Shapes of galactic satellites

Despite their name, globular clusters are not spherical, a fact well known to workers on the field, but not frequently raised in general astronomy text-books. The paper by Han and Ryden (1994) is probably the most recent contribution to the subject.

Possible causes of non - sphericity are: 1) Primordial, sustained by an-isotropic velocity distributions, in which case the satellite can be oblate, prolate, or (more likely) triaxial; 2) Rotation, which leads to oblate systems; 3) Tidal deformation, which results in almost prolate systems in spherical galaxies, or triaxial ones in disk galaxies. A primordial origin may be due either to the collapse of a non - spherical cloud (e.g., Aarseth and Binney 1978), or to the collapse of a spherical cloud subject to the effect of the radial orbit instability (e.g., Aguilar and Merritt 1990, and Carpintero and Muzzio 1995). Rotation may be primordial or acquired, as shown in the previous section, but it should be recalled that measured rotational velocities of globular clusters are usually small. Tidal effects are important for radii comparable to the tidal radius, while observers rarely have gone beyond 2 or 3 core radii in the past, the core radius being much smaller that the tidal radius for globular clusters. Nevertheless, large radii can now be reached with modern techniques (as mentioned above about Grillmair's 1992 work) and, besides, the deformation of the outermost parts, albeit difficult to observe, may affect what happens in the inner regions of the satellite.

Han and Ryden (1994) favor oblate spheroids for the globular clusters of the Milky Way and M31, and triaxial ellipsoids for those of the Magellanic Clouds. A possible scenario is that the clusters are born triaxial, supported by velocity anisotropy, are isotropized later on and become rounder, while tidal effects make them rotate and finally become oblate spheroids. Differ-ent histories would result in different ellipticities. The ages of the globular clusters (less than 20 Gy for the Milky Way and M31, but less than only 2.5 Gy for the Magellanic Clouds) lends support to that scenario.

What is the contribution of tidal deformation? While, as indicated before, it directly affects the outermost (difficult or impossible to observe) regions, it should be noticed that usual proofs that it is irrelevant assume that the elongation takes place along the line that joins the centers of the galaxy and the satellite (e.g., White and Shawl 1987), but that is simply not true.

Fig. 1 of Miller (1986), for example, clearly shows that there is a substantial angular difference between the elongation and the line that joins the centers. In a recent paper, Heggie and Ramamani (1995) built King models using the Jacobi integral instead of the energy, and got tidally deformed triaxial models (quasi prolate when orbiting around spherical galaxies). Bassino et al. (1995) performed numerical simulations of tidally truncated King models. They used 50,000 particles which allowed them to investigate the shapes with reasonable detail, and found clear ellipticities that should significantly affect the inner dynamics of the satellites.

6.   A new kind of galactic satellites?

Bassino et al. (1994) investigated the possibility that some globular clusters were just the nuclei of nucleated dwarf galaxies, captured by a giant galaxy and tidally stripped of their outer regions. They found that, while regular dwarf galaxies quickly became completely destroyed, the nuclei of the nucleated dwarfs managed to survive for periods longer than the Hubble time. Their unexpected result was that remnants about ten times more massive than globular clusters could be formed, i.e., galactic satellites intermediate between clusters and dwarf galaxies. Soon thereafter, Harris et al. (1995) found about three dozen objects around the central galaxy of the cluster of galaxies A2107 which could well be the satellites predicted by Bassino et al. Their luminosities are an order of magnitude brighter than those of globular clusters, suggesting a similarly larger mass and, moreover, a large fraction of them showed a non - stellar appearance, suggesting larger radii as well. It will be very interesting if an observational means could be found to decide whether those objects were captured by the galaxy, rather than being born in it.

7.   Acknowledgements

I am very grateful to L.P. Bassino, S.A. Cora, M.M. Vergne and F.C. Wachlin for their help. The referee, D. Merritt, was very helpful, spotting typos and a relevant missing reference. The technical assistance of S.D. Abal de Rocha, M.C. Fanjul de Correbo, R.E. Martínez and H.R. Viturro is gratefully acknowledged. This work was supported with grants from the Consejo Nacional de Investigaciones Científicas y Técnicas de la República Argentina and from the Fundación Antorchas.

References

Aarseth, S. J. and Binney, J.: 1978, *Mon. Not. Roy. Ast. Soc.* **185**, 27

Aguilar, L. and Merritt, D.: 1990, *Astrophys. J.* **453**, 33

Aguilar, L. and White, S. D. M.: 1986, *Astrophys. J.* **307**, 97

Allen, A. J. and Richstone, D. O.: 1988, *Astrophys. J.* **325**, 583

Bassino, L. P., Muzzio, J. C. and Rabolli, M.: 1994, *Astrophys. J.* **431**, 634

Bassino, L. P., Muzzio, J. C. and Wachlin, F. C.: 1995, 'Structure of galactic satellites' in J. C. Muzzio and S. Ferraz-Mello, ed(s)., *Chaos in Gravitational N-Body Systems*, Kluwer Acad. Publ.: Dordrecht, xxx

Bontekoe, Tj. R. and van Albada, T. S.: 1987, *Mon. Not. Roy. Ast. Soc.* **224**, 349

Binney, J. and Tremaine, S.: 1987, *Galactic Dynamics*, Princeton Univ. Press, Princeton

Carpintero, D. D. and Muzzio, J. C.: 1995, *Astrophys. J.* **440**, 5

Chandrasekhar: 1943, *Astrophys. J.* **97**, 255

Cincotta, P. M., Muzzio, J. C. and Núñez, J. A.: 1995, *Chaos, Solitons and Fractals* **6**, 89

Cora, S. A., Muzzio, J. C. and Vergne, M. M.: 1995, 'Structure of galactic satellites' in J. C. Muzzio and S. Ferraz-Mello, ed(s)., *Chaos in Gravitational N-Body Systems*, Kluwer Acad. Publ.: Dordrecht, xxx

Grillmair, C. J.: 1992, *Thesis*, The Australian National University, Canberra, Australia

Han, C. and Ryden, B. S.: 1994, *Astrophys. J.* **433**, 80

Harris, W. E., Pritchet, C. J., and McClure, R. D.: 1995, *Ap. J.* **441**, 120

Heggie, D. C. and Ramamani, N.: 1995, *Mon. Not. Roy. Ast. Soc.* , , in press

Hénon, M.: 1966a, *Bull. Astron. 3e. ser.* **T. 1, Fasc. 1**, 57

Hénon, M.: 1966b, *Bull. Astron. 3e. ser.* **T. 1, Fasc. 2**, 49

Hénon, M.: 1970, *Astron. Astrophys.* **9**, 24

Hernquist, L. and Weinberg, M. D.: 1989, *Mon. Not. Roy. Ast. Soc.* **238**, 407

Innanen: 1979, *Astron. J.* **84**, 960

Jefferys, : 1966, *Astron. J.* **71**, 306

Kalnajs, A.: 1972, 'Polarization clouds and dynamical friction' in M. Lecar, ed(s)., *Gravitational N - Body Problem*, D. Reidel Publ. Co.: Dordrecht, 13

King, I.: 1962, *Astron. J.* **67**, 471

Lin, D. N. C. and Tremaine, S.: 1983, *Astrophys. J.* **264**, 364

McGlynn, T. A. and Borne, K. D.: 1991, *Astrophys. J.* **372**, 31

Merritt, D.: 1988, 'Internal dynamics of galaxy clusters' in J.M. Dickey, ed(s)., *The Minnesota Lectures on Clusters of Galaxies and Large - Scale Structure*, Astronomical Society of the Pacific: San Francisco, 175

Miller, R. H.: 1986, *Astron. Astrophys.* **167**, 41

Muzzio, J. C.: 1994, 'Interacting spherical systems' in V.G. Gurzadyan and D. Pfenniger, ed(s)., *Ergodic Concepts in Stellar Dynamics*, Springer Verlag: Berlin, 243

Oh, K. S. and Lin, D. N. C.: 1992, *Ap. J.* **386**, 519

Oh, K. S., Lin, D. N. C. and Aarseth, S. J.: 1992, *Ap. J.* **386**, 506

Oh, K. S., Lin, D. N. C. and Aarseth, S. J.: 1995, *Ap. J.* **442**, 142

Ostriker, Spitzer and Chevalier: 1972, *Astrophys. J. Lett.* **176**, L51

Pesce, E., Capuzzo - Dolcetta, R. and Vietri, M.: 1992, *Mon. Not. Roy. Ast. Soc.* **254**, 466

Piatek, S. and Pryor, C.: 1995, *Astron. J.* **109**, 1071

Pfenniger, D.: 1986, *Astron. Astrophys.* **165**, 74

Prugniel, Ph. and Combes, F.: 1992, *Astron. Astrophys.* **259**, 25

Richstone, D. O.: 1975, *Astrophys. J.* **200**, 535

Rix and White, S. D. M.: 1989, *Mon. Not. Roy. Ast. Soc.* **240**, 941

Spitzer: 1958, *Astrophys. J.* **127**, 17

Stewart, P.: 1993, *Astron. Astrophys.* **269**, 135

Tremaine, S.: 1981, 'Galaxy mergers' in S. M. Fall and D. Lynden-Bell, ed(s)., *The Structure and Evolution of Galaxies*, Cambridge Univ. Press: Cambridge, 67

Tremaine, S. and Weinberg, M. D.: 1984, *Mon. Not. Roy. Ast. Soc.* **209**, 729
Weinberg, M. D.: 1986, *Astrophys. J.* **300**, 93
Weinberg, M. D.: 1989, *Mon. Not. Roy. Ast. Soc.* **239**, 549
Weinberg, M. D.: 1994a, *Astron. J.* **108**, 1398
Weinberg, M. D.: 1994b, *Astron. J.* **108**, 1403
Weinberg, M. D.: 1994c, *Astron. J.* **108**, 1414
White, R. E. and Shawl, S. J.: 1987, *Astrophys. J.* **317**, 246
White, S. D. M.: 1983, *Astrophys. J.* **274**, 53
Winter, O. C.: 1995, 'The Liapunov exponent as a tool for exploring phase space' in J.C. Muzzio and S. Ferraz-Mello, ed(s)., *Chaos in Gravitational N-Body Systems*, Kluwer Acad. Publ.:Dordrecht, xxx
Zaritsky, D. and White, S. D. M.: 1988, *Mon. Not. Roy. Ast. Soc.* **235**, 289

ON THE SATELLITE CAPTURE PROBLEM

Capture and stability regions for planetary satellites

ADRIÁN BRUNINI

Facultad de Ciencias Astronómicas y Geofísicas Universidad Nacional de La Plata Paseo del bosque s/n, La Plata (1900). ARGENTINA, and PROFOEG, CONICET.

Abstract.

The stability and capture regions in phase space for retrograde and direct planetary satellites are investigated in the frame of the Circular Restricted Three-Body Problem. We show that a second integral of motion furnishes an accurate description for the stability limit of retrograde satellites.

The distribution of heliocentric orbital elements is studied, and possible candidates to be temporary Jovian satellites are investigated.

Previous results, limited to orbits satisfying the Mirror Theorem, are extended in order to give a complete set of capture conditions in phase-space.

Key words: Restricted Three-Body Problem – Planetary Satellites

1. Introduction

The outer satellites of Jupiter and Phoebe, the outermost satellite of Saturn, share common physical characteristics, resembling asteroids rather than natural planetary satellites. Furthermore, most of them are in non equatorial, elongated and retrograde orbits, being thus unlikely a similar formation process as that of the innermost regular satellites.

Based upon these facts, astronomers have long thought that irregular planetary satellites have a capture origin (Kuiper 1961). Triton, the large moon of Neptune, is perhaps the largest example of satellite capture in the Solar System (Goldreich et al. 1989).

Another interesting application of satellite capture in cosmogony has been recently explored by Brunini (1995), who has shown that capture of planetesimals as planetary satellites has largely increased the efficiency of the accretion process of the outer planets.

Recent numerical studies have shown that short-period comets of the Jupiter family can be temporarily trapped as jovian satellites. Examples revealing the importance of the dynamics of capture are furnished by P / Gehrels 3 (Rickman 1979), P / Oterma which underwent a close encounter with Jupiter in 1934-1939 (Carusi et al. 1985), P / Helin - Roman - Crockett which will perform a long lasting capture of five revolutions about Jupiter (Tancredi et al. 1990) and the recent collision of P / Shoemaker - Levy 9 after being a Jupiter's satellite for several orbital periods.

Celestial Mechanics and Dynamical Astronomy **64**: 79–92, 1996.

© 1996 *Kluwer Academic Publishers. Printed in the Netherlands.*

The first rather complete description of the stability of irregular satellites was done by Hénon (1970) in the frame of Hill's approximation to the Restricted Three-Body Problem (hereafter RTBP). He treated only those orbits satisfying the Mirror Theorem: at the initial instant the minor body is at conjunction, and its velocity vector is perpendicular to the Sun-Planet line. The initial conditions as stated above produce a symmetric motion about both the x axis and the time (Roy and Ovenden 1955). Therefore, in this situation a capture is always followed by an escape.

In a purely gravitational scenario, captures are only of temporary nature. It has been proven by Hopf (1930) that each particle must pass arbitrarily close to its initial position in phase space, making zero the probability of permanent capture. In fact, there exist some initial conditions yielding to permanent capture, but they form a set of Lebesgue measure zero. Nevertheless, the "lifetimes" as temporary satellites may be very large, as the initial conditions approach arbitrarily close to stable periodic orbits (Murison 1989).

Since at present Jupiter's irregular satellites are stable, some other mechanism is required to transform temporary captures into permanent ones. Several theories involving dissipation of orbital energy have been proposed in order to facilitate permanent captures. Nevertheless, all the non gravitational theories require a temporary capture as a first step. Because dissipative diffusion is in general a slow process, long capture times enlarge the probability of decayment to the contition of eprmanent satellite. Is in this context were long-lasting temporary captures are of particular relevance. Longer capture times are in the regions near to stable periodic orbits around the planet, that lie necessarily close to the boundaries of the stability region in phase space, i. e. the region of permanent satellites.

In the frame of the Circular Restricted Three-Body Problem, libration-point capture may be interpreted as the transfer of a body from one mode of motion (heliocentric motion) to another mode (planetocentric motion). Thus, satellite capture can be thought as the competition between the primary masses for the possession of the particle. As expected in such a boundary regime, the structure of phase space is exceedingly complex.

Carusi and others have studied the phenomena of close encounters and temporary captures by means of numerical simulations (Carusi and Pozzi 1978; Carusi, Pozzi and Valsecchi 1979; Carusi and Valsecchi 1979; 1980a,b), whereas the stability and capture regions for retrograde satellites have been numerically explored by Huang and Innanen (1983).

The Jacobi constant C has been frequently used as a criterion for stability (Bailey 1972; Heppenheimer 1975; Brunini 1995) as stated in the theory of forbidden and permitted regions of motion (Szebehely 1967). Hunter (1967) and Hénon (1970) have shown that in fact this is true for direct orbits, whereas retrograde orbits have been found to be stable for much smaller values of C than the one corresponding to the inner libration point.

In this paper, special attention is paid to the question of the stability limit and the origin of chaos for retrograde satellites. Possible candidates to be captured as jovian satellites are also discussed in section 3 and the distribution of capture times is briefly analyzed in section 4.

The generalization of the previous results characterizing completely the phase-space covered with capture-escape orbits is outlined in section 5. In section 6 we discuss the consequence of considering the problem in the frame of the elliptic RTBP which is more suitable to be applied in Solar System problems.

The last section is devoted to conclusions.

2. Topology of the phase-space

Hénon (1970) studied the stability problem in satellite motion by means of Hill's formulation, which is an approximation to the circular RTBP when the mass ratio is near zero, i. e. well suited to satellites in the Solar System. The equations of the Hill's problem in rotating coordinates may be found in Hénon (1969). They allow a Jacobi integral whose expression is:

$$T = (C - 3)\mu^{-3/2}, \tag{1}$$

where C is the usual Jacobi constant of the RTBP (Moulton 1914) and μ is the mass ratio. In Hill's equations, the planet is in $x = y = 0$.

The circular RTBP, when limited to the plane case, is a system of two degrees of freedom with one integral of motion. The motion is thus confined to a three-dimensional manifold immersed in the four dimensional phase space, suitable to be investigated by the surfaces of section method, which is one of the most powerful tools for revealing the properties of the phase space. The conventional choice for the surface (Hénon 1970) is: $y = 0$, that is the $(x, \dot{x})$ plane, considering only intersections in the negative direction, i. e. with $\dot{y} < 0$.

Even for the plane circular case, the space of initial conditions to explore is fourth dimensional. Thus, restricting the initial conditions to satisfy the Mirror Theorem reduces it to two.

For $T = 5$ (Figure 2 in Hénon 1970) the transfer between heliocentric orbits and planetocentric orbits is not possible. Points with $x < 0$ represent retrograde satellites, whereas direct orbits are in the $x > 0$ region. At the resolution of the figure, quasi periodic trajectories appear to exist almost in all the accessible region. Retrograde and direct orbits are all stable. For values of $T > 5$ the picture is qualitatively similar.

For Jacobi constants smaller that the one corresponding to the inner libration point (i.e. for $T = 4.326749...$), however, the situation changes drastically. For the sake of clarity, the cases of direct and retrograde satellite orbits will be analyzed separately in the next sections.

2.1. DIRECT ORBITS

Surfaces of section for direct orbits were computed by Hénon for a sequence of the Jacobi constant in the range $T = 5$ to $T = 4.35$. For $T = 4.4$ (Fig. 4 in Hénon 1970) we see that the elliptic fixed point has given birth to a period 8 cycle, manifested by 8 stability islands. The main periodic family bifurcates, and consequently, a hyperbolic fixed point appears. This unstable periodic orbit gives rise to the chaotic sea exhibited in Figure 5 of Hénon's paper, which displays the surface of section for the case $T = 4.35$. Nevertheless, direct orbits can not escape from the planet sphere of influence on account that the Hill's curve is closed near the second libration point. For $T < 4.326749...$ the central region of the accessible phase space is open in the x direction, and the third body can escape. This is what happens in general, as the chaotic sea occupies almost all the available phase space. Regular orbits are found to fill a small region of phase space. We therefore conclude that escape (and therefore capture) of direct satellites is intimately associated with chaotic motion, and that the Jacobi constant is a useful stability criterion for direct orbits.

Murison (1989) has paid special attention to the sequence of bifurcations of periodic orbits. He has computed very accurate surfaces of section for the RTBP in the case $\mu = 0.01$. The most interesting surface of section is shown in his Figure 8c, where the oscillations between the stable and unstable manifolds and a subsequent homoclinic tangle are shown. Chaotic regions associated with hyperbolic points are interspersed between island elliptic points. The main feature of the satellite island structure is that they are self-similar.

2.2. RETROGRADE ORBITS

We have computed surfaces of section for retrograde orbits in the RTBP, in the case of μ corresponding to Jupiter's value.

For Jacobi constants greater than the one corresponding to the first collinear point ($C = 3.03844...$), regular trajectories fill practically all the accessible phase space. In addition, a new phenomenon is noted: the possibility of collisions and close encounters with the planet.

A frequent end-state of captured satellites is a collision onto the planet surface. Therefore, we have paid special attention to the conditions for collision orbits in the case of Jupiter. The region of phase-space full of collision

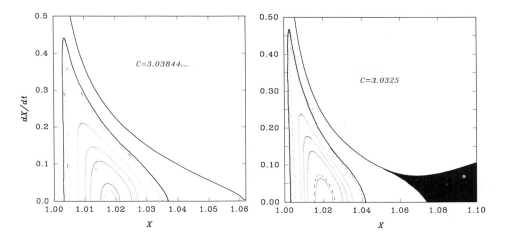

Figure 1. Surfaces of section for retrograde satellite orbits

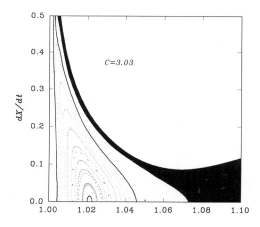

Figure 2. Surface of section for $\mu = 9.54 \times 10^{-4}$. Dark zones represent chaotic orbits. The zone between the chaotic region and the solid line corresponds to collision orbits.

orbits ($r_j \leq R_p$, where r_j is the distance between the particle and Jupiter, and R_p is Jupiter's radius) is represented as the white area, between the solid lines, in the surfaces of section of Figure 1b and Figure 2. They are all retrograde orbits and are placed just in the boundary of the stability region. It thus suggests that collision is the mechanism preventing retrograde orbits from escape, forming a separatrix between the two types of regime: heliocentric and planetocentric motion.

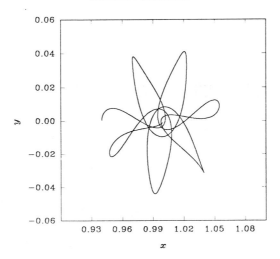

Figure 3. A typical direct-retrograde chaotic orbit. This satellite plows alternatively into the chaotic seas of direct and retrograde motion about the planet.

For C less than 3.025, KAM tori are surrounded by a chaotic region, and escape is then possible, as the chaotic regions extends up to initial conditions corresponding to heliocentric orbits. Collision orbits separate regular and chaotic regions in phase space. It suggests, at first sight, that chaos in retrograde motion would be originated by close encounters. Nevertheless, the chaotic region merges with the chaotic sea corresponding to direct orbits. The orbits of this region are alternatively direct and retrograde. Such a class of orbits is shown in Figure 3.

The stability region for retrograde orbits was explored in the space of initial conditions belonging to the Mirror Theorem.

As in some previous papers (Huang and Innanen 1983), we have defined a satellite as being stable if it is able to survive for more than 1000 orbits around the planet without escaping from the planet sphere of influence. The stable regions are shown in Figure 4.

Dark points represent regular regions, whereas blank spaces correspond to capture-escape regions. We see that regular orbits exist at almost all values of the Jacobi constant. Open circles are initial conditions corresponding to collision orbits. This diagram is somewhat different that the one computed by Huang and Innanen (1983) because they have erroneously used the diameter of Jupiter, rather than its radius, as the collision criterion.

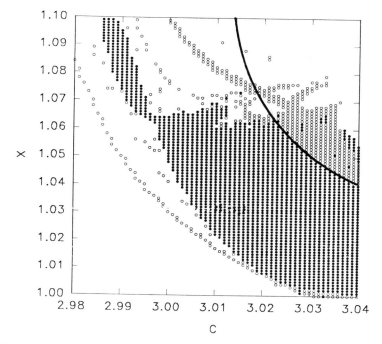

Figure 4. Dark dots: stable region. Open circles: collision orbits.
Full line: theoretical stability limit.

2.3. ANALYTIC DERIVATION OF THE STABILITY LIMIT

A first approximation to the limit of stability for retrograde orbits has been
given by Huang and Innanen (1983). They have shown that at the distance
to the planet

$$r \sim 0.84\mu^{1/3},\tag{2}$$

the total acceleration in the rotating frame falls to zero.

For the case of Jupiter $r = 0.084$, that should be compared with the
stability boundary of Figure 4.

A more accurate description of the stability limit for retrograde orbits
may be stated on rigorous dynamical grounds as follows: For regular orbits,
a second uniform integral confines the motion to an invariant manifold,
equivalent to a two dimensional torus in phase space. Such a second integral
of motion, besides the Jacobi constant, was formally found by Contopoulos

(1965) and Bozis (1966). Proceeding as in the two-body problem, we may write the angular moment integral as

$$\dot{x}y - \dot{y}x - r^2 = C_2 - \mu(1-\mu)\int_0^t y\left(\frac{1}{r_1^3} - \frac{1}{r_2^3}\right)dt. \tag{3}$$

The last integral in this expression exhibits small oscillations around a mean value, whose main frequency is half a planet year. The amplitude of the oscillations is in general small (Benner and McKinnon 1995). Neglecting this term, we may write, in usual polar coordinates:

$$C_2 = r^2\dot{\theta} + r^2. \tag{4}$$

The jacobi integral may also be written in polar coordinates (Moulton, 1914):

$$C = \dot{r}^2 + r^2\dot{\theta}^2 - (1-\mu)(r_1^2 + 1/r_1) + \mu(r_2^2 + 1/r_2). \tag{5}$$

Combining these two equations, we can eliminate $\dot{\theta}$ getting thus an expression for the limiting radius for regular orbits (we shall not write this rather complex expression here, which is only an algebraic exercise). In Figure 4, the theoretical limit (solid line) may be compared with the empirical stability limit for retrograde orbits. The agreement is remarkably good up to $C \sim 3.024$. The disagreement for $C \geq 3.024$ is originated by the presence of collision orbits. These orbits have a small angular momentum, and the integral term in the right hand side of eq. (3) can no longer be neglected. A first order approximation of this integral, might furnish a better description of the stability limit for retrograde orbits. Such an approximation may be found in the paper by Bozis (1966).

3. Heliocentric orbital elements

The results of the previous sections, suggest that captured satellites, such as Jupiter's irregular ones, originated in regions of chaotic motion about the Sun. In this section, we briefly discuss the possible candidates to be the progenitors of Jovian irregular satellites.

Sufficiently long time after a particle has escaped from Jupiter, its heliocentric orbital elements might be taken as the final elements (i.e.stables for a long time). We have thus computed heliocentric (a, e) elements for a sample of 5000 test escaping particles in the circular RTBP, Jupiter and the Sun being the primaries. The elements were computed at the instant 100 jovian years after escape. They are shown in Figure 5.

This figure should be compared with those obtained for the elliptic RTBP (Figure 1 of Huang and Innanen; 1983). The small eccentricity of Jupiter's

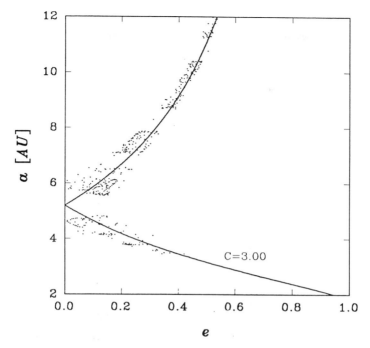

Figure 5. Heliocentric orbital elements of escape-capture bodies for the Jupiter-Sun case.

orbit is the reason for the fairly good concordance of both two distributions. The points in Fig. 5 are approximately distributed along two curves.
The motion being chaotic, no other uniform integral of motion exists and
therefore, these curves must necessarily be curves of $C = constant$.

Histograms of a and e are shown in Figure 6a and 6b respectively. We see
that the semiaxes distribute in three well definite spikes. The first one corresponds to $a = 3.95$ AU, which is very near the $3:2$ mean motion resonance
with Jupiter, whereas the clusters at 7.3 and 8.6 AU are associated, though
not exactly, with the $2:3$ and $1:2$ exterior mean motion resonances. The
reason of these accumulations remains to be investigated.

The histogram of eccentricities reveals that the preferred eccentricity of
capture is ~ 0.3. The reason on this fact is that this eccentricity corresponds
to close approaches with Jupiter of particles whose semiaxes are near the
accumulations in 3.95 AU and 7.3 UA.

It is instructive to compare the distribution of the semiaxes and eccentricities shown in Fig. 5 with those of the actual asteroids and short-period and
intermediate-period comets. They are plotted in Figure 7 where the curve
for $C = 3$ is also plotted. Among asteroids, some Hilda's overlap with this
curve. However, they are protected against capture because of the nature

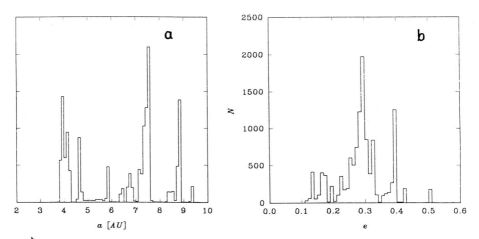

Figure 6. Distribution of semiaxes and eccentricities of escape-capture bodies in the Jupiter-Sun case.

of their resonant motion. We may conjecture that the present asteroids of the Hilda group are survivors of a more dense primordial population. Most of them were probably captured by Jupiter in the past the only steroids remaining are those protected by their libration motion.

Short period comets of the Jupiter family are plotted as open circles in Fig. 7. We observe that most of them are distributed very near the curve corresponding to capture conditions. Nevertheless, this diagram should not be taken too rigorously, as we have neglected the orbital inclination, which in fact decreases the probability of capture (Carusi and Valsecchi 1979).

4. Capture times

Capture times are defined as the number of orbits around the planet before the particle escapes into motion around the Sun. If the initial conditions are those satisfying the Mirror Theorem, the motion is symmetric with respect to the time and the x axis. In this situation, if the particle starts its motion in the sphere of influence of the smallest mass μ, the capture time is then twice the time elapsed before escape.

Initial conditions causing long captures are near the stability boundaries of phase space, and are associated with families of periodic orbits. As a consequence of this fact, the small-scale structure of capture times exhibits the same bifurcation sequence and self-similarity pattern as the periodic orbits (Murison, 1989). It is worth noting that there exist short lived orbits that are also periodic about both masses. Interesting examples of these class

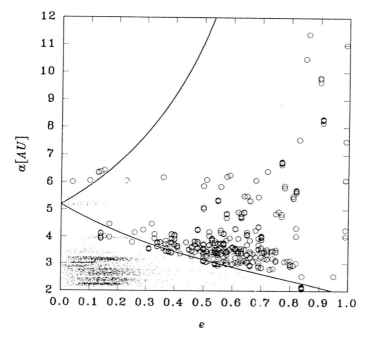

Figure 7. Distribution of actual asteroids (dots) and short period comets (circles)

are shown in Szebehely (1967, Fig. 9.24(b)) for the particular case of $\mu =$ 1/81.45, i. e. very near the Earth-Moon system. These class of periodic orbits are strongly unstable.

5. Complete set of capture conditions

As far as we know, all previous papers on satellite capture deal with particular sets of initial conditions, i. e. those satisfying the Mirror Theorem (Roy and Ovenden 1955). However, in order to estimate the probability of satellite capture, the complete set of conditions leading to capture is required.

A first attempt to resolve this problem has been carried out by Brunini et al. (1995).

In this paper, the massless particle has been initially placed on a straight line normal to the Sun-planet direction

$$x = x_0 = 1 - \mu - 1.1\rho \tag{6}$$

C=3.030

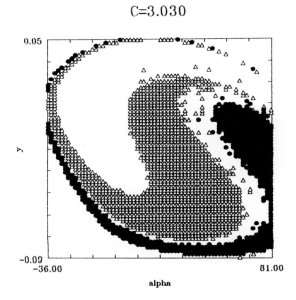

Figure 8. Capture initial conditions for the Jupiter-Sun case. Open triangles: direct orbits. Dark triangles: retrograde orbits. Collision orbits separate direct and retrograde orbits. In this figure $C = 3.024$.

where the value 1.1 - though arbitrary - guaranties the particle being initially outside the planetary sphere of influence (Danby 1988), since ρ is the distance between second libration point L_2 and the planet:

$$\rho = (\frac{\mu}{3})^{\frac{1}{3}} - \frac{1}{3}(\frac{\mu}{3})^{\frac{2}{3}} + ..., \tag{7}$$

In fact every heliocentric particle to be captured by the planet must necessarily pass through $x = x_0$. Given C, the only degree of freedom left is that of the direction of the velocity vector, defined by the angle with respect to the Sun-planet line, α, which may take values in the interval $(-90°, 90°)$.

For given C and μ, the initial conditions leading to capture cover the interior of a region in the (y, α) plane as the one shown in Figure 8, referred to as the "capture domain". Open circles correspond to direct orbits, whereas dark circles represent retrograde satellites. The empty area between retrograde and direct satellites are collision orbits. Brunini et al. (1995) have performed a great number of numerical simulations, characterizing capture conditions in terms of μ and C, and also generalizing these results to the 3D case.

6. Capture conditions in the elliptic RTBP

The main difficulty to study the problem in the frame of the elliptic RTBP is that an integral of motion, equivalent to Jacobi's, does no longer exist, and the convenient reduction of the degrees of freedom of the system is not allowed anymore. Since the Jacobi integral is not conserved, a particle crossing the plane $x = x_0$ with conditions belonging to the "capture domain" for a given value of C, could never be captured because of the shrinking of the neck around the second lagrangian point. Conversely, particles that would not be captured in the circular case could be captured when the planetary orbital eccentricity is considered.

As expected, these particles come from a thin bounding ring in the capture domain.

The net effect of the eccentricity of the primaries is the erosion and consequently the reduction of the stability region. Some numerical simulations (Brunini et al. 1995) have also revealed that the size of the stability region is properly scaled by the pericentric distance of the secondary rather that the orbital semimajor axis.

These results, though not conclusive, may be useful from a statistical point of view, guarantying that capture domains in the circular and actual cases will not differ much in the Solar System. A similar argument may be applied to capture times.

7. Conclusions

In this paper, we have made a revision of the main known aspects of the problem of satellite capture. In addition, we have shown that chaos for retrograde satellites is not originated in close encounters, but to an overlapping of capture conditions with those corresponding to direct motion about the planet. The chaotic regime at large values of the Jacobi constant is populated by a mixture between direct and retrograde chaotic orbits. Collisions play the formal role of separating direct and retrograde satellites.

Regular retrograde satellites exist for almost any value of the Jacobi constant, and the persistence of the angular momentum integral is the effective mechanism preventing from escape in this case.

We have also shown possible ways to generalize the results in order to characterize complete sets of capture conditions in phase space.

The effect of the eccentricity, in the range of interest for Solar System, does not affect much the results, at least from the statistical point of view.

References

Bailey, J. M., 1972, *J. Geophys. Res.* **76**, 7827.

Benner, L. A. M., McKinnon, W. B., 1994, *Icarus* **114**, 1.

Bozis, C., 1966, *Astron. J.* **71**, 404.

Brunini, A., 1995, *Earth. Moon and Planets* (in press).

Brunini, A. Giordano, C. M., Orellana, R. B., 1995, *Astron. Astrophys.* (submitted).

Carusi, A. Pozzi, E., 1978, *Moon Planets* **19**, 71.

Carusi, A. Pozzi, E. Valsecchi, G.B., 1979, in *Dynamics of the Solar System* (R. L. Duncombe ed.), D. Reidel, Dordrecht, 185.

Carusi, A. Valsecchi, G.B., 1979, in *Asteroids* (T.Gehrels ed.), Univ. Arizona Press, Tucson, 391.

Carusi, A. Valsecchi, G.B., 1980a, Moon Planets **22**, 113.

Carusi, A. Valsecchi, G.B., 1980b, Moon Planets **22**, 133.

Carusi, A. Kresák, L. Perozzi, E. Valsecchi, G.B., 1985, *Long-term Evolution of Short-period Comets*, Adam Hilger, Bristol.

Contopoulos, G., 1965, *Ap J* **142**, 802.

Danby, J.M.A., 1988, *Fundamentals of Celestial Mechanics*, The Macmillan Company, New York.

Goldreich, P., Murray, N., Longaretti, P. Y., Bandfield, D., 1989, *Science* **245**, 500.

Hénon, M., 1969, *Astron. Astrophys.* **1**, 223.

Hénon, M., 1970, *Astron. Astrophys.* **9**, 24.

Heppenheimer, T. A., 1975, *Icarus* **24**, 172.

Hopf, E., 1930, *Math. Ann.* **103**, 710.

Huang, T.-Y., Innanen K.A., 1983, Astron. J. **88**,1537.

Hunter, R. B., 1967, *Mon. Not. Roy. Astron. Soc.* **136**, 245.

Kuiper, G.P., 1961, in *Planets and Satellites*, (eds. Kuiper, G.P. & Middlehurst, B.M.) University of Chicago Press, 575.

Moulton, F.R., 1914, *An Introduction to Celestial Mechanics*, The Macmillan Company, New York.

Murison, M., 1989, Astron. J. **98**, 2346.

Rickman, H., 1979, in *Dynamics of the Solar System* (R.L. Duncombe ed.), D. Reidel, Dordrecht, 293.

Roy, A.E., Ovenden, M.W., 1955, Mon. Not. R. Astron. Soc. **115**, 296.

Szebehely, V., 1967, *Theory of Orbits*, Academic Press, New York.

Tancredi, G., Lindgren, M., Rickman, H.: 1990, Astron. Astrophys. **239**, 375.

CHAOTIC TRANSITIONS IN RESONANT ASTEROIDAL DYNAMICS

S. FERRAZ-MELLO, J.C. KLAFKE, T.A. MICHTCHENKO and
D. NESVORNÝ
Universidade de São Paulo
Instituto Astronômico e Geofísico
Caixa Postal 9638, São Paulo, Brasil.
sylvio @ vax.iagusp.usp.br

Abstract. The utilization of chaotic dynamics approaches allowed the identification of many modes of motion in resonant asteroidal dynamics. As these dynamical systems are not integrable, the motion modes are not separated and one orbit may transit from one mode to another. In some cases, as in the 3/1 resonance, these transitions may lead, in a relatively short time scale, to eccentricities so high that the asteroid may approach the Sun and be destroyed. In the 2/1 and 3/2 resonances these transitions are much slower and only indirect estimations of the time which is needed for a generic asteroid to leave the resonance are possible. It may reach hundreds of million years in the more robust regions of the 2/1 resonance and a time of the order of billions of years in those of the 3/2 resonance. These values are consistent with the observed depletion of the 2/1 resonance (only a few asteroids known while almost 60 asteroids are known in the 3/2 resonance).

Key words: asteroids – resonance – chaos – Kirkwood gaps

1. Introduction

The dynamics of resonant asteroids is a research domain that has experienced a rapid progression in recent years. After decades of belief on the insufficiency of gravitation alone to explain the peculiarities of the asteroids distribution, the pioneer work of Wisdom (1982) shed a first light on this problem showing that the fate of the missing asteroids in the resonance 3/1 could have been ruled by chaotic diffusion of the orbits and mode transitions (intermittences) increasing the eccentricity and leading the orbit to intersect many times the orbit of Mars. Because of the short time-scale of these intermittences ($10^5 - 10^6$ years), sooner or later, a close approach may have occurred and the asteroid was scattered from the resonance. Since then, this scenario was enriched. The study of the very 3/1 resonance has shown intermittences still larger and giving rise to orbits diving deeply in the inner Solar Systems while the study of the outer resonances 2/1 and 3/2 revealed the importance of close approaches to Jupiter. We review some of these new results in this paper.

If the mechanisms acting in the 3/1 and other inner belt resonances are well known, the main resonances of the outer belt are still a source of problems to be solved. The most puzzling situation is the difference of the asteroid distribution in the 3/2 and 2/1 resonances. The three-body models of these resonances show very similar dynamics (Henrard and Lemaître, 1987; Lemaître and Henrard, 1988; Ferraz-Mello *et al.*, 1995; Michtchenko and Ferraz-Mello, 1995, 1996). However, the distribution of the asteroids shows a gap, the *Hecuba gap* at the place of the 2/1 resonance, with only a few asteroids observed inside the resonance boundaries while it shows a group, the *Hilda group*, with almost 60 known asteroids, in the 3/2 resonance. We cannot say that we understand this difference but taking into account that global stochasticity is observed when more realistic models are considered and that the maximum Lyapunov exponents in the Hilda group are much smaller than those of the 2/1 resonance, we have proposed (Ferraz-Mello, 1994b; Ferraz-Mello *et al.*, 1995) that the same mechanisms that depleted the 2/1 resonance are acting in the 3/2 resonance but with such a slower pace that the age of the Solar System was not large enough to allow the depletion to be completed. To have a direct confirmation of this fact we should be able to construct analytical or numerical theories valid for some billion years, what is not yet the case.

2. The 3/1 resonance

The 3/1 resonance is the best known one. Until 1982, the only mode of motion known in this resonance was the ordinary low-eccentricity regime where, typically, the perihelion retrogrades and the eccentricity has a small periodic variation (for very low eccentricities, the asteroid's perihelion may oscillate about Jupiter's one). Some numerical experiments had shown anomalous increases of the eccentricity (Scholl and Froeschlé, 1974), but its dynamics started to be unravelled only when Wisdom (1982, 1983, 1985) discovered the mid-eccentricity mode of motion, in which the asteroid's perihelion oscillates about the position of Jupiter's perihelion and the eccentricity has large oscillations approaching values as high as 0.4. Wisdom showed also that orbits having a seemingly regular low-eccentricity motion for long times and suddenly transiting to the mid-eccentricity regime are common. Generic orbits alternate between these two modes of motion in a short timescale (much less than 1 Myr). This bifurcation is, in fact, a very complex one since the separatrix between the two modes of motion of the perihelion intersects the separatrix of the two modes of variation of the critical angle

$$\sigma = (r+1)\lambda_{\text{Jup}} - r\lambda - \varpi$$

(λ and ϖ denote, respectively, the asteroid's mean longitude and longitude of the perihelion, r is a rational number and $(r+1)/r$ is the commensurability ratio; in this resonance $r = \frac{1}{2}$). σ librates outside this separatrix and circulates inside it. Neishtadt (1987) has shown that dynamics of the

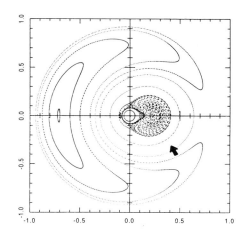 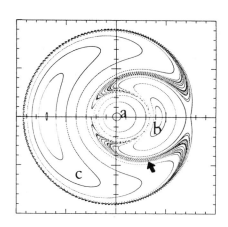

Figure 1. Poincaré maps ($\sigma = \pi/2$, $\dot{\sigma} < 0$) of the resonance 3/1 in the frame of the planar averaged Sun-Jupiter-asteroid problem at two different energy levels. Coordinates are $x = e.\cos(\varpi - \varpi_{\text{Jup}})$, $y = e.\sin(\varpi - \varpi_{\text{Jup}})$. On the left side, the chaotic domain found by Wisdom is seen in the inner part of the figure. It is confined by a bunch of almost regular motions (arrow). On the right side, these motions no more exist and a heteroclinic bridge (arrow) allows the transitions between (b) and (c) (after Ferraz-Mello and Klafke, 1991).

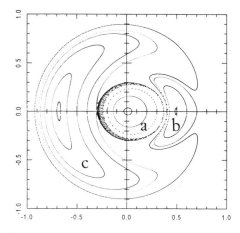 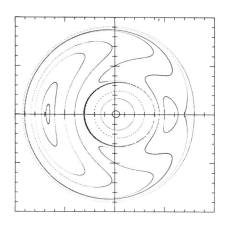

Figure 2. Same as fig.1 for lower energies. On the right side, (a) and (b) are separated and transitions may occur directly between (a) and (c).

transitions between the two modes of motion may be given by a mapping involving some random quantities.

From the cosmogonic point of view, the important fact is that in the mid-eccentricity regime, the asteroid orbit crosses the orbit of Mars and the asteroid may, eventually, have a close approach to the planet. When this happens, the orbit suffers an important energy change and leaves the resonance. The possibility of this scattering could explain the observed absence of permanent asteroids in the 3/1 resonance. The only doubts concerning this scenario come from the fact that Mars is a small planet and we do not know if this process could have been efficient enough to expel all asteroids expected to be there after the formation of the asteroid belt, 4.5 billion years ago.

Wisdom's hypothesis set a new paradigm in asteroidal dynamics and inspired all subsequent investigations which generalized his results in several directions. The work of Wisdom has been extended by Ferraz-Mello and Klafke (1991) through the use of new models for the averaged potentials allowing to study high-eccentricity orbits. They have shown the existence of a very-high-eccentricity mode of motion, associated with the secular resonance $\varpi - \varpi_{\mathrm{Jup}}$, in which the eccentricity oscillates in a wide range reaching values close to 1 and the perihelion may perform one complete revolution before the eccentricity decreases again. In the energy range considered by Ferraz-Mello and Klafke, this regime is almost always separated from the inner ones by some regular motions (fig.1 left) able to avoid transitions between them, at least in a timescale as short as the one observed in the transitions between the low- and the mid-eccentricity regimes. However, Ferraz-Mello and Klafke have shown that, decreasing the energy, these seemingly regular orbits cease to exist and a heteroclinic bridge appears allowing the solutions to go from one regime to another (fig.1 right) and the eccentricity to grow up to values close to 1. Decreasing the energy still more, the modes of motion studied by Wisdom become parted (fig. 2) and direct transitions between low and high-eccentricities become possible (Klafke *et al.* 1992).

The intermittences involving this new mode of motion can, now, drive the asteroidal eccentricity to values close to 1 and back. In this case, the asteroid will not only cross the orbit of Mars, but also those of the Earth and Venus, planets which are 10 times more massive than Mars.

When the effects of the outer planets are considered, excursions to very high eccentricity are the rule (see Moons and Morbidelli, 1995). Farinella *et al.* (1993) and Morbidelli and Moons (1995) have shown orbits whose eccentricity is high enough to allow the asteroid to fall in the Sun. In fact, the asteroid becomes liable to be disrupted and transformed into meteoroids even if the eccentricity is not enough close to 1 to allow it to reach the Sun. The Roche limit, in this case, is only 2 solar radii, but this value considers a scenario purely hydrostatic; the heating of an heterogeneous asteroid gives rise to internal forces, to be added to the tidal ones, and it may be disrupted even it does not reach the Roche limit.

3. The 2/1 resonance

The study of the several modes of motion, in the case of the 2/1 resonance, has been done by means of numerical integrations and the smoothing of the outputs with low-pass digital filters (Michtchenko, 1993; Michtchenko and Ferraz-Mello, 1993, 1995). In the case of the planar Sun-Jupiter-asteroid model, these smoothed integrations may be interpreted as solutions of an averaged dynamical system with 2 degrees of freedom and allow the construction of Poincaré maps (Ferraz-Mello, 1994b). Figure 3 shows some of these maps obtained from 1 Myr numerical integrations. In these maps, the low- and mid-eccentricity modes of motion of the 3/1 resonance are almost absent; they are only seen in maps corresponding to orbits with very large libration amplitudes ($\Delta\sigma > 200°$). For lower libration amplitudes, these regimes are substituted by a single chaotic low-eccentricity zone (Giffen, 1973; Froeschlé and Scholl, 1981; Murray, 1986). Lemaître and Henrard (1990) have shown that this chaotic zone has its origin in the existence of resonances between the libration of the critical angle σ and the perihelion motion. There is, in fact, a succession of microregimes of motion corresponding to these secondary resonances whose overlaps allow an orbit to transit through them. This chaotic zone is confined to low-eccentricities by apparent regular motions visible between $e \sim 0.2$ and $e \sim 0.5$ and is only slightly affected by the consideration of the inclinations and the web of secondary resonances involving also the motion of the node (Michtchenko and Ferraz-Mello, 1996). However, when these apparently regular regions are studied with the fine tools of wavelet and frequency analyses, the chaoticity associated with secondary resonances of higher orders becomes apparent and, when Saturn is added to model, transitions between neighbouring secondary resonances occur (Michtchenko and Nesvorný, 1996; Nesvorný and Ferraz-Mello, 1996).

The high-eccentricity mode of motion of the 3/1 resonance has, in the 2/1 resonance, two counterparts: One on the left side of the Poincaré map (asteroid's perihelion librating about Jupiter's aphelion) and another on the right side (asteroid's perihelion librating about Jupiter's perihelion). The motion in these regimes is such that both the critical angle σ and the perihelion are librating. The mode of motion in the right-side lobe is similar to the high-eccentricity regime of the 3/1 resonance: both grow up from the exact stable corotation existing at the maximum possible energy, that is, from an orbit where both the critical angle and the perihelion longitude are

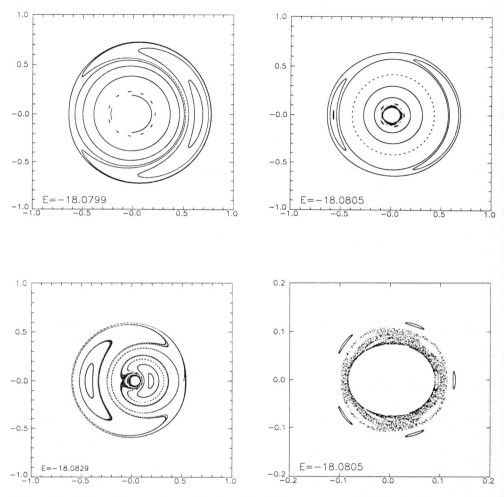

Figure 3. Poincaré maps ($\sigma = 0$, $\dot{\sigma} > 0$) of the resonance 2/1 in the frame of the planar averaged Sun-Jupiter-asteroid problem and magnification of the inner chaotic zone. Coordinates as in fig. 1. (from Ferraz-Mello, 1994b)

constant (see Ferraz-Mello 1989; Yoshikawa, 1989; Morbidelli and Giorgilli, 1990; Ferraz-Mello *et al.*, 1993).

The separation between the high-eccentricity lobes and the low-eccentricity chaotic zone persists even when the long-period perturbations of the orbit of Jupiter are considered (Morbidelli and Moons, 1993). The robustness of this separation led many authors to find impossible to explain the Kirkwood 2/1 gap by chaotic diffusion followed of gravitational scattering, as the 3/1 gap. However, when the full action of Saturn is considered all solutions become clearly chaotic. Maximum Lyapunov exponents were

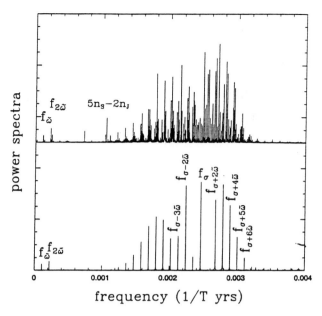

Figure 4. FFT power spectra of solutions starting at $e_0 = 0.3$ in the frame of two different models. *Bottom*: Sun-Jupiter-asteroid. *Top*: Sun-Jupiter-Saturn-asteroid.

estimated from some fifty 5–7 Myr integrations, for initial eccentricities in the interval $0.1 < e < 0.4$ and initial semi-major axes ranging from the middle to the border of the resonance region. The logarithmic median of the values found is $10^{-4.4\pm0.4}$ yr^{-1}, (Ferraz-Mello, 1994b). The disturbances due to Saturn are also clearly noted when the results of integrations are Fourier analyzed. Figure 4 shows FFT spectra of the semi-major axis of solutions whose initial conditions are $a_0 = 3.283$ AU, $e_0 = 0.30$, $I_0 = 1°$, $\sigma_0 = \Delta\varpi_0 = \Delta\Omega_0 = 0$. The power spectrum calculated over 5 Myrs in the frame of the Sun-Jupiter-asteroid model (fig. 4 *bottom*) shows well defined spectral lines associated with two independent modes: the libration, with frequency f_σ, and the circulation of the perihelion, with frequency f_ϖ, as well as their linear combinations. The action of Saturn introduces many additional spectral lines, mainly associated with g_6, creating multiplets of linear combinations with other independent modes (fig. 4 *top*). This clearly indicates chaotic motion. These results show that Saturn triggers the destruction of the almost regular structures separating the low-eccentricity chaotic region and the high-eccentricity lobes. In fact, KAM tori are not expected to exist in this system, since Jupiter's and Saturn's masses are much larger than the most optimistic evaluations of the small-parameter values of KAM and Nekhoroshev theories (Henon, 1967; Niederman, 1995).

We may repeat Guckenheimer (1991) and say that "labelling a system as chaotic does little good if that is the end of the story". It is now necessary to explain the origin of this chaoticity. Current researches (Henrard et al., 1995; Michtchenko and Ferraz-Mello, 1996) are devising intermittent mechanisms associated with secular resonances, secondary resonances and Kozai's ω-resonance able to produce chaotic behaviour and, eventually, lead to important transition between very different modes of motion. Recent investigations by Ferraz-Mello (1996) using a symplectic mapping have shown that the short-period perturbations of Jupiter's orbit associated with the angles $5\lambda_{Sat} - 2\lambda_{Jup}$ and $2\lambda_{Sat} - \lambda_{Jup}$ play a decisive role at least in accelerating the diffusion processes. One may note that the spectral line associated with the frequency $5n_{Sat} - 2n_{Jup}$ is clearly seen in the spectrum of fig. 4 (*top*).

The Hecuba gap is the cosmogonic consequence of this global stochasticity. Exploratory numerical integrations show the slow diffusion and, some of them, the final scattering of the orbit by Jupiter. The problem we face is the slowness of the diffusion. Simulations with a planar symplectic mapping have shown characteristic times of hundreds of Myrs, but some orbits remained in the resonance even after 1 Gyr (Ferraz-Mello, 1996). Some predictions can be done using the Lecar-Franklin formula which associates the Lyapunov maximum exponents and the so-called "crossing time" ($T_c \propto L_{max}^{-1.8\pm0.1}$) (see Murison *et al.,* 1994). The validity of this formula is still a controversial subject, but recent investigations have shown that when the perturbations are large enough to make the diffusion to be determined by the transitions between neighbouring secondary resonances (the so-called *Chirikov regime*), an exponential law of this kind is expected to hold even if the exponent -1.8 is not universal (Morbidelli and Froeschlé, 1996). We have used a planar symplectic mapping to study this relation in the case of this resonance: values in the range -1.7 to -1.9 were generally found (but in 25% of the experiments, the simulation time – 100 Myr – was not sufficient to have any macroscopic instability). This means that the most frequent "crossing times" are expected to be around 10^8 years. We have made a dozen exploratory numerical integrations of the exact equations in the central part of the resonance. Only 2 of them did not lead to escape from the resonance. Usually a jump in eccentricity to 0.9, or more, occurs in a time shorter than expected (from 15 to 80 Myr) followed by the escape from the resonance. These integrations confirm the stochasticity of the 2/1 asteroidal resonance and its role in the formation of the Hecuba gap.

4. The 3/2 resonance

The 3/2 resonance has been studied using the same techniques used to study the 2/1 resonance. Figure 5 shows two Poincaré maps of this resonance. They

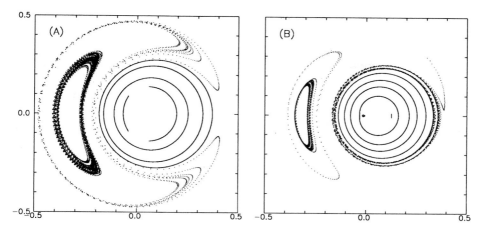

Figure 5. Poincaré maps ($\sigma = 0$, $\dot{\sigma} > 0$) of the resonance 3/2 in the frame of the planar averaged Sun-Jupiter-asteroid problem. Coordinates as in fig. 1. Orbits in the perihelion libration lobe are highly chaotic and are bound to close approaches to Jupiter in short times. Orbits in the innermost part remain regular for long times even when inclinations and perturbations of Saturn are taken into account. The actual Hildas are in the inner region (from Ferraz-Mello, 1994a).

differ from those of fig. 3 in several aspects. For instance, we see only two modes of motion: the inner domain of perihelion circulation and the perihelion libration lobe on the left side. At variance with the 2/1 resonance, no apparent chaotic activity exists in the region of perihelion circulation (confirmed by discrete spectra and Lyapunov exponents which tend to zero in numerical integrations over 10 Myr); these solutions show the same kind of secondary resonances web responsible for the inner chaotic region in the 2/1 resonance, but the chaotic regions associated with every secondary resonance in the web seem to be very narrow and without apparent overlap (Michtchenko, 1993; Michtchenko and Ferraz-Mello, 1995). The only source of appreciable chaoticity is the bifurcation between the two modes of oscillation of the perihelion. Strong chaos is visible spreading out over a large part of the perihelion libration lobe and in the outer region around the two regimes. The non-existence of observed orbits with mean eccentricities larger than ~ 0.3 is due to the fact that the outer orbits are scattered by approaches to Jupiter itself. The inclusion of Saturn accelerates the above described phenomena and allows the high-eccentricity orbits to be scattered in less than 1 Myr (see Morbidelli and Moons, 1993).

The spectral analysis of the output of numerical integration also shows a complex of chaotic orbits with eccentricities lower than 0.1 associated with secondary resonances involving the libration frequency f_σ and twice the frequency of the argument of the perihelion motion $f_{2\omega}$ ($\omega = \varpi - \Omega$) (Michtchenko and Ferraz-Mello, 1996). The rapid transitions between these

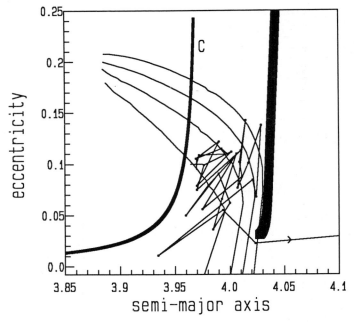

Figure 6. Chaotic diffusion of a solution through the secondary resonances $f_\sigma/f_{2\omega}$ from 1/1 to 1/4. C is the pericentric branch of libration centers in the 3/2 resonance. Initial position of the asteroid is shown by a cross.

resonances drive the orbit away from this region and thus explains the non-existence of orbits with small e among the Hildas.

As one example, we show the chaotic diffusion of the solution with initial conditions ($a_0 = 3.970$ AU, $e_0 = 0.10$, $I_0 = 15°$, $\sigma_0 = \Delta\varpi_0 = \Delta\Omega_0 = 0$) inside the zone of overlap of the secondary resonances of the type $f_\sigma/f_{2\omega}$. Figure 6 shows the (a, e)-plane for $I_0 = 15°$, $\sigma_0 = \Delta\varpi_0 = \Delta\Omega_0 = 0$, $\dot{\sigma} < 0$. The curves inside the resonance region indicate the locus of the secondary resonances, from 1/1 to 1/4. The rapid transition across the overlapping zone produces the slow chaotic diffusion of the orbit along a band of overlapping resonances (the so-called modulational diffusion) and the escape after 10 Myrs, when the orbit approaches the resonance border indicated by the thick curve.

The few existing low-eccentricity *Hildas* are outside this overlapping zone. The spreading of this chaotic region when the inclination increases is also in agreement with the observed distribution of the Hildas. The Lyapunov maximum exponents of the inner regular orbits calculated with this more complete model are in the range $10^{-5.5} - 10^{-7}\text{yr}^{-1}$. The cosmogonic implication of these small values is that the chaotic processes at work in the domain where the Hildas are found, which are the same acting in the 2/1 resonance are, now, about one hundred times slower. We may conjecture that the same process responsible for the depletion of the 2/1 resonance are depleting the 3/2 resonance, but at a slower pace and the time required to complete this depletion is some order of magnitude larger than the Solar System age.

5. Conclusion

We reviewed the main asteroidal mean-motion resonances emphasizing the several modes of motion possible in each case and the chaotic transitions between these modes. The results given are based on maps and simulations extending over $10^6 - 10^7$ years.

In what concerns the 3/1 resonance the main results refer to the possibility of orbits having their eccentricities raised to values very close to 1 and having catastrophic approaches to the Sun.

In what concerns the 2/1 and 3/2 resonances, the main qualitative result is the enormous similarity of their modes of motion. When the effects of Saturn are not taken into account both resonances shows very regular regions. When Saturn is taken into account the calculation of the Lyapunov maximum exponents shows huge differences in these two cases pointing to a slowness of the chaotic diffusion in the regular domain of the 3/2 resonance. However, we are dealing with slow processes and the necessary timespan to unravel the dynamical mechanisms at work in these resonances is of the order of $10^8 - 10^9$ years. The role of the global stochasticity of the 2/1 resonance in the formation of the Hecuba gap is just a first result. More can be expected in the near future.

References

Farinella, P., Froeschlé, Ch. and Gonczi, R.: 1993, 'Meteorites from the asteroid 6 Hebe', *Celest. Mech. Dyn. Astron.* **56**, 287-305.

Ferraz-Mello, S.: 1989, 'A semi-numerical expansion of the averaged disturbing function for some very-high-eccentricity orbits' *Celest. Mech. Dyn. Astron.* **45**, 65-68.

Ferraz-Mello, S.: 1994a, 'Kirkwood Gaps and Resonant Groups', in A.Milani *et al.*, (eds.), *Asteroids, Comets, Meteors 1993* Kluwer, Dordrecht, 175-188.

Ferraz-Mello, S.: 1994b, 'Dynamics of the asteroidal 2/1 resonance', *Astron. J.* **108**, 2330-2338.

Ferraz-Mello, S.: 1996, 'A symplectic mapping approach of the 2/1 asteroidal resonance', submitted.

Ferraz-Mello, S., Dvorak, R. and Michtchenko, T.A.: 1995, 'Depletion of the asteroidal belt at resonances', in A.E.Roy and B.A.Steves (eds.) *From Newton to Chaos*, Plenum Press, New York, 1995, pp. 157-169.

Ferraz-Mello, S. and Klafke, J.C.: 1991, 'A model for the study of very-high-eccentricity asteroidal motion. The 3:1 resonance', in A.E.Roy, (ed.), *Predictability, Stability and Chaos in N-Body Dynamical Systems*, Plenum Press, New York, 177-184.

Ferraz-Mello, S., Tsuchida, M. and Klafke, J.C.: 1993, 'On symmetrical planetary corotations', *Celest. Mech. Dyn. Astron.* **55**, 25-46.

Franklin, F., Lecar, M. and Murison, M.: 1993, 'Chaotic orbits and long-term stability: an example from asteroids of the Hilda group', *Astron. J.* **105**, 2336-2343.

Froeschlé, C. and Scholl, H.: 1981, 'The stochasticity of peculiar orbits in the 2/1 Kirkwood gap', *Astron. Astrophys.* **93**, 62-66.

Giffen, R.: 1973, 'A study of commensurable motion in the asteroidal belt', *Astron. Astrophys.* **23**, 387-403.

Guckenheimer, J.: 1991, 'Chaos: Science or Non-science', *Nonlinear Science Today* **1**(2), 6-9.

Henon, M.: 1967, 'Exploration Numérique du Problème Restreint - IV', *Bulletin Astronomique* **1**, 125-142.

Henrard, J. and Lemaître, A.: 1987, 'A perturbative treatment of the 2/1 Jovian resonance', *Icarus* **69**, 266-279.

Henrard, J., Watanabe, N. and Moons, M.: 1995, 'A bridge between secondary and secular resonances inside the Hecuba gap', *Icarus* **115**, 336-346.

Klafke, J.C., Ferraz-Mello, S. and Michtchenko, T.: 1992, 'Very-high-eccentricity librations at some higher-order resonances', in S.Ferraz-Mello (ed.), *Chaos, Resonance and Collective Dynamical Phenomena in the Solar System*, Kluwer, Dordrecht, 153-158.

Lemaître, A. and Henrard, J.: 1988, 'The 3/2 resonance', *Celest. Mech.* **43**, 91-98.

Lemaître, A. and Henrard, J.: 1990, 'Origin of the chaotic behaviour in the 2/1 Kirkwood gap', *Icarus* **83**, 391-409.

Michtchenko, T.A.: 1993, *Dr. Thesis*, University of São Paulo.

Michtchenko, T.A. and Ferraz-Mello, S.: 1993, 'The high-eccentricity libration of the Hildas. II. Synthetic-theory approach', *Celest. Mech. Dynam. Astron.* **56**, 121-129.

Michtchenko, T.A. and Ferraz-Mello, S.: 1995, 'Comparative study of the asteroidal motion in the 3:2 and 2:1 resonances with Jupiter-I', *Astron. Astrophys.* **303**, 945.

Michtchenko, T.A. and Ferraz-Mello, S.: 1996, 'Comparative study of the asteroidal motion in the 3:2 and 2:1 resonances with Jupiter-II', *Astron. Astrophys.* (in press).

Michtchenko, T.A. and Nesvorný, D.: 1996, 'Wavelet analysis of the asteroidal resonant motion', *Astron. Astrophys.* (submitted).

Moons, M. and Morbidelli, A.: 1995, 'Secular resonances in mean-motion commensurabilities. The 4/1, 3/1, 5/2 and 7/3 cases', *Icarus* **114**, 33-50.

Morbidelli, A. and Froeschlé, C.: 1996, 'On the relationship between Lyapunov times and macroscopic instability times', *Astron. J.* (submitted).

Morbidelli, A. and Giorgili, A.: 1990, 'On the dynamics of the asteroids belt. Part II: Detailed study of the main resonances', *Celest. Mech. Dynam. Astron.* **47**, 173-204.

Morbidelli, A. and Moons, M.: 1993, 'Secular resonances in mean-motion commensurabilities. The 2/1 and 3/2 cases', *Icarus* **102**, 316-332.

Morbidelli, A. and Moons, M.: 1995, 'Numerical evidences of the chaotic nature of the 3/1 mean-motion commensurability', *Icarus* **115**, 60-65.

Murison, M.A., Lecar, M. and Franklin, F.A.: 1994, 'Chaotic motion in the outer asteroid belt and its relation to the age of the solar system', *Astron. J.* **108** 2323-2329.

Murray, C.D.: 1986, 'Structure of the 2:1 and 3:2 Jovian Resonances', *Icarus* **65**, 70-82.

Neishtadt, A.I.: 1987, 'Jumps in the adiabatic invariant on crossing the separatrix and the origin of the 3:1 Kirkwood gap', *Dokl. Akad. Nauk SSSR* **295**, 47-50.

Nesvorný, D. and Ferraz-Mello, S.: 1996, 'The chaotic diffusion in the 2/1 asteroidal resonance', *Astron. Astrophys.* (submitted).

Niederman, L.: 1995, 'Stability over exponentially long times in the planetary problem', in A.E.Roy and B.A.Steves (eds.) *From Newton to Chaos*, Plenum Press, New York, 1995, pp. 109-118.

Scholl, H. and Froeschlé, C.: 1974, 'Asteroidal motion at the 3/1 commensurability', *Astron. Astrophys.* **33**, 455-458.

Wisdom, J.: 1982, 'The origin of Kirkwood gaps: A mapping for asteroidal motion near the 3/1 commensurability', *Astron. J.* **85**, 1122-1133.

Wisdom, J.: 1983, 'Chaotic behaviour and the origin of the 3/1 Kirkwood gap', *Icarus* **56**, 51-74.

Wisdom, J.: 1985, 'A perturbative treatment of motion near the 3/1 commensurability', *Icarus* **63**, 279-282.

Yoshikawa, M.: 1989, 'A survey of the motions of asteroids in commensurability with Jupiter', *Astron. Astrophys.* **213**, 436-458.

A NOTE CONCERNING THE 2:1 AND THE 3:2 RESONANCES IN THE ASTEROID BELT

JACQUES HENRARD
Département de mathématique FUNDP
8, Rempart de la Vierge, B-5000 Namur, Belgique

Abstract. The dynamical behavior of asteroids inside the 2:1 and 3:2 commensurabilities with Jupiter presents a challenge. Indeed most of the studies, either analytical or numerical, point out that the two resonances have a very similar dynamical behavior. In spite of that, the 3:2 resonance, a little outside the main belt, hosts a family of asteroids, called the Hildas, while the 2:1, inside the main belt, is associated to a gap (the Hecuba gap) in the distribution of asteroids.

In his search for a dynamical explanation for the Hecuba gap, Wisdom (1987) pointed out the existence of orbits starting with low eccentricity and inclination inside the 2:1 commensurability and going to high eccentricity, and thus to possible encounters with Mars. It has been shown later (Henrard *et al.*), that these orbits were following a path from the low eccentric belt of secondary resonances to the high eccentric domain of secular resonances. This path crosses a bridge, at moderate inclination and large amplitude of libration, between the two chaotic domains associated with these resonances.

The 3:2 resonance being similar in many respects to the 2:1 resonance, one may wonder whether it contains also such a path. Indeed we have found that it exists and is very similar to the 2:1 one. This is the object of the present paper.

Key words: Asteroids – Kirkwood gap – chaotic motion

1. The Planar Problem

The Hamiltonian of the restricted three body problem (Sun-Jupiter-Asteroid) can be written

$$H = -\frac{1-\mu}{2a} - \mu \left[\frac{1}{|\mathbf{r} - \mathbf{r}'|} - \frac{(\mathbf{r} \mid \mathbf{r}')}{r'^3} \right]$$

where $(1 - \mu)$ and μ are the reduced mass of the Sun and Jupiter, a (resp. a') the semi-major axis of the asteroid (resp. Jupiter) and $\mathbf{r}$ (resp. $\mathbf{r}'$) the position vectors (relative to the Sun). We are considering here that the Hamiltonian function is expressed implicitly in terms of the usual modified Delaunay's elements:

$$
\begin{aligned}
\lambda \;\; &= \;\; \text{mean longitude of the asteroid} & L &= \sqrt{(1-\mu)a}\,, \\
-p \;\; &= \;\; \text{longitude of the pericenter} & P &= L(1 - \sqrt{1 - e^2})\,, \\
-q \;\; &= \;\; \text{longitude of the node} & Q &= (L - P)(1 - \cos I)\,.
\end{aligned}
$$

The Hamiltonian function is also a periodic function of the time through its dependence upon the mean longitude λ' of Jupiter. In models more

Celestial Mechanics and Dynamical Astronomy **64**: 107–114, 1996.
© 1996 *Kluwer Academic Publishers. Printed in the Netherlands.*

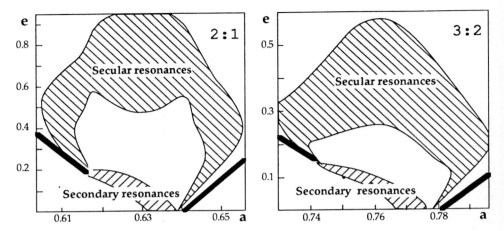

Figure 1. Localization of the chaotic domains in the planar problem for the 2:1 commensurability (to the left) and the 3:2 commensurability (to the right). The diagrams map the initial values of semi-major axis a and eccentricity e of orbits starting at $\sigma = 0$ and $\nu = \pi$ or $\nu = 0$. The shaded area corresponds to initial condition of orbits in the chaotic layers. The bold lines represent the position in these diagrams of the separatrices marking the boundaries of the mean motion resonances as explained in the text. The Figure summarizes the results of semi-analytical perturbation theories (Henrard and Lemaître, 1986, Lemaître and Henrard, 1988; Morbidelli and Moons, 1993).

sophisticated than the restricted three body problem, one can assume that Jupiter itself is quasi-periodically perturbed by the other planets and then the Hamiltonian function is a quasi-periodic function of the time.

In the case where there exists a mean motion resonance

$$(i + j)n' - in \simeq 0 ,$$

between the unperturbed mean motion of the asteroid, $n = [(1 - \mu)/a^3]^{1/2}$ and the mean motion n' of Jupiter, one introduces the Poincaré's resonance variables (Poincaré, 1902; Schubart, 1966):

$$
\begin{aligned}
\sigma &= [(i + j)\lambda' - i\lambda + jp]/j , & S &= P , \\
\sigma_z &= [(i + j)\lambda' - i\lambda + jq]/j , & S_z &= Q \\
\nu &= -[(i + j)\lambda' - i\lambda - j\varpi']/j , & N &= jL/i + P + Q .
\end{aligned}
$$

and average the Hamiltonian over the remaining "fast variable", the mean longitude of Jupiter. The fast frequency being much faster than the remaining ones, this operation does not oversimplify the system.

In the case of the *planar* problem, when Jupiter and the asteroid move in the same inertial plane, the momentum S_z is zero and the corresponding

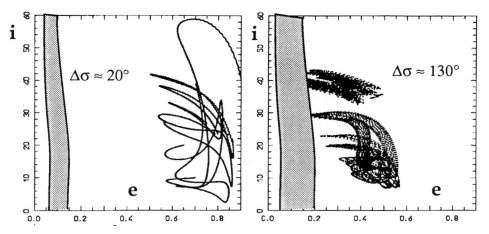

Figure 2. Localization of Kozai resonance in the 2:1 mean motion commensurability. Orbits were numerically computed in the general area of the pericentric chaotic complex for two different values of the amplitude of libration of the resonant angle σ. Segments exhibiting libration of the angle $\sigma - \sigma_z$, characteristic of the Kozai resonance were selected and projected on the plane (a, e). The localization of secondary resonances is roughly indicated as the shaded area to the left of the Figures. These results are extracted from (Henrard *et al.*, 1995).

angle disappears from the Hamiltonian. The problem is reduced to a two-degree-of-freedom problem.

A further simplification, taking Jupiter on a circular orbit, reduces the problem to a one degree of freedom problem as the angular variable ν disappears then from the Hamiltonian. Of course this problem is too simple to modelize in any detail the motion of an asteroid, but it allow us to define with surprising accuracy the region of phase space governed by the mean motion resonance. The one-degree-of-freedom problem exhibit doubly asymptotic orbits to unstable equilibria. The location in phase space of this family of separatrices marks, even for more sophisticated problems taking into account the eccentricity of Jupiter and the slow evolution of its orbit, an abrupt change in the behavior of the resonant angle σ and thus the boundary of the mean motion resonance. In Figure 1 we have indicated the position of these separatrices.

The belt of chaotic motion found numerically in the planar elliptic problem, (Froeschlé and Scholl, 1981; Murray, 1986; Wisdom, 1987) at low to moderate eccentricity in the 2:1 resonance has been shown to be related to *secondary resonances* between the frequency of libration of the resonant angle σ and the frequency of ν (Henrard and Lemaître 1986; Lemaître and Henrard, 1990). It was already known by Froeschlé and Scholl and later

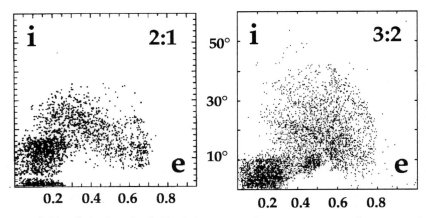

Figure 3. Orbits "crossing the bridge" from secondary resonances at low eccentricity to secular resonance at high eccentricity in the 2:1 mean motion resonance (Figure 3a) and in the 3:2 mean motion resonance (Figure 3b).

confirmed by Moons and Morbidelli (1993) that, in the planar restricted problem this chaotic domain is "bottled up" by invariant tori from the high eccentricity region.

On the other hand, Moons and Morbidelli (1993) have shown how to identify *secular resonances* inside the mean motion resonances. The secular resonances are resonances between the mean frequency of the longitude of the pericentrer of the asteroid and one of the frequencies of the longitude of the pericenters of the planets. They pointed out that the overlap of the ν_5 and ν_6 resonances (resonances with the longitude of pericenter of Jupiter and Saturn respectively) were responsible for another large zone of chaotic motion at large eccentricity, a region of phase space which had not been explored previously. The localization of these two chaotic domains in the plane of initial conditions (a, e) is shown in Figure 1.

2. The Third Dimension

One of the most prominent features in the third dimension is the so-called "Kozai resonance", a 1:1 commensurability between the unperturbed frequencies of σ and σ_z. Inside the first order mean motion resonances we are dealing with, the Kozai resonance appears at rather large eccentricities and, of course, it becomes important only for moderate to high inclinations as the restoring torque of the resonance term is proportional to S_z. As it overlaps with the above mentioned secular resonances ν_5 and ν_6 it enlarges the chaotic domain until it becomes, at moderate inclination, the largest feature of the dynamics inside the mean motion commensurabilities. We call this chaotic domain the pericentric complex of overlapping resonances.

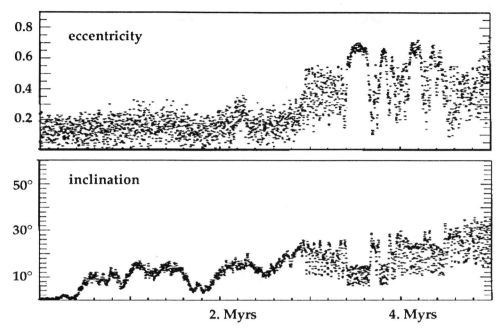

Figure 4. Time evolution of the orbit of Figure 3a. The initial conditions of this orbit are: $a = 0.64, e = 0.05, i = 0°, \omega = 90°, \Omega = \sigma = 0°$

This pericentric complex appears at lower eccentricity for large amplitude of libration. This is already the case in the planar problem (see Figure 1), but it is still more evident in the three-dimensional problem as can be seen in Figure 2. Figure 2 reproduces projections on the (e, i) plane of segments of orbits (computed numerically) caught in the Kozai resonance of the 2:1 Kirkwood gap. They are arranged according to the value of the amplitude of libration of the resonant angle σ. It is only a projection but it gives a fair idea of the extend of the volume of phase space affected by the Kozai resonance and of its localization.

The position of the layer of overlapping secondary resonances is not very much affected by the value of the inclination as can be seen in Figure 2. But, because the pericentric complex goes down in eccentricity with larger inclinations and larger amplitudes of libration, the two domains of chaotic motion meet for large amplitude of libration at about 30° of inclination in the 2:1 case.

There is thus a path from the low eccentricity, low inclination chaotic domain to the high eccentricity chaotic domain going "over" a seemingly

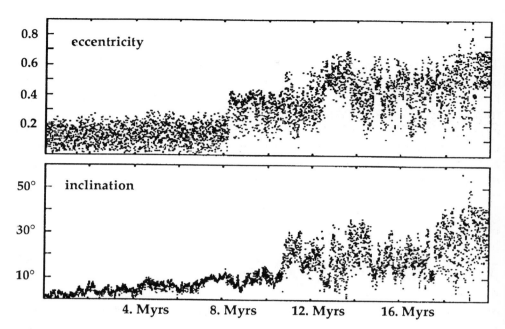

Figure 5. Time evolution of the orbit of Figure 3b. The initial conditions of this orbit are: $a = 0.7425, e = 0.15, i = 0°, \omega = 90°, \Omega = \sigma = 0°$

regular low inclination, moderate eccentricity region. An orbit taking this path was exhibited by Wisdom (1987) as an indication of how the regular region "bottling up" the secondary resonance layer in the planar problem could be bypassed. A similar orbit computed in (Henrard *et al.*, 1995) is shown in Figure 3a and Figure 4. From these Figures one can see how the orbit remains confined for some time in the low eccentricity region before finding, at moderate inclination, the bridge that will take it to high eccentricity. In the times series of the same orbit, shown in Figure 4, the chaotic character of the orbit is clearly revealed; the two passages at very high eccentricity correspond to temporary capture in the ν_5 secular resonance.

In Figure 3b and Figure 5, is displayed an orbit following a very similar orbit but this time inside the 3:2 mean motion resonance. Like the orbit in figure 4, this orbit was computed by means of an "averaging integrator" *à la Schubart* (Moons, 1994), with an orbit of Jupiter affected by all the main frequencies of the solar system. This orbit shows that the path from low eccentricity to large eccentricity is not a special feature of the 2:1 mean motion resonance (the Hecuba gap) but exists also in the 3:2 mean motion resonance (the Hilda family) and thus cannot serve as an explanation for the absence of asteroids in the Hecuba gap.

3. Conclusion

The two orbits displayed in the previous section are typical of the behaviour of orbits starting in the secondary resonance zone with large amplitude of libration of the resonant angle σ. They confirm that indeed there is a path from low eccentricity to large eccentricity through the "third dimension" as announced by Wisdom (1987). At the same time they show that this path is not a special feature of the 2:1 mean motion resonance (the Hecuba gap) but exists also in the 3:2 mean motion resonance (the Hilda family) and thus cannot serve as an explanation for the absence of asteroids in the Hecuba gap.

The dynamics inside the 2:1 and 3:2 mean motion resonance are so similar that one may wonder why one resonance corresponds to a gap in the asteroid belt and the other one to a group. Figures 4 and 5 hint at one difference which may play a role. The time scales are different. The orbit displayed in Figure 4 takes about 3 Myrs to find the door to the high eccentricity regime; for the orbit in Figure 5, it is 10 Myrs. These time scales are typical for other orbits we have computed, all starting in the zone of secondary resonances with large amplitude of libration. The time scales would be much larger for orbits starting with smaller amplitude of libration, but it is not unreasonable to conjecture that the ratio between the time scales would remain.

Ferraz-Mello (Ferraz-Mello, 1994; Ferraz-Mello et al., 1996) also finds difference of this type in time scales between the 2:1 and 3:2 resonances, both for Lyapunov time scales and for diffusion time scales.

A priori one would have expected the time scales to be shorter for the 3:2 resonance than for the 2:1 as the period of libration is between 200 and 250 years in the 3:2 resonance and between 400 and 500 years in the 2:1 resonance.

It remains to investigate whether this difference in time scales is enough to explain the formation of the Kirkwood gap, and to understand why the time scales are different.

References

Ferraz-Mello, S.: 1994, 'Kirkwood gaps and resonant groups'. In *Asteroids, Comets, Meteors 1993* (Milani *et al.* eds.), pages 175–188, Kluwer A.P.

Ferraz-Mello, S., Klafke, J., Michtchenko, T., and Nesvorny, N.: 1996, 'Chaotic transitions in resonant asteroidal dynamics', *Celest. Mech.*, this issue.

Froeschlé, C. and Scholl, H.: 1981, 'The stochasticity of peculiar orbits in the 2/1 Kirkwood gap', *Astron. Astrophys.*, **93**, 62–66.

Henrard, J. and Lemaître, A.: 1986, 'A perturbation method for problems with two critical arguments', *Celest. Mech.*, **39**, 213–238.

Henrard, J., Watanabe, N., and Moons, M.: 1995, 'A bridge between secondary and secular resonances inside the Hecuba gap', *Icarus*, **115**, 336–346.

Lemaître, A. and Henrard, J.: 1990, 'On the origin of chaotic behaviour in the 2/1 Kirkwood gap', *Icarus*, **83**, 391–409.

Moons, M.: 1994, 'Extended Schubart averaging', *Celest. Mech.*, **60**, 173–286.

Moons, M. and Morbidelli, A.: 1993, 'The main mean motion commensurabilities in the planar circular and elliptic problem', *Celest. Mech.*, **57**, 99–108.

Morbidelli, A. and Moons, M.: 1993, 'Secular resonances in mean motion commensurabilities: the 2/1 and 3/2 cases', *Icarus*, **102**, 316–332.

Murray, C.: 1986, 'Structure of the 2:1 and 3:2 Jovian resonances', *Icarus*, **65**, 70–82.

Poincaré, H.: 1902, 'Sur les planètes du type d'Hécube', *Bull. Astron.*, **19**, 289–310.

Schubart, J.: 1966, 'Special cases of the restricted problem of three bodies'. In *Proc. of the I.A.U. Symp. No. 25*, pages 187–193.

Wisdom, J.: 1987, 'Urey prize lecture : Chaotic dynamics in the Solar system', *Icarus*, **72**, 241–275.

LARGE SCALE CHAOS AND MARGINAL STABILITY
IN THE SOLAR SYSTEM

JACQUES LASKAR*

CNRS, Astronomie et Systèmes Dynamiques, Bureau des Longitudes,
3 rue Mazarine, 75006 Paris, France
E-mail: laskar@bdl.fr

Abstract. Large scale chaos is present everywhere in the solar system. It plays a major role in the sculpting of the asteroid belt and in the diffusion of comets from the outer region of the solar system. All the inner planets probably experienced large scale chaotic behavior for their obliquities during their history. The Earth obliquity is presently stable only because of the presence of the Moon, and the tilt of Mars undergoes large chaotic variations from 0° to about 60°. On billion years time scale, the orbits of the planets themselves present strong chaotic variations which can lead to the escape of Mercury or collision with Venus in less than 3.5 Gyr. The organization of the planets in the solar system thus seems to be strongly related to this chaotic evolution, reaching at all time a state of marginal stability, that is practical stability on a time-scale comparable to its age.

Contents

* This lecture was given at the XIth International Congress of Mathematical Physics, Paris, july 1994

Celestial Mechanics and Dynamical Astronomy **64**: 115–162, 1996.
© *1996 Kluwer Academic Publishers. Printed in the Netherlands.*

1. Introduction

Since the seminal work of Poincaré (1892-99) on the non integrability of the three-body problem and the existence of heteroclinic intersections, completed by the results of Kolmogorov (1954) and Arnold (1963a-b), it is known that in general, in celestial bodies dynamics, many non regular motions will appear besides the quasiperiodic orbits confined on invariant tori in the phase space. In practice, positive Lyapunov exponents, reflecting the exponential divergence of nearby orbits will be detected during long time numerical integrations of these non regular orbits lying in the chaotic zones.

As a result of this exponential divergence, a practical limit will arise for the possibility of making precise predictions for the motion of these celestial bodies. If this limit is much larger than the age of the solar system, the motion still could be very well approximated by a regular solution and the chaotic behavior will not be sensible.

On the other case, one needs to give up the program of Laplace which was to determine with the ultimate precision the motion of all the objects of the heaven. This may be of no physical consequences if the chaotic region where the motions wanders is practically confined in a narrow region over the age of the solar system. For a planet, it would just mean for example that the orientation of the orbit or its position on this orbit is not known. Much more important are the cases of extended chaos, when the diffusion of the action like variables is sensible over the considered period. In this case, the orbit will explore a large portion of the phase space, and significant physical changes may occur. It could mean changes in semi major axis, eccentricity, or inclination.

Despite the pioneered work of (Hénon and Heiles, 1964) in galactic dynamics, the exhibition of these large scale chaotic behaviors in the real solar system is very recent, and most of the work reported here was done in the last few years. This leads to a completely new vision of celestial mechanics which in the previous decade was considered in astronomy as an old and dusty field, uniquely concerned by the determination of more and more precise paths for already well known objects. Indeed, until very recently, most people assumed that everything was regular and smooth in the solar system, and the motion of the planets was considered as the paradigm of regularity.

In fact, many objects in the solar system present large scale chaotic behavior, and the analysis of their possible evolution over long times changed profoundly the understanding of the evolution of the solar system, inducing also many new elements for the understanding of its formation.

2. Minor bodies in the solar system

The solar system is crowded with a multitude of small objects : asteroids, comets, small satellites. Many of them are supposed to be the remains of the material of the primitive solar system which did not contributed to the formation of the planets. They are of great interest for the understanding of the formation of the solar system, as their material should not have changed much since the primordial state of the solar system. This leads in the recent years to numerous studies on their dynamics, aiming to the understanding of their dynamical evolution since the formation of the solar system.

From a practical point of view, these small bodies can usually be studied with simplified models. Their very small masses do not perturb the remaining part of the solar system and, as we are more interested by their collective behavior than by the very precise orbit of a single of them. On the other hand, for the same reason, it will be necessary to understand the considered dynamics in a global way, in the large part of the phase space which will correspond to the numerous possible initial conditions of these small bodies.

Apart from the review of (Wisdom, 1987b), the papers of (Ferraz-Mello, 1994, Farinella et al., 1994) can be consulted for an overview of the recent developments in the understanding of the Kirkwood gaps in the asteroid belt, and the delivery of meteorites to the Earth. The dynamical studies on comets have been reviewed in (Fernández, 1994).

2.1. THE CHAOTIC MOTION OF HYPERION

The first striking example of chaotic behavior in the solar system was given by the chaotic tumbling of Hyperion, a small satellite of Saturn which strange rotational behavior was detected during the encounter of the Voyager spacecraft with Saturn (Wisdom, Peale, Mignard, 1984). This example, which dynamics can be reduced to the perturbed pendulum one, provides a simple illustration of the arising of chaotic behavior in the vicinity of a resonance. This study will apply more generally to any satellite of irregular shape in the vicinity of spin-orbit resonance (Wisdom, 1987a).

The equations of motion for the orientation of a satellite S orbiting around a planet P on a fixed elliptical orbit of semi major axis a and eccentricity e are given by the Hamiltonian

$$H = \frac{y^2}{2} - \frac{3}{4}\frac{B-A}{C}\left(\frac{a}{r(t)}\right)^3 \cos 2(x - v(t))$$

where $r(t)$ is the distance from the planet to the satellite, x gives the orientation of the satellite with respect to a fixed direction (here the direction of periapse), $y = dx/dt$ is its conjugate variable, v is the true anomaly of

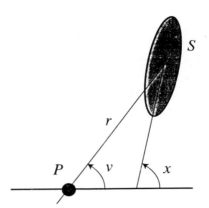

Figure 1. The position of the satellite S around the planet P is defined by the distance r and the angle v from the direction of periapse (true anomaly). The angle x provides the orientation of the satellite.

the satellite, and $A < B < C$ are the principal moments of inertia of the satellite (Fig.1). The associated equations of motion are

$$\frac{dy}{dt} = -\frac{\partial H}{\partial x}; \qquad \frac{dx}{dt} = \frac{\partial H}{\partial y} .$$

The unit of time is taken such that the mean motion $n = 1$. When expanding the Hamiltonian with respect to the eccentricity (e) which is supposed to be small, and retaining only the terms of first order in eccentricity, one obtains

$$H = \frac{y^2}{2} - \frac{\alpha}{2} \cos 2(x - t) + \frac{\alpha e}{4} [\cos(2x - t) - 7 \cos(2x - 3t)] \qquad (1)$$

with $\alpha = 3(B - A)/2C$.

If S has a rotational symmetry, $\alpha = 0$, and the hamiltonian is reduced to $H_0 = y^2/2$. The satellite rotates with constant velocity $dx/dt = y_0$. When the orbit is circular, the problem is also integrable as H_0 is reduced to the first two terms of Eq. (1).

$$H_0 = \frac{y^2}{2} - \frac{\alpha}{2} \cos 2(x - t) \qquad (2)$$

This motion will be similar to the simple pendulum motion, with possibility of libration of the satellite around the direction of the planet (spin-orbit

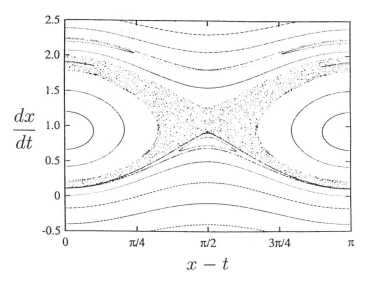

Figure 2. Surface of section in the phase space of Deimos, a small satellite of Mars. $x - t$ defines the orientation of the satellite and dx/dt its rotational velocity. A small chaotic zone appears in the vicinity of the separatrix of the unperturbed problem ($e = 0$) (Wisdom, 1987b).

resonance occurs), or circulation motion for large values of the initial angular rotational velocity of the satellite.

In the general case, $\alpha e \neq 0$, and the hamiltonian H_0 of (2) is perturbed by the remaining terms of (1). At the transition between librational motion and rotational motion of the satellites, appears a small chaotic zone. This can be observed in a section of the phase space portrait of Phobos orientation motion when orbiting around Mars (Fig.2).

When the size of the perturbation αe increases, resonant zones corresponding to the various possible resonant terms $\cos 2(x - t), \cos(2x - t), \cos(2x - 3t)$ will overlap (Chirikov, 1979), giving rise to large scale chaotic motion.

This is the case for Hyperion (Fig.3), where $\alpha e \approx 0.039$. The resulting effect is that the rotational motion of Hyperion is not regular, and it becomes impossible to adjust any periodic or quasiperiodic model to its lightcurve (Klavetter, 1989). The consideration of this chaotic motion was necessary to explain the observations of Hyperion, and this example demonstrated that significant physical phenomenon on the solar system could result from this complicated chaotic dynamics.

It should be noticed that the models used in these computations are of two degrees of freedom. In this case, according to KAM theory, invariant tori of dimension 2 may subsist and will divide the phase space. Thus, although

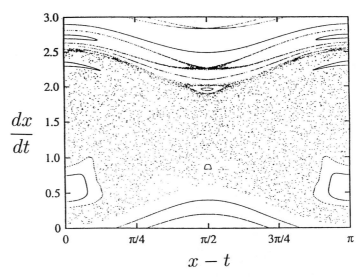

$$\frac{dx}{dt}$$

$$x - t$$

Figure 3. Surface of section in the phase space of Hyperion, an outer satellite of Saturn. When the perturbation parameter αe is large, like in the case of Hyperion, many of the resonances overlap, giving rise to a large chaotic zone of irregular motion (Wisdom, 1987b).

an orbit may be chaotic, it can be bounded for all time by these invariant surfaces (Celletti, 1990a,b). In the complete model, the addition of the extra degrees of freedom leaves place for the possibility of diffusion although this diffusion may be extremely small.

2.2. THE KIRKWOOD GAPS

The distribution of the asteroids, the minor planets which orbits lay primarily between Mars and Jupiter, has puzzled astronomers for many decades, since Kirkwood observed in 1867 that they are not randomly distributed. Indeed, when plotting the number of asteroids against their semi major axis, one can observe gaps and accumulations (Fig.4). Kirkwood noticed that these gaps coincide with commensurabilities of mean motion with Jupiter, and the extend of the gaps also coincide with the libration zones of the resonances (Dermott and Murray, 1983). It was thus thought that these gaps result from the effect of these resonances, but although the first numerical integrations of asteroids placed inside the resonance lead to some increase of their eccentricities (Froeschlé and Scholl, 1977), no satisfying explanation was given. Later on, using a simplified model of two degrees of freedom (an averaged planar problem where the asteroid is uniquely submitted to the perturbation of Jupiter orbiting on a fixed ellipse), Wisdom (1983, 1985), inspired by the work of Chirikov (1979), managed to integrate the orbits

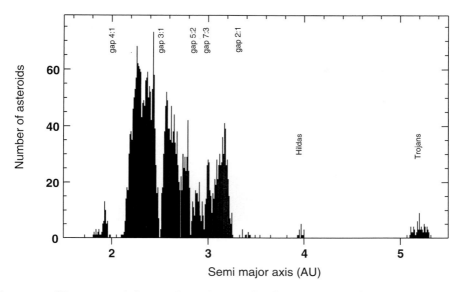

Figure 4. Histogram of the number of asteroids plotted against their semi-major axis showing the location of gaps and accumulations.

approximately over much extended time and showed that in the vicinity of the 3/1 resonance, there exists a chaotic zone which can be easily observed in a Poincaré surface of section of the trajectories, corresponding to the successive intersections with a plane of fixed argument of perihelion. An orbit starting in this chaotic zone can have a moderate eccentricity for a long time, but it could enter in some other branch of this chaotic zone, which would lead to a large increase of the eccentricity, sufficient to cross Mars orbit. A possible close encounter with this planet can then expel the asteroid from its primitive orbit. The location and extent of the chaotic zone related to the 3/1 resonance is in good agreement with the 3/1 asteroid gap. The understanding of this complex dynamics, which differs very much from the ordered motion of the integrable problems, thus allowed to obtain convincing explanation for one famous problem of celestial mechanics.

Since the work of Wisdom, which applies more specifically to the 3/1 gap, many other studies analyzed the possible chaotic behavior in the vicinity of other asteroidal resonances involving more complicated interactions, and models of many degrees of freedom. In particular, in their analysis of the 2/1 and 3/2 resonances, Morbidelli and Moons (1993) needed to take into account the spatial problem and the secular resonances due to the slow

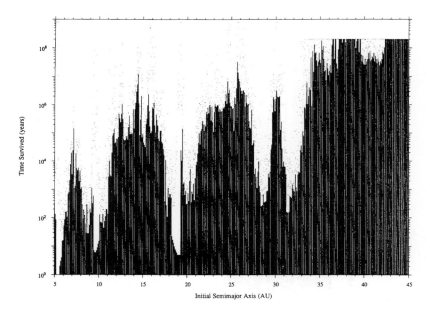

Figure 5. The time survived by each test particle as a function of initial semimajor axis. For each semimajor axis bin, six test particles were started at different longitudes. The vertical bars mark the minimum of the six termination times. The spikes at 5.2, 9.5, 19.2, and 30.1 AU, at the semimajor axes of the planets (Jupiter, Saturn, Uranus and Neptune), correspond to test particle in librating in Trojan or horseshoelike orbits before close encounter. Interior to Neptune the integration extends to 800 Myr; exterior to Neptune to 200 Myr. Beyond about 43 AU all the test particles survive the full integration (Holman and Wisdom, 1993).

precession of Jupiter's orbit under planetary perturbations. Moreover, using the same model, they demonstrated that the overlap of the secular resonances inside the 3/1 libration region provides a more efficient mechanism for the depletion of this gap than the one originally proposed by Wisdom (Moons and Morbidelli, 1994).

2.3. THE CHAOTIC MOTION OF THE COMETS AND THE DYNAMICS OF THE KUIPER BELT

The asteroids are not the only small bodies of the solar system which can be subject to chaotic motion. Indeed, many cometary orbits are chaotic.

When Halley's comet came to visit us in 1985, several numerical integrations were carried out to retrace its orbit over all the extent of the observations, that is beyond 163 BC, date of the most ancient observation of this comet (Stephenson *et al.*, 1984). After such a long time, all the different numerical integrations showed different behavior, and their accuracy was questioned. In fact, these divergences were later on explained by the analysis of Chirikov and Vecheslavov (1989), which demonstrated that the motion of Halley's comet could be chaotic, and thus practically unpredictable after 29 revolutions. Indeed, it can be shown that, due to the perturbation of Jupiter, there exists a large chaotic zone for nearly parabolic comets orbits, which extends up to the Oort cloud (Petrosky, 1986, Sagdeev and Zaslavsky, 1987, Natenzon et al., 1990). This chaotic behavior of Halley's comet was later on confirmed by direct numerical integration (Froeschlé and Gonczi, 1988). More generally, most of the long period comets have chaotic orbits, where the chaotic behavior results from repeated close encounters with the planets.

The existence of the Oort's cloud could explain the observation of the long period comets, but the distribution of the inclination of the short period comet lead to the hypothesis of the existence of an other source of comets, the Kuiper belt, located beyond Neptune close to the planetary plane (Kuiper, 1951, Fernandez, 1980)). In order to study this hypothesis, and as the integration of the outer solar system becomes accessible to long time computations, many efforts have been conducted recently to understand the dynamics of small particles in the outer solar system. From these studies, which consist mainly in the numerical integration of thousand of massless particle in the outer solar system, it was found that apart from some special locations, like the trojans Lagrangian points of Jupiter (fig. 4), there was practically no stable orbits which could last for more than 1 Gyr among the outer planets (Duncan et al., 1989, Gladman and Duncan, 1990, Holman and Wisdom, 1993, Levison and Duncan, 1993). On the contrary, there exist some stable orbits at about 40 AU and further, where some planetesimal could last for a long time (Levison and Duncan, 1993). Close to them, unstable regions exist which will provide from time to time, by chaotic diffusion, planetesimal which would enter a more internal part of the solar system, and could be captured temporarily into resonance, as a short period comet (Torbett and Smoluchovski, 1990). During some recent observation campaign, several of these transneptunian objects where observed, at the location of the supposed Kuiper belt, at about 40 AU (see Luu, 1994 for a review of this search). In this case again, the understanding of the non regular orbits, which can thus explore a large part of the solar system, gave some insight of the observed distribution of the short period comets.

These findings are also in good agreement with the scenario of formation of the solar system including a phase where planetesimals are present everywhere (Safronov, 1969). Indeed, as it was forecasted by Kuiper, in the outer solar system the removal of the planetesimals non accreted to form the planets can probably be explained by the gravitational perturbations of the large planets while some remaining bodies are actually present in the stable regions of the outer solar system where these perturbations decrease.

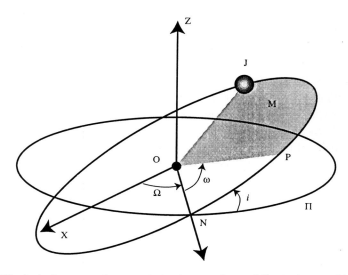

Figure 6. Elliptical elements. At any given time, a planet (J) can be considered to move on an elliptical orbit, with semimajor axis a and eccentricity e, with the sun at one focus (O). The orientation of this ellipse with respect to a fixed plane Π, and a direction of reference OX, is given by three angles: The inclination i, the longitude of the node Ω, and the longitude of perihelion $\varpi = \Omega + \omega$, where ω is the argument of perihelion (P). The position of the planet on this ellipse is given by the mean longitude $\lambda = M + \varpi$, where M (mean anomaly) is an angle which is proportional to the area OPJ (third Kepler's law).

3.   The chaotic motion of the planets

The first studies of chaotic motion in the solar system concerned small objects, with simplified dynamical models which could often be reduced to two degrees of freedom. With these simplifications, it was possible to describe their global dynamics, and to study their chaotic zones, which gave rise to new insight in the organization and evolution of the solar system. But chaotic behavior is not confined to the small bodies of the solar system, and concern also the main celestial objects, the planets. Despite the outstanding work of Poincaré, the discovery of the non regular behavior of the actual planets is very recent, as it requested the possibility to study the evolution of the actual solar system over a very long time, which was only achieved in the last few years.

3.1. Historical introduction

The problem of the stability of the solar system has fascinated astronomers and mathematicians since antiquity, when it was observed that among the seemingly fixed stars, there were also "wandering stars"—the planets. Efforts were first focused on finding a regularity in the motion of these wanderers, so their movement among the fixed stars could be predicted. For Hipparcus and Ptolemy, the ideal model was a combination of uniform circular motions, the epicycles, which were adjusted over the centuries to conform to the observed course of the planets. Astronomy had become predictive, even if its models were in continual need of adjustment.

From 1609 to 1618, Kepler fixed the planets' trajectories: having assimilated the lessons of Copernicus, he placed the Sun at the center of the universe and, based on the observations of Tycho Brahe, showed that the planets describe ellipses around the Sun. At the end of a revolution, each planet found itself back where it started and so retraced the same ellipse. Though seductive in its simplicity, this vision of a perfectly stable solar system in which all orbits were periodic would not remain unchallenged for long.

In 1687 Newton announced the law of universal gravitation. By restricting this law to the interactions of planets with the Sun alone, one obtains Kepler's phenomenology. But Newton's law applies to all interactions: Jupiter is attracted by the Sun, as is Saturn, but Jupiter and Saturn also attract each other. There is no reason to assume that the planets' orbits are fixed invariant ellipses, and Kepler's beautiful regularity is destroyed.

In Newton's view, the perturbations among the planets were strong enough to destroy the stability of the solar system, and divine intervention was required from time to time to restore planets' orbits to their place. Moreover, Newton's law did not yet enjoy its present status, and astronomers wondered if it was truly enough to account for the observed movements of bodies in the solar system.

The problem of solar system stability was a real one, since after Kepler, Halley was able to show, by analyzing the Chaldean observations transmitted by Ptolemy, that Saturn was moving away from the Sun while Jupiter was moving closer. By crudely extrapolating these observations, one finds that six million years ago Jupiter and Saturn were at the same distance from the Sun. In the 18th century, Laplace took up one of these observations, which he dated March 1st, 228 BC: *At 4:23 am, mean Paris time, Saturn was observed "two fingers" under Gamma in Virgo.* Starting from contemporary observations, Laplace hoped to calculate backward in time using Newton's equations to arrive to this 2000 year-old observation.

The variations of planetary orbits were such that, in order to predict the planets' positions in the sky, de Lalande was required to introduce artificial "secular" terms in his ephemeris tables. Could these terms be accounted for by Newton's law?

The problem remained open until the end of the 18th century, when Lagrange and Laplace correctly formulated the equations of motion. Lagrange started from the fact that the motion of a planet remains close, over a short duration, to a Keplerian ellipse, and so had the notion to use this ellipse as the basis for a coordinate system (Fig.6). Lagrange then wrote the differential equations that govern the variations in this elliptic motion under the effect of perturbations from other planets, thus inaugurating the methods of classical celestial mechanics. Laplace (1772) and Lagrange (1776), whose work converged on this point, calculated secular variations, in other words long-term variations in the planets' semi-major axes under the effects of perturbations by the other planets. Their calculations showed that, up to first order in the masses of the planets, these variations vanish. Poisson (1809) and Haretu (1885) later showed that this result remains true through second order in the masses of the planets, but not through third order.

This result seemed to contradict Ptolemy's observations from antiquity, but by examining the periodic perturbations between Jupiter and Saturn, Laplace discovered a quasi-resonant term $(2\lambda_{Jupiter} - 5\lambda_{Saturn})$ in their longitudes. This term has an amplitude of $46'50''$ in Saturn's longitude, and a period of about 900 years. This explains why observations taken in 228 BC and then in 1590 and 1650 could give the impression of a secular term.

Laplace (1785) then calculated many other periodic terms, and established a theory of motion for Jupiter and Saturn in very good agreement with 18th century observations. Above all, using the same theory, he was able to account for Ptolemy's observations to within one minute of arc, without additional terms in his calculations. He thus showed that Newton's law was in itself sufficient to explain the movement of the planets throughout known history, and this exploit no doubt partly accounted for Laplace's determinism.

Laplace showed that the planets' semi-major axes undergo only small oscillations, and do not have secular terms. At the same time, the eccentricity and inclination of planets' trajectories are also very important for solar system stability. If a planet's eccentricity changes appreciably, its orbit might cut through another planet's orbit, increasing the chances of a close encounter which could eject it from the solar system.

Laplace (1784) revisited his calculations, taking into account only terms of first order in the perturbation series, and showed that the system of equations describing the mean motions of eccentricity and inclination in a planetary system with k planets may be reduced to the system of linear differential equations with constant coefficients

$$
\frac{d}{dt}
\begin{bmatrix} z_1 \\ \vdots \\ z_k \\ \zeta_1 \\ \vdots \\ \zeta_k \end{bmatrix}
= \sqrt{-1}
\begin{bmatrix} A_k & 0_k \\ 0_k & B_k \end{bmatrix}
\begin{bmatrix} z_1 \\ \vdots \\ z_k \\ \zeta_1 \\ \vdots \\ \zeta_k \end{bmatrix}
\tag{3}
$$

where $z = e \exp \sqrt{-1}\varpi$, $\zeta = \sin(i/2) \exp \sqrt{-1}\Omega$, A_k and B_k are (k, k) matrices with real coefficients which depends only on the planetary masses and semi major axis; 0_k is the (k, k) null matrix. Using the invariance of the angular momentum

$$C = \sum_{i=1}^{k} m_i \sqrt{\mu_i a_i (1 - e_i^2)} \cos i_i \tag{4}$$

and retaining only the terms of degree 2 in eccentricity and inclination, and arguing that the eccentricity and inclination evolutions are decoupled in the linear equations, Laplace deduced that the quantities

$$\sum_{i=1}^{k} m_i \sqrt{a_i} e_i^2$$

$$\sum_{i=1}^{k} m_i \sqrt{a_i} \sin^2 i_i/2$$

should be constant, and thus there cannot exist polynomial or exponential terms in the solutions of these linear equations. Therefore, he deduced that all the eigenvalues g_i of A and s_i of B are real and distinct, and the solutions of this linear secular system are quasiperiodic expressions of the form

$$z_i = \sum_{j=1}^{k} \alpha_{ij} \, e^{ig_j t}$$

$$\zeta_i = \sum_{j=1}^{k} \beta_{ij} \, e^{is_j t}$$

where α_{ij} and β_{ij} are complex quantities. The values of the secular frequencies g_i and s_i, computed in the more complete semi-analytical solution of (Laskar, 1990) are given in table I.

The variations in eccentricity thus reduce to a superposition of uniform circular motions (Fig.7) of frequencies g_i and s_i. The inclinations and eccentricities of the orbits are therefore subject to only small variations about their mean values (in fact, this was really established by Le Verrier for the whole solar system). It must be stressed that Laplace's solutions are very different from Kepler's, because the orbits are no longer fixed. They are subject to a double precessionary motion with periods ranging from about 45 000 to a few million years (table I): precession of the perihelion, which is the slow rotation of the orbit in its plane, and precession of the nodes, which is the rotation of the plane of the orbit in space.

Table I

Fundamental frequencies of the precession motion of the
solar system (excluding Pluto). These values are taken as the
mean values over 20 million years of from the recent solution
La90. For the inner planets, due to chaotic diffusion, the frequencies can change significantly with time (Laskar, 1990).

	ν (" /yr)	period (yr)
g_1	5.596	231 000
g_2	7.456	174 000
g_3	17.365	74 600
g_4	17.916	72 300
g_5	4.249	305 000
g_6	28.221	45 900
g_7	3.089	419 000
g_8	0.667	1 940 000
s_1	- 5.618	230 000
s_2	- 7.080	183 000
s_3	-18.851	68 700
s_4	-17.748	73 000
s_5	0.000	
s_6	-26.330	49 200
s_7	- 3.005	431 000
s_8	- 0.692	1 870 000

The work of Laplace concerns only the linear approximation of the secular motion of the planets. In modern language, one can say that Laplace
demonstrated that the origin (planar and circular motions) is an elliptical
fixed point in the secular phase space which is obtained after averaging over
the mean longitudes. Later, Le Verrier (1856), famed for his discovery in
1846 of the planet Neptune through calculations based on observations of
irregularities in the movement of Uranus, took up Laplace's calculations
and considered the effects of higher order terms in the series. He showed
that these terms produce significant corrections and that Laplace's and
Lagrange's calculations "could not be used for an indefinite length of time."
He then challenged future mathematicians to find exact solutions, without
approximations. The difficulty posed by "small divisors" showed that the
convergence of the series depended on initial conditions, and the proof of
the stability of the solar system remained an open problem.

But Poincaré (1892–99) formulated a negative response to Le verrier's
question. In so doing he rethought the methods of celestial mechanics along
the lines of Jacobi's and Hamilton's work. Poincaré showed that it is not
possible to integrate the equations of motion of three bodies subject to
mutual interaction, and not possible to find an analytic solution representing

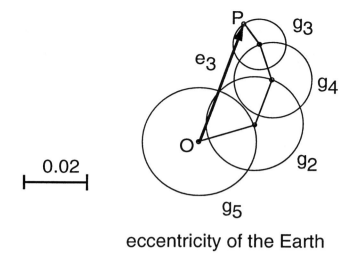

eccentricity of the Earth

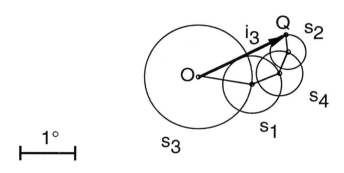

inclination of the Earth

Figure 7. The solutions of Laplace for the motion of the planets are combinations of circular and uniform motions with frequencies the precession frequencies g_i and s_i of the solar system (Table I). The eccentricity e_3 of the Earth is given by OP, while the inclination of the Earth with respect to the invariant plane of the solar system (i_3) is OQ (Laskar, 1992b).

the movement of the planets valid over an infinite time interval, since the perturbation series used by astronomers to calculate the movement of the planets are not convergent on an open set of initial conditions.

Kolmogorov (1954) reexamined this problem and demonstrated that in a perturbed non degenerated Hamiltonian system, among the non regular solutions described by Poincaré, there still exist some quasiperiodic trajectories lying on isolated tori in the phase space. This result was completed by Arnold (1963a) who demonstrated that for a sufficiently small perturbation, the set of invariant tori foliated by quasiperiodic trajectories is of strictly positive measure, tending to unity when the perturbation decreases to zero. Moser (1962) established the same kind of results for less stronger conditions which did not require the analyticity of the Hamiltonian. These theorems are known generically as KAM theorems, and have been employed in various fields. Unfortunately, they do not apply directly to the planetary problem which presents proper degeneracy (the unperturbed Hamiltonian depends only on the semi major axes, and not on the other action variables related to eccentricity and inclination). This led Arnold to extend the proof of existence of invariant tori, taking into account this phenomenon of proper degeneracy. He then applied his theorem explicitly to a planar planetary system of two planets with a ratio of the semi major axis close to zero, demonstrating the existence of quasi periodic trajectories for sufficiently small values of the planetary masses and eccentricities (Arnold 1963b). This result was recently extended to more general planetary systems of two planets (Robutel, 1995).

Arnold's results motivated many discussions; indeed, as the quasiperiodic KAM tori are isolated, an infinitely small change in the initial conditions can change the solution from being stable for all time, to a chaotic orbit. Moreover, as the planetary system is of more than two degrees of freedom, none of the KAM tori separates the phase space, leaving the possibility for chaotic trajectories to travel large distances in the phase space. In fact, several subsequent results demonstrated that very close to a KAM tori, the diffusion of the solutions is very slow (Nekhoroshev, 1977, Giorgilli et al., 1989, Lochak, 1993, Morbidelli and Giorgilli, 1995), and can be negligible over very long time, eventually as long as the age of the universe.

Although the actual masses of the planets are much too large for these results to apply directly to the solar system, it was generally supposed that the scope of these mathematical results extends much further than their actually proven bounds, and until very recently it was generally assumed that the solar system was stable over its lifetime, "by any reasonable acceptation of this term".

In the past few years, the problem of solar system stability has advanced considerably, due largely to the help provided by computers which allow extensive analytic calculations and numerical integrations over model time scales approaching the age of the solar system, but also due to a better understanding of the underlying dynamics, resulting from the expansion of the overall field of dynamical systems theory.

3.2. NUMERICAL INTEGRATIONS

The motion of the planets of the solar system has a very privileged status; indeed, it is one of the best modelized problems in physics, and its study can be practically reduced to the study of the behavior of the solutions of its gravitational equations (Newton's equation completed with relativistic corrections for the most inner planets), neglecting all dissipation, and treating the planets as point masses, except in the case of the Earth, where for more precise results, one likes to take into account the perturbation introduced by the existence of the Moon.

The mathematical complexity of this problem, despite its apparent simplicity is daunting, and has been a challenge for mathematicians and astronomers since its formulation three centuries ago. Since the work of Poincaré, it is also known that the analytical perturbative methods which were used in planetary computations for nearly two centuries cannot provide good approximations of the solutions over infinite time. Moreover, as stated above the stability results obtained by Arnold (1963) do not apply to realistic planetary systems.

Since the introduction of computers, numerical integration of the planetary equations appeared as a straightforward way to overcome this complexity of the solutions, but has always been bounded until now by the available computer technology. The first long time numerical studies of the solar system were limited to the motion of the outer planets, from Jupiter to Pluto (Cohen *et al.*, 1973, Kinoshita and Nakai, 1984). Indeed, the more rapid the orbital movement of a planet, the more difficult it is to numerically integrate its motion. To integrate the orbit of Jupiter, a step-size of 40 days will suffice, while a step-size of 0.5 days is required to integrate the motion of the whole solar system using a conventional multistep integrator.

The project LONGSTOP (Carpino *et al.*, 1987, Nobili *et al.*, 1989) used a CRAY computer to integrate the system of outer planets over 100 million years. At about the same time, calculations of the same system were carried out at MIT on a parallel computer specially designed for the task over even longer periods, corresponding to times of 210 and 875 million years (Applegate *et al.*, 1986, Sussman and Wisdom, 1988). This latter integration showed that the motion of Pluto is chaotic, with a Lyapunov time (the inverse of the Lyapounov exponent) of 20 million years. But since the mass of Pluto is very small, (1/130 000 000 the mass of the Sun), this does not induce macroscopic instabilities in the rest of the solar system, which appeared relatively stable in these studies.

3.3. CHAOS IN THE SOLAR SYSTEM

The numerical integrations can give very precise solutions of the trajectories, but are limited by the short stepsize necessary for the integration of the whole solar system and it should be stressed, that until 1991, the only available numerical integration of a realistic model of the full solar system was the numerically integrated ephemeris DE102 of JPL (Newhall *et al.*, 1983) which spanned only 44 centuries.

My approach to to this problem was different, and more in the spirit of the analytical works of Laplace and Le Verrier. Indeed, since these pioneered works, the *Bureau des Longitudes**, has traditionally been the place for development of analytical planetary theories based on classical perturbation series (Brumberg and Chapront, 1973, Bretagnon, 1974, Duriez, 1979). Implicitly, these studies assume that the motion of the celestial bodies is regular and quasiperiodic. The methods used are essentially the same ones which were used by Le Verrier, with the additional help of the computers for symbolic computations. Indeed, such methods can provide very satisfactory approximations of the solutions of the planets over several thousand years, but they will not be able to give answers to the question of the stability of the solar system over time span comparable to its age. This difficulty which is known since Poincaré is one of the reasons which motivated the previously quoted long time direct numerical integrations of the equations.

However, the theoretical results of Arnold (1963) supported the idea that it may have been possible with the help of computer algebra to extend very much the scope of the classical analytical planetary theories, but this revealed to be hopeless when considering the whole solar system, because of severe convergence problems encountered in the Birkhoff normalization of the secular system of the inner planets (Laskar, 1984). This difficulty which revealed to be inherent to this complicated system led me to proceed in two very distinct steps: a first one, purely analytical, consists on the averaging of the equations of motion over the rapid angles, that is the motion of the planets along their orbits. Indeed, from all the achievements of classical celestial mechanics obtained since the XIXth century, it could be forecasted that no severe problems would occur during this first step involving only possible resonances among the orbital motion of the planets.

This averaging process was conducted in a very extensive way, without neglecting any term, up to second order with respect to the masses, and through degree 5 in eccentricity and inclination, conducting to the truncated secular equations of the solar system on the form

---

* The *Bureau des Longitudes* was founded the 7 messidor year III (june 25, 1795) in order to develop Astronomy and Celestial Mechanics. Its founding members were Laplace, Lagrange, Lalande, Delambre, Méchain, Cassini, Bougainville, Borda, Buache, Caroché.

$$\frac{d\alpha}{dt} = \sqrt{-1}\,(\mathcal{A}\,\alpha + \Phi_3(\alpha,\bar{\alpha}) + \Phi_5(\alpha,\bar{\alpha})) \tag{5}$$

where $\alpha = (z_1,\ldots,z_8,\zeta_1,\ldots,\zeta_8)$, and $\mathcal{A}$ is similar to the linear matrix of Laplace (eq. 3); $\Phi_3(\alpha,\bar{\alpha})$ and $\Phi_5(\alpha,\bar{\alpha})$ gather the terms of degrees 3 and 5.

The system of equations thus obtained comprises some 150000 terms, but it can be considered as a simplified system, as its main frequencies are now the precessing frequencies of the orbits of the planets, and no longer comprises their orbital periods. The full system can thus be numerically integrated with a very large stepsize of about 500 years. Contributions due to the Moon and to the general relativity are added without difficulty (Laskar, 1985, 1986).

This second step, i.e. the numerical integration, is then very efficient because of the symmetric shape of the secular system, and was conducted over 200 millions years in just a few hours on a super computer. The main results of this integration was to reveal that the whole solar system, and more particularly the inner solar system (Mercury, Venus, Earth, and Mars), is chaotic, with a Lyapunov time of 5 million years (Laskar, 1989). An error of 15 meters in the Earth's initial position gives rise to an error of about 150 meters after 10 million years; but this same error grows to 150 million km after 100 million years. It is thus possible to construct ephemerides over a 10 million year period, but it becomes practically impossible to predict the motion of the planets beyond 100 million years.

This chaotic behavior essentially originates in the presence of two secular resonances among the planets: $\theta = 2(g_4 - g_3) - (s_4 - s_3)$, which is related to Mars and the Earth, and $\sigma = (g_1 - g_5) - (s_1 - s_2)$, related to Mercury, Venus, and Jupiter (the g_i are the secular frequencies related to the perihelions of the planets, while the s_i are the secular frequencies of the nodes (table I)). The two corresponding arguments change several times from libration to circulation over 200 million years, which is also a characteristic of chaotic behavior (fig. 8). It should be stressed that these two combinations of frequencies were not chosen in a random way. In fact, the frequency analysis (Laskar, 1990) of the numerical solutions of the secular system showed that these combinations appear with a large amplitude in the very first terms of the inner planets solutions. Indeed, as soon as one goes further than the linear model, they need to be taken into account.

When these results were published, the only possible comparison was the comparison with the 44 centuries ephemeris DE102, which already allowed to be confident on the results by comparing the slopes of the solutions at the origin (Laskar, 1986, 1990). At the time, there was no possibility to obtain similar results with direct numerical integration. In fact, partly due to the very rapid advances in computer technology, and in particular to the development of workstations, only two years later, Quinn *et al.* (1991) were

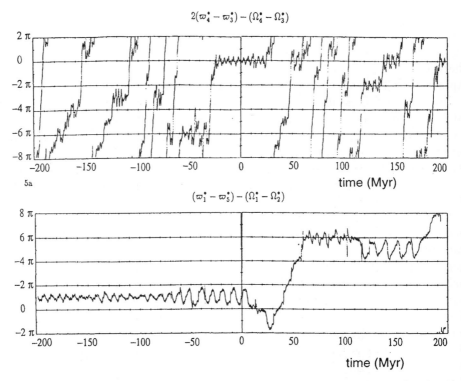

$$2(\varpi_4^\bullet - \varpi_3^\bullet) - (\Omega_4^\bullet - \Omega_3^\bullet)$$

5a

$$(\varpi_1^\bullet - \varpi_5^\bullet) - (\Omega_1^\bullet - \Omega_2^\bullet)$$

time (Myr)

Figure 8. The secular resonances $\theta = 2(g_4 - g_3) - (s_4 - s_3)$ and $\sigma = (g_1 - g_5) - (s_1 - s_2)$. From -200 Myr to $+200$, the corresponding argument present several transitions from libration to circulation (Laskar, 1992a).

able to publish a numerical integration of the full solar system, including the effects of general relativity and the Moon, which spanned 3 million years in the past (completed later on by an integration from -3Myrs to +3Myrs). Comparison with the secular solution of (Laskar, 1990) shows very good quantitative agreement (fig. 9), and confirms the existence of secular resonances in the inner solar system (Laskar *et al.*, 1992a). Later on, using a symplectic integrator directly adapted towards planetary computations which allowed them to use a larger stepsize of 7.2 days, Sussman and Wisdom (1992) made an integration of the solar system over 100 million years which confirmed the existence of the secular resonances as well as the value of the Liapunov exponent of about 1/5 Myrs for the solar system.

3.4. PLANETARY EVOLUTION OVER MYR

The planetary eccentricities and inclinations present variations which are clearly visible over a few million of years (fig. 9). Over 1 million years, the perturbation methods of Laplace, and Le verrier (see section 3.1) will give a

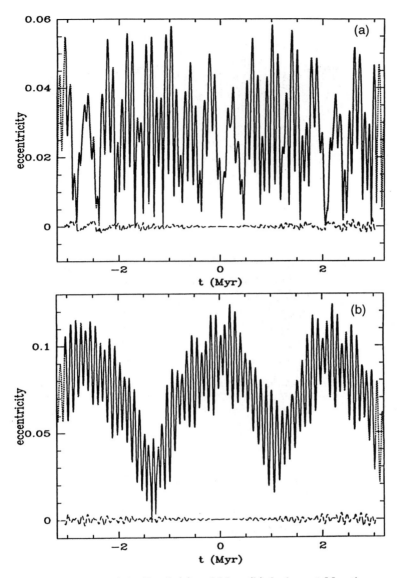

Figure 9. The eccentricity of the Earth (a) and Mars (b) during a 6 Myr timespan centered at the present. The solid line is the numerical solution QTD (Quinn *etal.* 1991), and the dotted line the integration La90 of the secular equations (Laskar, 1990). For clarity, the difference between the two solutions is also plotted (from Laskar, Quinn, Tremaine, 1992).

good account of these variations which are mostly due to the linear coupling present in the secular equations. They involve the precessional periods of the orbits, ranging from 45 000 years to a few million years (Table I). Over sev-

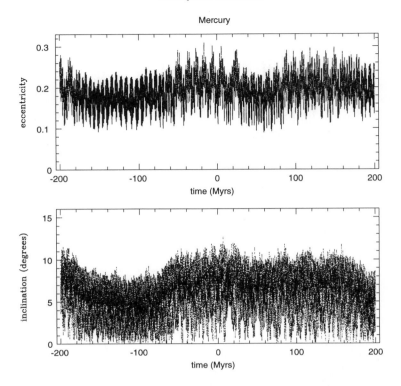

Figure 10. Computed evolution of the eccentricity and inclination of Mercury with respect to time from -200 to +200 million years. On each curves one can see a rapid variation, with periods of about 100 000 years which should correspond to the regular part of the solution as described by Laplace, and a slow variation which reveals the diffusion resulting from the chaotic dynamics (adapted from Laskar, 1992a).

eral hundred million years, the behavior of the solutions for the outer planets (Jupiter, Saturn, Uranus and Neptune) are very similar to the behavior over the first million years, and the motion of these planets appears to be very regular, which was also shown very precisely by means of frequency analysis (Laskar, 1990).

For the Earth, over such time span, the chaotic effect will induce a lost of predictability for the orbit. The additional change of eccentricity resulting from the chaotic diffusion is moderate and may be estimated to about 0.01 for the Earth (Laskar, 1992a,b). The most perturbed planet is Mercury, the effects of its chaotic dynamics being clearly visible over 400 million years (Laskar, 1992a,b) (Fig.10).

It should be stressed that the exponential divergence of the orbits revealed by the computation of the Lyapunov exponent result mostly from the change from libration to circulation of the resonant precession angles, which induce after some time a total indeterminacy of the precessional angles of the orbit, that is its orientation in space. The eccentricity and inclination (which are action like variables) variations due to the chaotic diffusion are much less rapid, and an important question is to estimate their wandering over the life time of the solar system.

3.5. THE CHAOTIC OBLIQUITY OF THE PLANETS

Instabilities of another sort also manifest themselves in the motion of the solar system's planets. These motions are not present in the orbits, but rather in the orientation of the planets' axes of rotation. Because of their equatorial bulge, the planets are subject to torques arising from the gravitational forces of their satellites and of the Sun. This causes a precessional motion, which in the Earth's case has a period of about 26,000 years. Moreover, the obliquity of each planet—the angle between the equator and the orbital plane—is not fixed, but suffers a perturbation due to the secular motion of the planet's orbit. Over one million year period, this variation is only ±1.3 degrees around the mean value of 23.3 degrees. This may not seem like much, but it is enough to induce variations of nearly 20 percent in the summer insolation received at 65 degrees north latitude (fig. 11). According to Milankovitch theory (see Imbrie 1982), the amount of additional heat received during the summer at high latitudes is an important factor in climate studies, as it melts ice accumulated over the winter and prevents the ice caps from extending their reach. When this insolation is not sufficient, the ice cap extends, inducing a general cooling of the temperature on Earth, and eventually leading to an ice age. Weak variations in the Earth's obliquity are therefore a determining factor in regulating the climate enjoyed by the Earth over the last several million years. The quaternary ice ages constitute significant climatic changes, but were not so severe as to permanently change the conditions for life on the Earth's surface.

The full equations of precession are presented in (Laskar *et al.*, 1993a, b, Laskar and Robutel, 1993). In fact, in order to understand the dynamics of the problem, the very small terms of these equations can be neglected, although they are taken into account in the numerical computations. In the following simplified equations, we shall also neglect the eccentricity of of the Earth and the Moon as well as the inclination of the Moon. This will provide a simple but realistic form for the equations of precession which will allow the curious reader to check directly most of the computations. Using the action variable $X = \cos\varepsilon$, where ε is the obliquity, and the precession angle ψ, the hamiltonian reduces to

$$H(X, \psi, t) = \frac{1}{2}\alpha X^2 + \sqrt{1 - X^2}(\mathbf{A}(t)\sin\psi + \mathbf{B}(t)\cos\psi) \qquad (6)$$

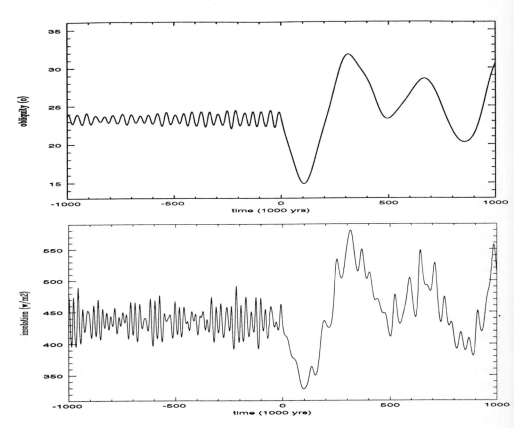

Figure 11. Changes in obliquity (a) and insolation at 65N ($\lambda_d = 120°$) (b) computed in the presence of the Moon from -1Myr to 0, and for 1 Myr after its suppression at $t = 0$ (Laskar *et al.*, 1993a).

where

$$\alpha = \frac{3}{2}\frac{C - A}{C}\frac{1}{\nu}(n_M^2 m_M + n_\odot^2 m_\odot) \tag{7}$$

A and C are the momentum of inertia of the planet (we assume that $B = A$), ν its rotational angular velocity, n_M and $n_\odot$ the mean motions of the Moon and the Sun (around a fixed Earth), m_M and $m_\odot$ their masses. The expression

$$\mathbf{A}(t) + i\mathbf{B}(t) \approx 2\frac{d\zeta}{dt}$$

where $\zeta = \sin i/2 \; e^{i\Omega}$ depends only on the change of inclination of the Earth with respect to a fixed plane and is given by the already computed

Table II

Quasiperiodic approximation of $\mathbf{A}+i\mathbf{B}$ obtained by frequency analysis of the Earth orbital solution over 18 Myr. The 12 Major terms are listed as well as a smaller, well isolated term, due to the perturbation of Jupiter and Saturn. $\mathbf{A}+i\mathbf{B} \approx \sum_{k=1}^{13} \alpha_k e^{i(\nu_k t+\phi_k)}$ (the α_k are expressed in yr^{-1}) (from Laskar et al., 1993b).

k		ν_k ("/yr)	$\alpha_k \times 10^6$	$\phi_k(°)$
1	s_3	-18.8504	1.616070	151.724
2	s_4	-17.7544	0.691588	199.002
3		-18.3016	0.478868	176.641
4	s_6	-26.3302	0.340738	37.294
5	s_1	-5.6128	0.274325	270.479
6		-19.3997	0.286930	305.514
7	s_2	-7.0772	0.237068	9.899
8		-19.1251	0.165838	46.398
9		-6.9564	0.132989	199.316
10		-7.2037	0.112089	176.470
11		-6.8283	0.108391	233.037
12		-5.4892	0.080168	289.422
13	$s_6 - g_6 + g_5$	-50.3021	0.001043	120.161

solution of the solar system from (Laskar, 1990). Although the motion of the solar system is chaotic, for qualitative understanding of the behaviour of the solution, it is convenient to use a quasiperiodic approximation of this quantities over a short time span of a few millions of years (table II):

$$\mathbf{A}(t) + i\mathbf{B}(t) \approx \sum_{k=1}^{N} \alpha_k e^{i(\nu_k t+\phi_k)} \ .$$

With this approximation, the hamiltonian now reads

$$H = \frac{1}{2}\alpha X^2 + \sqrt{1 - X^2} \sum_{k=1}^{N} \alpha_k \sin(\nu_k t + \psi + \phi_k) \qquad (8)$$

which is the hamiltonian of an oscillator of frequency αX_0, perturbed by a quasiperiodic external oscillation of small amplitude ($|\alpha_k| \ll 1$). We will thus obtain a resonance, when $\dot\psi \approx \alpha X_0 = \alpha \cos \varepsilon_0 = 50.47"/yr$ will be opposite to one of the frequency ν_k.

When limited to a single term ($N = 1$), this Hamiltonian is integrable (Colombo, 1966). On the contrary, when ($N > 1$), a simple application of Chirikov overlap criterion already allows to forecast the existence of chaotic zone for the obliquity.

In the frequency decomposition of the $\mathbf{A}(t) + i\mathbf{B}(t)$ planetary forcing term, there exists a periodic term of small amplitude related to perturbations exerted by Jupiter and Saturn and of frequency $s_6-g_6+g_5 = -50.30207"/yr$, which could enter into resonance with the precession frequency.

In order to see the effect of this resonance, we slightly changed the value of the dynamical ellipticity $(C - A/C)$ of the Earth, keeping fixed its angular momentum. This is what can happen for example during an ice age, where the redistribution of the ice changes a little the dynamical ellipticity of the Earth. The integration of the obliquity of the Earth in this small resonance showed and increase of the maximum obliquity of the Earth of about 0.5 degrees. This small term could thus be of great importance in the computation of the past insolation of the Earth (Laskar et al., 1993b).

After investigating the effect of this small resonance, we investigated the global dynamics of the obliquity of the Earth, by means of frequency map analysis (Laskar, 1990, 1993a, Laskar et al., 1992, Dumas and Laskar, 1993). Briefly speaking, for a Hamiltonian system with n degrees of freedom close to integrable $H(J_i, \theta_i) = H_0(J_i) + \varepsilon H_1(J_i, \theta_i)$, we shall construct numerically, the frequency map

$$F_T : \mathbf{R}^n \times \mathbf{R} \longrightarrow \mathbf{R}^n$$
$$(J, \tau) \longrightarrow \nu(J, \tau) \tag{9}$$

which associates to the action like variables $(J_i)_{i=1,n}$ and to the starting time τ, the frequency vector $(\nu_i)_{i=1,n}$ obtained numerically with a refined Fourier analysis of the solution of initial conditions (J_i, θ_{i0}) over the finite time interval $[\tau, \tau+T]$ (the initial phases $\theta_i(0) = \theta_{i0}$ are fixed to an arbitrary value).

The regularity of the trajectories can then be monitored by the analysis of the frequency map (9), which allows also to make refined estimates of the chaotic diffusion of the orbit in the phase space. Indeed, the frequencies $(\nu_i)_{i=1,n}$ can be thought as the "best" action variables obtained locally for the given initial condition.

In the case of the present $1 + 15$ degrees of freedom problem, we fix $\tau = 0$, and as the orbital motion of the solar system is not supposed to be perturbed by the orientation of the planets, the frequency map will reduce to a $\mathbf{R} \longrightarrow \mathbf{R}$ map. The regularity of the motion will then be studied directly by looking to the regularity of the frequency curve giving the numerically determined precession frequency for various values of the initial obliquity.

This analysis, which was performed for every 0.1° in obliquity over 18 million years shows immediately that the obliquity of the Earth is presently stable, but reveals also the existence of a very large chaotic region, ranging from 60° to 90° (fig. 12) (Laskar et al., 1993b).

We are far from this chaotic region, and the changes of obliquity of the Earth remains small (23.3° ± 1.3°), but if the Moon were not present, the value of the precession constant α would be roughly divided by 3 (as for the

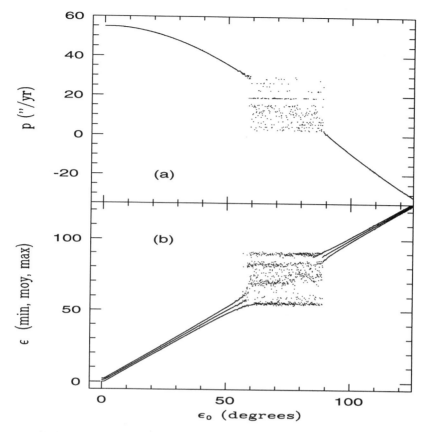

Figure 12. Stability of the rotation axis of the Earth in presence of the Moon. A different numerical integration of the precession equations is made over 18 Myr for each value of the initial obliquity of the Earth ranging from 0 to 125 degrees by 0.1 degree stepsize. For each integration, the minimum, mean, and maximum values of the obliquity are retained (b). A frequency analysis is also performed in order to determine precisely the averaged precession frequency of the Earth over this 18 Myr time span. The regularity of the frequency map (a) reflects the regularity of the motion. In particular, an extended chaotic zone is clearly visible from 60° to 90°. (Laskar *et al.*, 1993a).

ocean tides, the Moon accounts for $\approx 2/3$ in α, and the Sun for $\approx 1/3$). The precession frequency will also be divided by 3, and will be in resonance with the perturbations due to the motion of the orbital plane of the Earth. Even more, many resonances overlap, giving rise to an extended chaotic zone.

We investigated the global stability of the precession of the Earth for many values of its rotation speed ν (it should be noted that the dynamical ellipticity $C - A/C$ is proportional to ν^2), and found that for all primordial rotation period ranging from 12 h to about 48 h, the Earth obliquity would

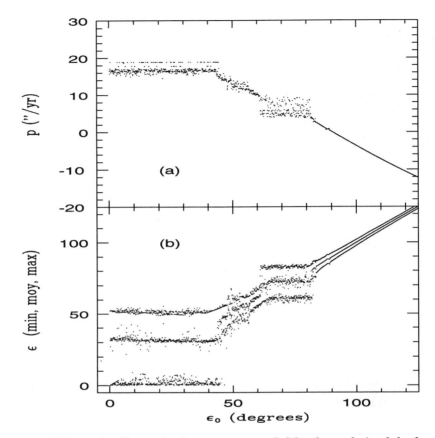

Figure 13. Without the Moon, the chaotic zone revealed by the analysis of the frequency map over 18 Myr (a) extends from 0° to ≈ 85°, for a period of rotation of the Earth of about 20 hours (Laskar *et al.*, 1993a).

suffer very large chaotic variations, from nearly 0° to about 85° (figs. 13, 14), which would probably lead to terrible climate variations on its surface (Laskar *et al.*, 1993a, Laskar 1993b) (typical changes from nearly 0° to about 60° can occur in less than 2 Myr, while transition to higher values of the obliquity should take a much longer time).

   In much the same way as described above for the Earth, we studied the stability of the axial orientation of all the principal planets of the solar system. Mercury and Venus are special cases, since—no doubt because of solar tides acting over time—their rotational speeds are now very slow. Venus also possesses a trait that has long intrigued astronomers: it does not rotate in the same direction as the other planets, or in other words, it is upside down.

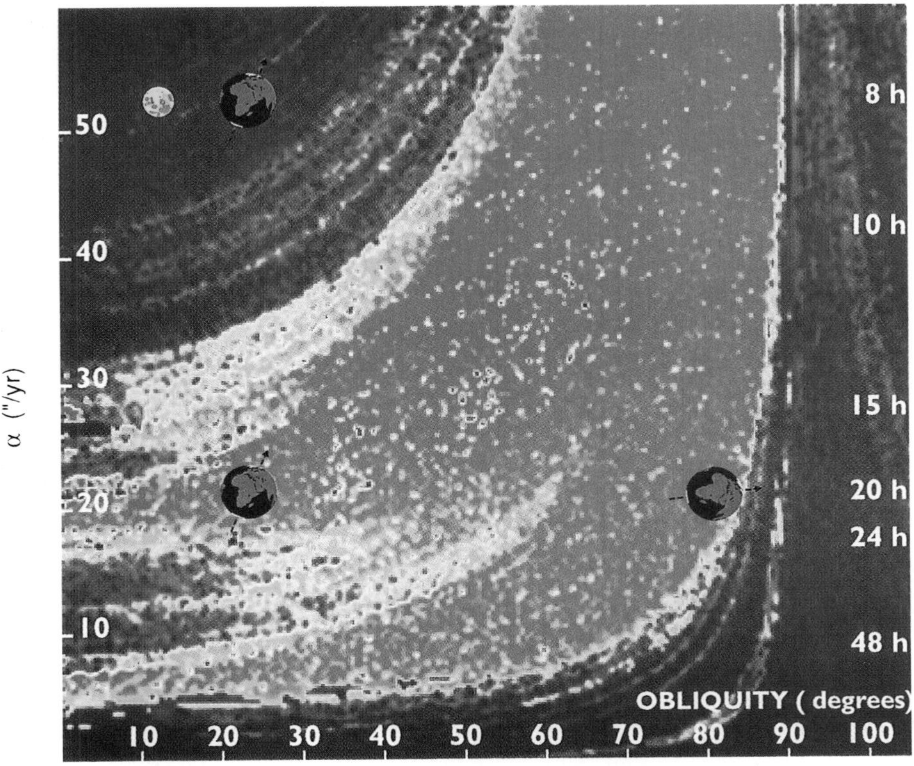

Figure 14. The zone of large scale chaotic behavior for the obliquity of The Earth (without the Moon) for a wide range of spin rate. The precession constant α is given on the left in arcseconds per year, and the estimate of the corresponding rotation period of the planet on the right, in hours. The blue region corresponds to the stable orbits, where the variations of the obliquity are moderate, while the orange and red zone is the chaotic zone. Indeed, the chaotic motion is estimated by the diffusion rate of the precession frequency measured for each initial condition (ε, α) via numerical frequency map analysis over 36 Myr. In the large chaotic zones visible here, the chaotic diffusion will occur on horizontal lines (α is fixed), and the obliquity of the planet can explore horizontally all the red orange zone. The extend of the chaotic zone should even be larger, when considering the diffusion of the orbits over longer time scale. With the Moon, one can consider that the present situation of the Earth can be represented approximately by the point of coordinates $\varepsilon = 23°, \alpha = 55\ ''/$yr, which is in the middle of a large zone of regular motion. Without the Moon, for spin period ranging from about 12h to 48h, the obliquity of the Earth would suffer very large chaotic variations ranging from nearly 0° to about 85°. This figure summarizes the results of about 250 000 numerical integrations of the Earth obliquity variations under the whole solar system perturbations for various initial conditions over 36 Myr. (Laskar and Robutel, 1993, Laskar, 1993b).

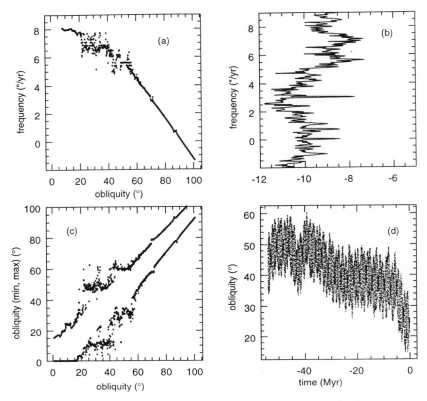

Figure 15. (a) Frequency map analysis of the obliquity of Mars over 56Myr. 1000 integrations of the obliquity of Mars have been conducted over 56 Myr for various initial conditions. A large chaotic zone is visible, ranging from 0° to 60°. In (b), the power spectrum of the orbital forcing term $\mathbf{A}(t) + i\mathbf{B}(t)$ is given in logarithmic scale, showing the correspondence of the chaotic zone with the main frequencies related to Venus and Mercury. (c) Maximum and minimum values of the obliquity reached over 56Myr. (d) Actual variations of the obliquity over 56 Myr for a selected orbit. (adapted from Laskar and Robutel, 1993).

It was generally assumed that Venus was formed upside down—or at least with its rotational axis in its orbital plane, since then dissipative effects arising from solar tides, core-mantle interactions, or from atmospheric tidal forces due to the Sun could bring it into an upside down position (Goldreich and Peale, 1970, Dobrovolski, 1980). Indeed, this was considered as a constraint on the models for the formation of the solar system, which would then require a "stochastic phase" at the end of the formation process, with a moderate number of large impacts by massive objects in order to obtain the desired orientation of this planet (e.g. Dones, and Tremaine, 1993a) We have shown instead that, even if Venus started with a rotational speed similar to the Earth's, and in the same direction, the presence of a large chaotic zone in its obliquity could subject it to severe tilting, bringing its rotational axis very nearly into its orbital plane. The dissipative effects just described could then bring it into its present position, where ultimately it might be stabilized as its rotation slowed further.

The situation for Mercury is slightly different. As is the case for Venus, we do not know Mercury's primordial rotational period, but it is enough to assume it was shorter than 300 hours to assure that, in the course of its history, Mercury underwent strongly chaotic variations in its obliquity, ranging from 0 to 90 degrees in the space of a few million years (Laskar and Robutel, 1993). As with Venus, the continued effects of tides could then slow its rotation, causing it to right itself and end up in its present position (Peale, 1974, 1976).

Mars is far from the Sun, and its satellites Phobos and Deimos have masses far too small to slow its rotation, so that its present rotational period of 24 hours 37 minutes is close to its primordial rotational period. Mars' equator is inclined 25 degrees with respect to its orbital plane, and its speed of precession, 7.26 seconds per year, is close to certain frequencies of motion of its orbit (Ward, 1974, Ward and Rudy, 1991). Moreover, variations in the inclination of Mars' orbit are considerably stronger than those of the Earth. It follows that variations in its obliquity over a period of one million years are also much stronger than the Earth's, and Ward has found obliquity variations on the order of ± 10 degrees about a mean value of 25 degrees. These variations bring about strong climatic changes on Mars' surface, and certain surface structures seem to bear witness to these changes.

Our computations (Laskar and Robutel, 1993), and numerical results obtained by Touma and Wisdom (1993), provide evidence that the motion of Mars' rotational axis is chaotic. This has two consequences. First, as it is also the case for the orbital motion of the inner planets, it is not possible to predict the orientation of Mars' axis for periods longer than a few million years.

But more important, the obliquity of Mars is subject to much larger variations than those predicted by Ward, ranging from about 0 to 60 degrees in less than 50 million years (fig. 15) (Laskar and Robutel, 1993). Models of the past climates of Mars need then to be reviewed in light of these new results. In particular, the large obliquity possibly reached for this planet will lead to higher temperature on its surface which may then allow the possibility of liquid water on its surface (Jakosky et al., 1993).

On figure 15, obtained by frequency map analysis, it is clear that the size of the chaotic zone of the obliquity ranges from 0° to about 60°, and these values are actually reached during numerical integrations over less than 50 Myrs, but it can also be seen that the chaotic zone is divided into two main boxes: one is essentially related to secular resonances with Venus, and the second one with Mercury. The diffusion in each of these boxes is rapid, while the passage from one box to the other one is more difficult. This explains why Touma and Wisdom (1993), as they performed only a very limited numbers of integrations, were not able to see this transition and found only limited variations of Mars obliquity from 11° to 49°.

The existence of this large chaotic zone in the orientation motion of Mars also removes some constraints from models of solar system formation, since Mars' obliquity cannot be considered primordial, and its present orientation which is very similar to the Earth's, is purely due to chance.

On the other hand, our investigations showed that the obliquities of the outer planets are essentially stable. It is thus not possible to explain like that the very large obliquity of Uranus (98°), but it should be investigated if a chaotic behavior sufficient to lead to such an obliquity could have occurred during the formation of the solar system, at a time when it was supposed to be much more massive.

These results show that the situation of the Earth is very particular. The common status for all the terrestrial planets is to have experienced very large scale chaotic behaviour for their obliquity, which, in the case of the Earth and in absence of the Moon, may have prevented the appearance of evoluted forms of life. It is presently difficult to say exactly what would be the climate on the Earth with very large values of the obliquity, and even more with the possibility of drastic changes in the Earth orientation within a few million years, as no realistic models have yet been constructed taking into account these new results. But it is important to realize that up to now, it was generally assumed that in a planetary system similar to our, the planet located not too close to the sun, in which case runaway greenhouse effect may occur, and not to far from it in order to prevent runaway glaciation (Hart, 1978), would be very similar to the Earth. Our study demonstrated that this "reasonable hypothesis" is wrong, and that in the case of the Earth, we owe our relative present climate stability to an exceptional event: the presence of the Moon. While many results since the acceptance of heliocentrism have tendency to show that our Earth should be very common in the Universe, the present findings go in the opposite direction.

Moreover, the presence for the Earth of such a large satellite as the Moon, still puzzle astronomers, and the currently mostly accepted scenario for its origin relies on a non generic event: a large body, of the size of Mars, formed at the same time as the other planets enter in collision with the Earth, the subsequent accretion of the resulting debris forming the Moon (see the review of Stevenson, 1987). Indeed, if we accept that our presence on the Earth is correlated to the existence of the Moon, there is no problem for accepting an improbable scenario for the formation of the Moon, as soon as it does agree with all other physical and chemical constraints. Moreover, we may accept even more improbable models, if they better agree with the present observations. As a result, this may reduce the chances of finding extraterrestrial civilizations similar to ours around the nearby stars.

3.6. PLANETARY EVOLUTION ON GYR TIME SCALES

If the motion of the solar system were close to quasiperiodic, that is close to a KAM tori in the phase space, then it could be expected that some bound on the possible diffusion of the orbit over 5 Gyr would result from a Nekhoroshev like theorem (e.g. Niederman, 1994). In fact, as it was shown in (Laskar, 1990), although the system reduced to the outer planets may be considered as close to a KAM tori, the full solar system evolves far from a KAM tori, and diffusion of the action like variables (eccentricity and inclination) can occur. The natural question is thus to estimate this diffusion. Let us remind that in contrast to two degrees of freedom systems, where the diffusion may be bounded, in such a many degrees of freedom system (15 independent degrees of freedom for the secular system), there exist no results on the existence of invariant set which will bound the evolution of the system on infinite time span.

One may be tempted to integrate the motion of the solar system over 5 Gyr, that is over its expecting time life. For direct numerical integrations, this can be considered as an interesting challenge as it is still out of reach of present computer technology, but it should be stressed, that by no means it can be considered as the description of the evolution of the solar system over 5 Gyr. Indeed, because of the exponential divergence with a Lyapunov time of 5 Myr, after about 100 Myr the computed solution will be very different from the real solution followed by the actual solar system. Such a solution still present some interest, as it gives one of the possible behavior of the solar system, but it is much more important to obtain some description of the chaotic zone where the solar system evolves. In particular, it is more interesting to estimate the speed of diffusion in this chaotic zone. For such a goal, a single integration of the solar system over 5 Gyr will not be sufficient.

Quite surprisingly, we can use integrations over even longer time span, which will act as scouts exploring this chaotic zone. We can also send multiple of these explorers with very close initial conditions, in order to reach a larger portion of the phase space which can be attained by the solar system in 5 Gyr.

Doing this kind of search, it becomes obvious that we need to be able to integrate very rapidly the equations of motion for the solar system, and the present work analyzed the results of many such numerical integrations, totalling an integration time larger than 200 Gyr.

In order to achieve this task, the secular equations of the solar system were used, after some simplifications (laskar, 1994). Indeed, the initial secular system consisted into about 150000 polynomial terms, but many of them

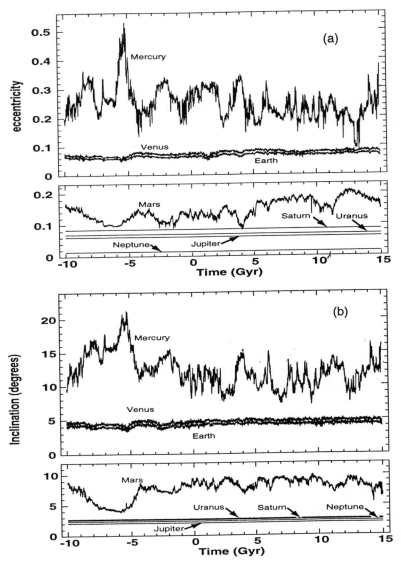

Figure 16. Numerical integration of the averaged equations of motion of the solar system 10 Gyr backward and 15 Gyr forward. For each planet, the maximum value obtained over intervals of 10 Myr for the eccentricity (a) and inclination (in degrees) from the fixed ecliptic J2000 (b) are plotted versus time. For clarity of the figures, Mercury, Venus and the Earth are plotted separately from Mars, Jupiter, Saturn, Uranus and Neptune. The large planets behavior is so regular that all the curves of maximum eccentricity and inclination appear as straight lines. On the contrary the corresponding curves of the inner planets show very large and irregular variations, which attest to their diffusion in the chaotic zone.(Laskar, 1994)

are of very small amplitude. It was thus possible to suppress them, and to reduce the system to only 50 000 terms, conserving the symmetries which were present in the equations. Doing that, only about 6000 terms need really to be computed during the evolution of the second hand member of the equations, and the computations could be achieved on an IBM RS6000/370 workstation at a rate of about 1 day of CPU time per Gyr, without any significant loss in the precision. Moreover, the numerical integration of the secular system has been improved, and the stepsize reduced to 250 years, which allowed the best precision. As we want to understand the dynamics of this secular system, it is actually necessary to make the integration with great accuracy. The secular system is an approximation of the real equations of motion, but the understanding of the global dynamical behavior of this system will provide a lot of information on the original system.

Some first integrations were conducted over 25 Gyr (-10 Gyr to + 15 Gyr) (fig. 16). It may seem strange to try to track the orbit of the solar system over such an extended time, longer than the age of the universe, but one should understand that it is done in order to explore the chaotic zone where the solar system evolves, and after 100 Myr, can give only an indication of what can happen. On the other hand, if there is a sudden increase of eccentricity for one planet after 10 Gyr, this still tells us that such an event could probably also occur over a much shorter time, for example in less than 5 Gyr. In the same way, what happens in negative time can happen as well in positive time.

In order to follow the diffusion of the orbits in the chaotic zone, one needs quantities which behave like action variables, that is quantities which will be almost constant for a regular (quasiperiodic) solution of the system. Such quantities are given here by the maximum eccentricity and inclination attained by each planet during intervals of 10 Myr (Fig. 16).

The behavior of the large planets is so regular that all the corresponding curves appear as straight lines (Fig. 16). On the contrary the maxima of eccentricity and inclination of the inner planets show very large and irregular variations, which attest to their diffusion in the chaotic zone. The diffusion of the eccentricity of the Earth and Venus is moderate, but still amounts to about 0.02 for both planets. The diffusion of the eccentricity of Mars is large and reaches more than 0.12, leading to values higher than 0.2 for the eccentricity of Mars. For Mercury, the chaotic zone is so large (more than 0.4) that it reaches values larger than 0.5 at some time. The behavior of the inclination is very similar.

Strong correlations between the different curves appear in figure 16. Indeed, as the solar system wanders in the chaotic zone, it is dominated by the linear coupling among the proper modes of the averaged equations (eq. 3), which induces a very similar behavior for the maximum eccentricity and inclination of Venus and the Earth. This coupling is also noticeable in the solution of Mars. On the other hand, an angular momentum integral exists in the averaged equations, and explains why when Mercury's maximum eccentricity and inclination increase, the similar quantities for Venus, the Earth and Mars decrease. Thus it appears that, despite the small values of the inner planets' masses, the conservation of angular momentum plays a decisive role in limiting their excursions in the chaotic zone. Thus the same argument which allowed Laplace to "prove" the stability of the solar system in the linear approximation (see section 3.1) appear to be indeed primordial for limiting the wandering of the Earth's orbit in the chaotic zone, and thus achieving practical stability over the age of the solar system.

3.7. ESCAPING PLANETS

At some time, Mercury suffered a large increase in eccentricity (fig. 16) rising up to 0.5. But this is not sufficient to cross the orbit of Venus. The question then arises whether it is possible for Mercury to escape from the solar system in a time comparable to its age. A first attempt to answer this was made by slightly changing the initial conditions for the planets. Indeed, because of the chaotic behavior, very small changes in the initial conditions lead to completely different solutions after 100 Myr. Using this, only one coordinate in the position of the Earth was changed, amounting to a physical change of about 150 meters (10^{-9} in eccentricity). The full system was integrated with several of these modified solutions, but led to similar (although different) solutions. In fact, it should not be too easy to get rid of Mercury, otherwise it would be difficult to explain its presence in the solar system.

I thus decided to guide Mercury towards the exit. A first experiment was done for negative time: for 2 Gyr, the solution is left unchanged, then, 4 different solutions are computed for 500 Myr, in each of which the position of the Earth is shifted by 150 meters, in a different direction (due to the exponential divergence, this corresponds to a change smaller than Planck's length in the original initial conditions).

The solution which leads to the maximum value of Mercury's eccentricity is retained up to the nearest entire Myr, and is started again. In 18 of such steps, Mercury attains eccentricity values close to 1 at about -6 Gyr when the solution enters a zone of greater chaos, with Lyapunov time ≈ 1 Myr, giving rise to much stronger variations of the orbital elements of the inner planets. A second solution was also computed in positive time, with changes in initial condition of only 15 meters instead of 150 meters. As anticipated, this led to a similar increase in Mercury's eccentricity, this time in only 13 steps and about 3.5 Gyr (fig. 17).

While the eccentricity increases, the inclination of Mercury can change very much but the computation of the relative positions of the intersection

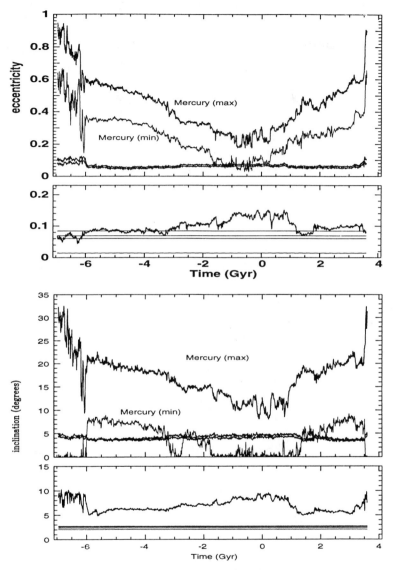

Figure 17. Orbit of the solar system leading to very large values for the eccentricity of Mercury, and possibility of escape at -6.6 Gyr and +3.5 Gyr. The plotted quantities are the same as in Fig. 16, except for Mercury, where minimum eccentricity and inclination over 10 Myr are also plotted. During all the integrations, the motion of the large planets is very regular (Laskar, 1994).

of the orbits of Mercury and Venus with their line of nodes demonstrated that the orbits effectively intersect at about 3.5 Gyr. At this time, the two planets can experience a close encounter which can lead to the escape of Mercury or to collision (Laskar, 1994).

For very high eccentricity of Mercury, the model used here no longer gives a very good approximation to the motion of Mercury, but in the real system the addition of the extra degrees of freedom related to semi-major axes and longitudes, will probably lead to even stronger chaotic behavior, as in general, addition of degrees of freedom increases the stochasticity of the motion. When one examines the results of direct numerical integrations of asteroids in secular resonances (see for example Farinella *et al.*, 1993), one can see very often that at the beginning, one can assist to a similar increase in eccentricity due to the secular resonance, but after some times, when the eccentricity is high enough, the perturbations related to mean motion resonances become very important, and large scale chaos related to mean motion resonances occurs, resulting from overlap of the mean motion resonances, which induces a large diffusion of the semi major axis. Most probably, the natural scenario for an escape of Mercury will be of this type and for high eccentricities the complete solar system should be much less stable than the secular system, where the semi-major axis are frozen.

Similar computations were made for Mars and the Earth, but did not lead up to now to an escaping solution. For the Earth, the maximum eccentricity reached after 5 Gyr is about 0.1, while for Mars, the eccentricity attained about 0.25 after 5 Gyr. With such a high eccentricity, Mars comes very close to the Earth, and it may be possible to find some escaping solution for Mars when considering the complete equations, but it should be noted that the search for the escaping solutions of Mars, in positive time uniquely, necessitated about 100 numerical integrations, each of them over 500 Myr. This is obviously out of reach of direct numerical integration with present computer technology, and a mixed solution should be envisaged, where most of the increase of eccentricity is made with the secular equations, and the final part with direct numerical integration.

3.8. Marginal stability of the solar system

The existence of an escaping orbit for Mercury does not mean that this escape is very likely to occur. In fact, the solution computed here which lead to an escape was very carefully tailored, by selecting at each step one solution among 4 or 5 equivalent ones. The result obtained here is a result of existence for an escaping orbit, but does not tell us the probability for this escape to occur. The computation of an estimate of this probability would require to follow more completely all the studied orbits, and also most probably to take into account the full equations in order to be accurate. From the present computation, it can be thought that this probability is small, but not null, which is compatible with the present existence of Mercury.

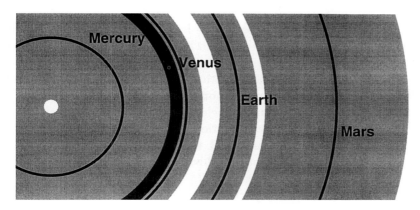

Figure 18. Estimates of the zones possibly occupied by the inner planets of the solar system over 5 Gyr. The circular orbits correspond to the bold lines, and the zones visited by each planet resulting from the possible increase of eccentricity are the shaded zones. In the case of Mercury and Venus, these shaded zones overlap. Mars can go as far as 1.9 AU, which roughly corresponds to the inner limit of the asteroid belt.

Without speaking of escaping orbits, the very large diffusion of the inner planets orbits is very striking. Even after the discovery of the chaotic behavior of the solar system, and despite the results of (Laskar, 1990) where estimates of the diffusion were already computed by means of frequency analysis, many people assumed that the chaotic diffusion in the solar system was very small. Here it is clearly demonstrated that for the inner planets, it is not the case. More, for the inner planets, the excursion of the eccentricity and inclination variables seems to be essentially constrained by the angular momentum conservation which explains that when the maximum eccentricity of Mercury increases, the maximum eccentricity of Venus, the Earth and Mars decreases. This is quite surprising, when considering that most of the angular momentum comes from the outer planets. In fact, the outer planets system is very regular, and practically no diffusion will take place among the degrees of freedom related to the outer planets.

On figure 16, it appears that the less massive planets are subject to the largest variation of eccentricity. This becomes obvious when considering that these variations are essentially bounded by the angular momentum conservation, which for each planets is proportional to $m\sqrt{a}$, where m is the mass of the planet, and a its semi major axis.

If for each planet, we consider the maximum diffusion of the eccentricity observed over 5 Gyr (fig. 18) during similar numerical experiments as for Mercury, we find that Mercury's eccentricity can go sufficiently high to allow Mercury's orbit to cross the orbit of Venus, Venus and the Earth's eccentricity can go up to 0.1, and Mars as high as 0.25. Apart from some small place in between Venus and the Earth, or the Earth and Mars, all the inner solar system is swept by the planetary orbits, and the small planets (Mercury and Mars) are the planets which present the largest excursions. Practically, we can conclude that the inner solar system is full. That is there is no room for any extra planet. Indeed, even if there are some place which seems not to be possibly reached in 5 Gyr, an additional planet orbit will most probably intersect one of the already existing ones. If we add a large planet, of the size of the Earth or Venus, its orbital elements will not vary much but it will induce strong short periods perturbations. On the contrary, a small object will suffer large orbital variations, as it will not be much constrained by the angular momentum conservation. In this case, encounters with the already existing planets is very probable.

The variations which are reported in fig. 17 are the maximum variations observed over 5 Gyr, and not the most probable variations, but the addition of an extra planet will most probably increase very much the diffusion by increasing the numbers of degrees of freedom, and these maximum possible variations can probably be considered as good estimates of the probable variations over 5 Gyr in the eventuality of this addition of an extra planet in the inner solar system. It thus becomes interesting to speak of marginal stability when considering the solar system. Maybe there were some extra planets at the early stage of formation of the solar system, and in particular in the inner solar system, but this led to so much instability that one of the planets (probably among the smallest ones, of the size of Mercury or Mars) suffered a close encounter, or a collision with the other ones. This lead eventually to the escape of this planet, and the remaining system gets more stable. Indeed, this is what was observed when Mercury was suppressed in the numerical simulation, after the crossing of Venus orbit. Quinlan (1993) also observed similar results on experiments conducted on examples of planetary systems with the full equations over shorter time scales. In this case, at each stage of its evolution, the system should have a time of stability comparable with its age, which is roughly what is achieved now, when ones finds that escape of one of the planets (Mercury) can occur within 5 Gyr.

4. Discussion

4.1. STABILITY OF THE SOLAR SYSTEM

The Lyapounov time of 5 Myr for the solar system (Laskar, 1989), as well as the existence of secular resonances of large amplitude in the inner solar system demonstrates that the motion of the solar system is not regular, and cannot be approximated by a quasiperiodic trajectory over more than 10 to 20 Myr. Moreover, it will be practically impossible to make any precise prediction for the evolution of the solar system beyond 100 Myr, due to the exponential divergence of the orbits. Thus, we are far from the regular solutions, whose existence was exhibited by Arnold.

Nevertheless, this result applies more specifically to the inner planets (Mercury, Venus, Earth and Mars). Although the outer planets (Jupiter, Saturn, Uranus and Neptune) are perturbed gravitationally by the inner planets, this perturbation is small, and the induced effect of their chaotic motion will only generate a small diffusion of their trajectories. For a planetary system restricted to the outer planets, and even more for the Jupiter -Saturn couple, it still should be possible to obtain rigorous (in the mathematical sense) stability results along the lines described by Arnold and Nekhoroshev, although this will necessitate specially adapted version of the theorems.

In their integrations of the outer planets system, Sussman and Wisdom (1992) have reported Lyapounov times ranging from 3 to 30 Myr. This result needs to be taken cautiously as it seems to be very dependent of the numerical procedure they used to integrate the equations. Moreover, as the Lyapounov time of the secular system seems to be much larger, these instabilities should be related to the fast orbital motion of the planets, and not to the slow precession of the orbits. They probably involve very high order mean motion resonances whose amplitude will be very small, and no physical consequence will result. The orbits of the outer planets should still be confined to very narrow regions over the age of the solar system.

For the secular system, the problem is very different. The main frequencies are of the order of 100 000 years. The Lyapounov time of 5 Myr (which is also of the same order as the libration period of the identified main secular resonance) is only equal to 50 times the fundamental periods of the motion, which explains why this can lead to large scale chaotic behavior. Indeed, we have seen that all the inner planets experience significant chaotic diffusion over billion years timescale, and the existence of an escaping orbit for Mercury demonstrates that the solar system in not stable, even when considering the strongest meaning of this word, that is the possibility of evasion or collision of the planets.

However, although the solar system is not stable, it can be considered as marginally stable, that is strong instabilities (collision or escape) can only occur on a time scale comparable to its age, that is about 5 Gyr. We have exhibited an escaping or collisional orbit for Mercury in less than 3.5 Gyr (Laskar, 1994). For Mars, the large diffusion of its orbit can drive the eccentricity to about 0.25, and it still should be checked, using the full equations of motion in order to add the possibility of instabilities related to the mean motion, whether this could also lead to a collisional orbit with the Earth.

On the other hand, the orbits of Venus and the Earth, because of their larger masses, their linear coupling, and because of the angular momentum conservation constraint, seem to be practically confined only to small deviations from their presents path. These two planets, although their orbits are not close to quasi periodic, can thus be considered as stable over the age of the solar system, without regard to the possibilities of collision with Mercury or Mars.

4.2. Constraints on the formation of the solar system

This new vision of the evolution of the solar system over its age also induces some changes in the dynamical constraints for the formation models of the solar system (Harris and Ward, 1982).

In particular, the recognition that the solar system is in a state of marginal stability suggests that the organization of the planets in the solar system (often quoted as the Titius-Bode law), and more particularly of its inner part, is most probably due to its long run orbital evolution, and not uniquely to its rapid (less than 100 million years) formation process. We have shown that the inner solar system is full from 0 AU to about 2 AU, which coincides with the inner edge of the asteroidal belt. Some extra inner planets may have existed, but their existence then gave rise to a much more instable system, leading to the escape or collision of one of the planets, the remaining part then being much more stabilized. Indeed, this is what was observed in our numerical computations, after the simulation of the escape of Mercury. In particular, these findings show that minor bodies in the inner solar system will probably not be able to survive for a very long time. This result is important for the understanding of the formation of the solar system, as it tells us that the solar system at the end of its formation process may have been significantly different from the present one, and has then evolved towards the present configuration because of the gravitationnal instabilities. It still should be very interesting to investigate this point further using simulations with the addition of an extra planet, but many features have already been deduced here from the present computations.

On the other hand, the outer system is very stable, but the long time recent numerical integrations (see section 2.5) also demonstrate that the outer solar system is full, that is most of the objects introduced in this system will escape on time scale much shorter than 5 Gyr. Apart from some special locations, like the Jupiter Lagrangian positions, stable zones only begin at about 40 AU, where several objects were recently founded.

Moreover, by showing that none of the obliquities of the inner planets are primordial (section 3.5), we have removed one of the constraint on the formation models for the solar system. We have also proven the stability of the obliquity of the outer planets, since the solar system is in its present state, but instabilities may have existed during the formation of the solar system, when the planetary disk was supposed to be more massive, and it is of great interest to study the possibility of such a scenario.

As was discussed in section 3.5, we have established the possibility of a strong correlation between our existence and the existence of the Moon, which should leave the possibility to accept an improbable scenario for its formation, if it properly accounts for the other physical and chemical observations. The models for the formations of the Moon thus need to be reevaluated in this scope.

4.3. GENERIC PLANETARY SYSTEMS

One may be now tempted to answer to the question of what will be a generic planetary system ?

Such a question is of course delicate to answer, after having only studied our solar system stability, but the observation that our solar system is in a state of marginal stability, that is practical stability on a time scale comparable to its age, can be a clue for answering this question.

Indeed, I would like to suggest that a planetary system will always be in this state of marginal stability, as a result of its gravitational interactions.

In particular, a planetary system with only one or two planets should be excluded, because it will then be much too stable*, or more precisely, if it does exist, it would be crowded with asteroids everywhere which would be the original remaining planetesimals, not ejected by planetary perturbations.

On the other hand, if the formation process is such that there exist some large outer planets, and some small inner planets, after 5 Gyr, the small inner planets will still be subject to instabilities similar to the present ones in the solar system, and thus so will be their obliquities.

It should indeed be noted that if a planet evolves at about 1 AU from a solar type star, that is in good condition to have liquid water on its surface, then its precession frequency will depend essentially on its rotation period and will thus be similar to the one of the Earth in absence of the Moon (fig. 14). Thus if the precessing frequencies of this planetary system are of the same order as those of our solar system, this planet will have a large probability to be subject to very large chaotic variations for its obliquity.

Moreover, in order to have an orbital stability comparable to the one of the Earth, a terrestrial planet probably needs to have a sufficiently large mass, otherwise it could be subject to orbital variations similar to the ones of Mercury or Mars, which would induce even larger variations of its obliquity.

These considerations show that it may not be so easy to find around a nearby star another planet with a similar orbital and rotational stability as the Earth, situated at a distance from the central star allowing the existence of liquid water.

---

* This results from some work in progress with P. Robutel

4.4. Epilogue

Many fundamental problems still remain in order to clarify the questions raised here on the genericity of our solar system and of the Earth orbital and rotational stability. Some concern the formation of the planetary system; in particular the understanding of the origin of the rotation of the planets (Dones and Tremaine, 1993b, Lissauer and Safronov, 1991) appears as a crucial point for the analysis of the stability of their orientations. As important will be possible improvements on the understanding of the response of the planets atmosphere behavior under insolation forcing. The direct observation of another planetary system, which may occur in the near future, should also provide important elements for answering these questions, but it should be stressed that improvements of the present theoretical knowledge of the global dynamics of planetary systems can also provide very important constraints on the possible organization of planetary systems.

Most of the results on the planetary orbits presented here rely on the analysis of the secular equations of the solar system, and not on the complete equations. This was the price to pay for allowing a more global approach on the problem of the stability and long time evolution of the solar system. Some integrations of the full equations are still welcome, but it is doubtful that these future integrations will change much the global landscape of the dynamics of the solar system portrayed here.

Acknowledgements

This work was partially supported by the Programme National de Planétologie. Computations were made partially with the help of IDRIS and on the SP2 of CNUSC.

References

Applegate, J.H., Douglas, M.R., Gursel, Y., Sussman, G.J. and Wisdom, J.: 1986, 'The solar system for 200 million years,' *Astron. J.* **92**, 176–194

Arnold V.: 1963a, Proof of Kolmogorov's theorem on the preservation of quasi-periodic motions under small perturbations of the hamiltonien, *Rus. Math. Surv.* **18**, N6 9–36

Arnold, V. I.: 1963b, 'Small denominators and problems of stability of motion in classical celestial mechanics,' *Russian Math. Surveys,* **18**, 6, 85–193

Bretagnon, P.: 1974, Termes à longue périodes dans le système solaire, *Astron. Astrophys* **30** 341–362

Brumberg, V.A., Chapront, J.: 1973, Construction of a general planetary theory of the first order, *Cel. Mech.* **8** 335–355

Carpino, M., Milani, A. and Nobili, A.M.: 1987, Long-term numerical integrations and synthetic theories for the motion of the outer planets, *Astron. Astrophys* **181** 182–194

Celletti, A.: 1990a, Analysis of resonances in the spin-orbit problem in celestial mechanics: the synchronous resonance (Part I), *J. Appl. Math. Phys. (ZAMP)* **41** 174-204

Celletti, A.: 1990b, Analysis of resonances in the spin-orbit problem in celestial mechanics: higher order resonances and some numerical experiments (Part II), *J. Appl. Math. Phys. (ZAMP)* **41** 453-479

Chirikov, B.V.: 1979, A universal instability of many dimensional oscillator systems, *Physics Reports* **52** 263-379

Chirikov, B.V., Vecheslavov, V.V.: 1989, Chaotic dynamics of comet Halley, *Astron. Astrophys.* **221**, 146-154

Cohen, C.J., Hubbard, E.C., Oesterwinter, C.: 1973, , *Astron. Papers Am. Ephemeris* **XXII** 1

Colombo,G.: 1966, *Astron. J.*, **71**, 891-896

Dermott, S.F., Murray, C. D.: Nature of the Kirkwood gaps in the asteroidal belt, *Nature*, **301**, 201-205

Dobrovolskis, A.R.: 1980, Atmospheric Tides and the Rotation of Venus :II.Spin Evolution, *Icarus*, **41**, 18-35

Dones, L., Tremaine, S.: 1993a, Why does the Earth spin forward?, *Science* **259** 350-354

Dones, L., Tremaine, S.: 1993b, On the origin of planetary spins, *Icarus* **103** 67-92

Dumas, H. S., Laskar, J.: 1993, Global Dynamics and Long-Time Stability in Hamiltonian Systems via Numerical Frequency Analysis, *Phys. Rev. Lett.* 70, 2975-2979

Duncan, M., Quinn, T., Tremaine, S.: 1988, The origin of short period comets *Astrophys. J. Lett*, **328**, L69-L73

Duncan, M., Quinn, T., Tremaine, S.: 1989, The Long-Term evolution of orbits in the solar system: a mapping approach *Icarus*, **82**, 402-418

Duriez, L.: 1979, Approche d'une théorie générale planétaire en variable elliptiques héliocentriques, *thèse* Lille

Farinella, P., Froeschlé, Ch.,Gonczi, R.: 1993, Meteorites from the asteroid 6 Hebe, *Cel. Mech.* **56** 287-305

Farinella, P. , Froeschlé, C., Gonczi, R.: 1994, Meteorite delivery and transport, in Symposium IAU 160, A. Milani, M. Di Martino, A. Cellino, eds, 205-222, Kluwer, Dordrecht

Fernández, J.A.: 1980, On the existence of a comet belt beyond Neptune, *Mon. Not. Roy. AStron. Soc.*, **192**, 481-492

Fernández, J.A.: 1994, Dynamics of comets: recent developments and new challenges, in Symposium IAU 160, A. Milani, M. Di Martino, A. Cellino, eds, 223-240, Kluwer, Dordrecht

Ferraz-Mello, S.: 1994, Kirkwood gaps and resonant groups, in Symposium IAU 160, A. Milani, M. Di Martino, A. Cellino, eds, 175-188, Kluwer, Dordrecht

Froeschlé, C., Scholl, H.: 1977 A qualitative comparison between the circular and elliptic Sun-Jupiter-asteroid problem at commensurabilities *Astron. Astrophys.* **57**, 33-59

Froeschlé, C., Gonzci, R.: 1988, On the stochasticity of Halley like comets, *Celes. Mech.* **43**, 325-330

Giorgilli A., Delshams A., Fontich E., Galgani L., Simo C.: 1989, Effective stability for a Hamiltonian system near an elliptic equilibrium point, with an application to the restricted three body problem, *J. Diff. Equa.* 77 167-198

Gladman, B., Duncan, M.: 1990, On the fates of minor bodies in the outer solar system *Astron. J.*, **100(5)**

Goldreich, P., Peale S.J.: 1970, The obliquity of Venus, *Astron. J.*, **75**, 273-284

Haretu, S.C.: 1885, Sur l'invariabilité des grands axes des orbites planétaires *Ann. Obs. Paris, XVIII*, I1-I139

Harris A.L., Ward, W.R.: 1982, Dynamical constraints on the formation and evolution of planetary bodies, *Ann. Rev. Earth Planet Sci.* **10** 61-108

Hart, M.H.: 1978, The evolution of the atmosphere of the Earth, *Icarus* **33** 23-39

Hénon, M. and Heiles, C.: 1964, The applicability of the third integral of motion: some numerical experiment, *Astron. J.*, **69**, 73-79

Holman, M.J., Wisdom, J.: 1993, Dynamical stability in the outer solar system and the delivery of short period comets *Astron. J.*, **105(5)**

Imbrie, J.: 1982, Astronomical Theory of the Pleistocene ice ages: a brief historical review, *Icarus* **50** 408-422

Jakosky, B.M., Henderson, B.G., Mellon, M.T.: 1993, Chaotic obliquity and the nature of the Martian climate, Bull.Am.Astron.Soc.,**25**,1041

Kinoshita, H., Nakai, H.: 1984, Motions of the perihelion of Neptune and Pluto, *Cel. Mech.* **34** 203

Klavetter, J.J.: 1989, Rotation of Hyperion. I. Observations *Astron. J.*, **97(2)**, 570–579

Kolmogorov, A.N.: 1954, On the conservation of conditionally periodic motions under small perturbation of the Hamiltonian *Dokl. Akad. Nauk. SSSR*, **98**, 469

Kuiper, G. P.: 1951, On the origin of the solar system, in *Astrophysics*, J.A. Hyneck Ed.), McGraw-Hill, New York, 357–427

Lagrange, J. L.: 1776, Sur l'altération des moyens mouvements des planètes *Mem. Acad. Sci. Berlin*, 199 Oeuvres complètes **VI** 255 Paris, Gauthier-Villars (1869)

Laplace, P.S.: 1772, Mémoire sur les solutions particulières des équations différentielles et sur les inégalités séculaires des planètes Oeuvres complètes **9** 325 Paris, Gauthier-Villars (1895)

Laplace, P.S.: 1784, Mémoire sur les inégalités séculaires des planètes et des satellites *Mem. Acad. royale des Sciences de Paris*, Oeuvres complètes **XI** 49 Paris, Gauthier-Villars (1895)

Laplace, P.S.: 1785, Théorie de Jupiter et de Saturne *Mem. Acad. royale des Sciences de Paris*, Oeuvres complètes **XI** 164 Paris, Gauthier-Villars (1895)

Laskar, J.: 1984, *Thesis*, Observatoire de Paris

Laskar, J.: 1985, Accurate methods in general planetary theory, *Astron. Astrophys.* **144** 133-146

Laskar, J.: 1986, Secular terms of classical planetary theories using the results of general theory,, *Astron. Astrophys.* **157** 59–70

Laskar, J.: 1988, Secular evolution of the solar system over 10 million years, *Astron. Astrophys.* **198** 341-362

Laskar, J.: 1989, A numerical experiment on the chaotic behaviour of the Solar System *Nature*, **338**, 237–238

Laskar, J.: 1990, The chaotic motion of the solar system. A numerical estimate of the size of the chaotic zones, *Icarus*, **88**, 266–291

Laskar, J.: 1992a, A few points on the stability of the solar system, in *Chaos, Resonance and collective dynamical phenomena in the Solar System, Symposium IAU 152*, S. Ferraz-Mello ed., 1–16, Kluwer, Dordrecht

Laskar, J.: 1992b, La stabilité du Système Solaire, in *Chaos et Déteminisme*, A. Dahan *et al.*, eds., Seuil, Paris

Laskar, J.: 1993a, Frequency analysis for multi-dimensional systems. Global dynamics and diffusion, *Physica D* **67** 257–281

Laskar, J.: 1993b, La Lune et l'origine de l'homme, *Pour La Science*, **186**, *avril 1993*

Laskar, J.: 1994, Large scale chaos in the solar system, *Astron. Astrophys.* **287** L9-L12

Laskar, J., Quinn, T., Tremaine, S.: 1992a, Confirmation of Resonant Structure in the Solar System, *Icarus*, **95**,148–152

Laskar, J., Froeschlé, C., Celletti, A.: 1992b, The measure of chaos by the numerical analysis of the fundamental frequencies. Application to the standard mapping, *Physica D,* **56,** 253-269

Laskar, J. Robutel, P.: 1993, The chaotic obliquity of the planets, *Nature*, **361**, 608–612

Laskar, J., Joutel, F., Robutel, P.: 1993a, Stabilization of the Earth's obliquity by the Moon, *Nature* **361** 615-617

Laskar, J., Joutel, F., Boudin, F.: 1993b, Orbital, Precessional, and insolation quantities for the Earth from -20Myr to +10Myr, *Astron. Astrophys* **270** 522–533

Le verrier U.J.J.: 1856, *Ann. Obs. Paris*, *II* Mallet-Bachelet, Paris

Levison, H.F., Duncan, M.J.: 1993, The gravitational sculpting of the Kuiper belt, *Astrophys. J. Lett.*, **406**, L35-L38

Lissauer J.J., Safronov V.S.: 1991, The random component of planetary rotation, *Icarus* **93** 288-297

Lochak, P.: 1993, Hamiltonian perturbation theory: periodic orbits, resonances and intermittency, *Nonlinearity*, **6**, 885-904

Luu,J.: 1994, The Kuiper belt, in Symposium IAU 160, A. Milani, M. Di Martino, A. Cellino, eds, 31-44, Kluwer, Dordrecht

Morbidelli, A., Moons, M.: 1993, Secular resonances in mean motion commensurabilities: the 2/1 and 3/2 cases, *Icarus* **102** 1-17

Morbidelli, A., Giorgilli, A.: Superexponential stability of KAM tori, *J. Stat. Phys.* **78**, (1995) 1607-1617

Natenzon, M.Y., Neishtadt, A.I., Sagdeev, R.Z., Seryakov, G.K., Zaslavsky, G.M.: 1990, Chaos in the Kepler problem and long period comet dynamics, *Phys. Lett.A*, **145**, 255-263

Nekhoroshev, N.N.: 1977, An exponential estimates for the time of stability of nearly integrable Hamiltonian systems, *Russian Math. Surveys*, **32**, 1-65

Newhall, X. X., Standish, E. M., Williams, J. G.: 1983, DE102: a numerically integrated ephemeris of the Moon and planets spanning forty-four centuries, *Astron. Astrophys.* **125** 150-167

Niederman, L., 1994, Résonance et stabilité dans le problème planétaire *Thesis*, Paris 6 Univ.

Nobili, A.M., Milani, A. and Carpino, M.: 1989, Fundamental frequencies and small divisors in the orbits of the outer planets, *Astron. Astrophys.* **210** 313-336

Peale S.J.: 1974, Possible History of the Obliquity of Mercury, *Astron. J.*, **6**, 722-744

Peale S.J.: 1976, Inferences from the Dynamical History of Mercury's Rotation, *Icarus*, **28**, 459-467

Petrosky, T.Y.: 1986, Chaos and cometary clouds in the solar system, *Phys. Lett.A*, **117**, 328-332

Poincaré, H.: 1892-1899, Les Méthodes Nouvelles de la Mécanique Céleste, tomes I-III, *Gauthier Villard, Paris*, reprinted by Blanchard, 1987

Poisson, S.D.: 1809, Sur les inégalités séculaires des moyens mouvements des planètes *Journal de l'Ecole Polytechnique*, **VIII**, 1

Quinlan, G.: 1993, *personal communication*

Quinlan, G.D.: 1992, Numerical experiments on the motion of the outer planets., *Chaos, resonance and collective dynamical phenomena in the solar system* **Ferraz-Mello, S.** 25-32 Kluwer Acad. Publ. IAU Symposium 152

Quinn, T.R., Tremaine, S., Duncan, M.: 1991, 'A three million year integration of the Earth's orbit,' *Astron. J.* **101**, 2287-2305

Robutel, P.: 1995, Stability of the planetary three-body problem. II KAM theory and existence of quasiperiodic motions, *Celes. Mech.* **62**, 219-261

Safronov: 1969, Evolution of the protoplanetary cloud and formation of the Earth and the planets, *Nauka, Moskva*

Sagdeev, R.Z., Zaslavsky, G.M.: 1987, Stochasticity in the Kepler problem and a model of possible dynamics of comets in the OOrt cloud, *Il Nuovo Cimento*, **97B**, 119-130

Stephenson, F.R., Yau, K.K.C., Hunger, H.: 1984, *Nature*, **314**, 587

Stevenson, D.J.: 1987, Origin of the Moon. The collision hypothesis, *Ann. Rev. Earth Planet. Sci.* **15** 271-315

Sussman, G.J., and Wisdom, J.: 1988, 'Numerical evidence that the motion of Pluto is chaotic.' *Science* **241**, 433-437

Sussman, G.J., and Wisdom, J.: 1992, 'Chaotic evolution of the solar system', *Science* **257**, 56-62

Torbett M.V., Smoluchovski, R.: 1990, Chaotic motion in a primordial comet disk beyond Neptune and comet influx to the solar system, *Nature*, **345**, 49-51

Touma, J., Wisdom, J.: 1993, 'The chaotic obliquity of Mars', *Science* **259**, 1294-1297

Ward W.R.: 1974, Climatic Variations on Mars: 1.Astronomical Theory of Insolation, *J. Geophys. Res.*, **79**, 3375-3386

Ward W.R., Rudy D.J.: 1991, Resonant Obliquity of Mars?, *Icarus,* **94,** 160-164

Wisdom, J.: 1983, Chaotic behaviour and the origin of the 3/1 Kirkwood gap, *Icarus* **56** 51-74

Wisdom, J.: 1985, A perturbative treatment of motion near the 3/1 commensurability, *Icarus* **63** 272-289

Wisdom, J.: 1987a, Rotational dynamics of irregularly shaped natural satellites, *Astron. J.* **94 (5)** 1350-1360

Wisdom, J.: 1987b, Chaotic dynamics in the solar system, *Icarus* **72** 241-275

Wisdom, J., Peale, S.J., Mignard, F.: 1984, The chaotic rotation of Hyperion, *Icarus* **58** 137-152

CHAOS IN THE SOLAR SYSTEM

MYRON LECAR

Harvard-Smithsonian Center for Astrophysics
60 Garden Street, Cambridge, Massachusetts 02138

Abstract. We have accumulated thousands of orbits of test particles in the Solar System from the asteroid belt to beyond the orbit of Neptune. We find that the time for an orbit to make a close encounter with a perturbing planet, T_c, is a function of the Lyapunov time, T_{ly}. The relation is $\log(T_c/T_o) = a + b \log(T_{ly}/T_o)$ where T_o is a fiducial period which we have taken as the period of the principal perturber or the period of the asteroid. There are exceptions to this rule interior to the 2/3 resonance with Jupiter. There, at least in the restricted problem, for sufficiently small Jupiter mass, orbits may have a positive Lyapunov exponent and still be blocked from having a close approach to Jupiter by a "zero velocity curve". Of more serious concern is whether the relation holds for purely secular resonances, and if it does, how to choose T_o. This is the case of interest for the planets in the solar system.

Key words: Chaotic motion – Lyapunov time – asteroids

My colleague, Fred Franklin, and I have been studying asteroidal orbits for more than two decades. We were joined by a student of mine, Paul Soper, in the 1980's. Paul took a seminar I taught on the origin of the solar system, and later, coincidentally I believe, decided to become a priest. He is now a priest in Lowell, Massachusetts, but continues to work with us. More recently, we were joined by Marc Murison who is now at the U.S. Naval Observatory. Our original motivation was to explain why the asteroid belt was so elaborately sculpted, and in particular, why the density of asteroids dropped off so sharply just exterior to the 2/1 resonance with Jupiter. We believe that the asteroids were the building blocks of the planets and were initially more uniformly distributed.

We also argued that the empty regions between all the planets were once populated by asteroids. We were able to show that an initial uniform distribution between Jupiter and Saturn would be cleaned out by perturbations by the two planets in about 10^8 years. In the course of that investigation, we discovered a relation between the Lyapunov time and the time for a test particle to make a close approach to one of the planets. The relation had the form:

$$\log(T_c/T_o) = a + b \log(T_{ly}/T_o)$$

where T_c is the time to a close encounter, T_{ly} is the Lyapunov time (the inverse of the Lyapunov exponent) and T_o is a fiducial time, which in that earlier study, we took to be the period of Jupiter. We now prefer the asteroid period. In fact, we do not have a reliable prescription for choosing To and

Celestial Mechanics and Dynamical Astronomy **64**: 163–166, 1996.
© 1996 *Kluwer Academic Publishers. Printed in the Netherlands.*

that is a loose end in the relation. The choice of T_o is folded into the value of constant a. The value of the exponent b does not depend on T_o and seems, in a variety of numerical experiments, to hover around 1.75.

So far, we have not been able to construct a model for this relation, and for that reason alone we do not claim that the relation is "universal". We are also aware of the fact that in the Restricted Three-Body Problem there can be orbits with a positive Lyapunov exponent which nevertheless are prevented from having a close encounter with Jupiter. For example, the region of "overlapping resonances" (Wisdom, 1980) extends inward from Jupiter, a distance $\Delta a_{res} = 1.49\mu^{2/7}$ using the coefficient found by Duncan, Quinn and Tremaine (1989). On the other hand, orbits further from Jupiter than $\Delta a_{Hill} = 2.04\mu^{1/3}$ are prevented from having a close approach to Jupiter by a "zero velocity curve" (Gladman, 1989; Birn, 1973). If $\Delta a_{res} > \Delta a_{Hill}$, then in the overlap region we have orbits with positive Lyapunov exponent that are nevertheless bounded away from Jupiter. This condition obtains when $\mu^{1/21} < (1.49/2.04)$ or when $\mu < 1/22,260 = \mu_{Jup}/21$. This shows rigorously that for sufficiently small mass ratios, there is a region where orbits with positive Lyapunov exponents never make a close approach to Jupiter, in violation of our relation. Froeschlé, at this meeting, provided a more general framework for similar occurrences.

This leaves open the question of the stability of the outer asteroids. In our 1994 paper (Murison, Lecar and Franklin) we looked at asteroidal orbits perturbed by a Jupiter with a mass too large by a factor of 10. This enhancement of Jupiter's mass brings the outer asteroids into the region of overlapping resonances and out of the region that is protected by a "zero velocity curve", and so our results in that paper cannot be simply applied to the outer asteroids. At least in the Restricted Three-Body Problem and in the Hill Problem, for the correct Jupiter mass, the outer asteroids are protected by a "zero-velocity-curve" and are outside the region of overlapping resonances. However, they do have rather short Lyapunov times. In the case of real solar system, we have not, as yet, identified the overlapping resonances that cause the short Lyapunov times. In some cases, it is hard to identify even a single resonance. Furthermore, we just don't know if "Hill Stability" (the analog to the "zero-velocity-curve") holds for a Jupiter with the varying eccentricity and rotating perihelion induced by Saturn, and for a perturber with Jupiter's mass. In short, we own up to not appreciating the effect of increasing Jupiter's mass by a factor of 10, but we don't know the right answer. The mere fact that asteroids with relatively short Lyapunov times are still there proves nothing! Mercury is still there yet Laskar has shown that it has a chaotic orbit with a Lyapunov time some 900 times shorter than the age of the solar system. Furthermore, Laskar demonstrated, at this meeting, an escape orbit for Mercury with time scale of about 3.5 billion years.

In our 1992 paper, we found for orbits between Jupiter and Saturn, that the exponent b in our relation was between 1.72 and 1.75 and the constant a was between 0.42 and 0.50 (Lecar, Franklin and Murison). Interior to Jupiter, the exponent b was 1.73 but the constant a was 1.53; a difference of 1.0 in the Log or a factor of 10 in the multiplier. This limited data indicates that even if the exponent b turns out to be robust, the constant a (or the fiducial time T_o) will have to be determined separately for each region between two planets. Practically, this has not been a serious limitation on the usefulness of the relation. If the region extends less than a factor of two or so in semi-major axis, one can use the asteroid period for T_o and determine a by numerical experiment. In order to determine a, one does have to integrate a few orbits to close encounter.

However, we have no (numerical) experimental data on systems with only secular resonances so we have no (rigorous) reason to believe that our relation, or a similar relation, holds for such systems. The planets in our solar system are just such a system. I suggest that it would be worthwhile to numerically investigate, say the inner planets of the solar system, with "beefed-up" planetary masses to speed up the evolution. If a relation like ours exists, I think that the enhanced masses will affect the Lyapunov time and the time for a close encounter similarly so as to preserve the relation. This would probably also be true if the higher order terms in the mutual perturbations were deleted. The resulting system would not accurately model the solar system, but I believe that if this toy model yielded a relation, a more accurate model would yield the same relation. Such a program might be useful to determine whether "Bode's Law" was a result of the physics of planetary formation or of dynamical evolution. Closer to home, could it be that there were additional planets in the early solar system that have since been ejected?

Acknowledgements

I thank Fred Franklin for his careful reading of this manuscript and for the pleasure he has provided me in our long association on this problem.

References

Birn, J.: 1973, On the Stability of the Planetary System, *Astron. & Astrophys.*, **24**, 283–293

Duncan, M., Quinn, T. and Tremaine, S.: 1989, The Long-Term Evolution of Orbits in the Solar System: A Mapping Approach, *Icarus*, **82**, 402–418

Gladman, B.: 1993, Dynamics of Systems of Two Close Planets, *Icarus*, **106**, 247–263

Lecar, M., Franklin, F. and Soper, P.: 1992, On the Original Distribution of the Asteroids IV. Numerical Experiments in the Outer Asteroid Belt, *Icarus*, **96**, 234–250

Lecar, M., Franklin, F. and Murison, M: 1992, On Predicting Long-Term Orbital Instabil-
    ity: A Relation between the Lyapunov Time and Sudden Orbital Transitions, *Astron.
    Journal*, **104**, 1230–1236
Murison, M., Lecar, M. and Franklin, F.: 1994, Chaotic Motion in the Outer Asteroid Belt
    and its Relation to the Age of Solar System, *Astron. Journal*, **108**, 2323–2329
Soper, P., Franklin, F. and Lecar, M.: 1990, On the Original Distribution of the Asteroids
    III. Orbits Between Jupiter and Saturn, *Icarus*, **87**, 265–284
Wisdom, J.: 1980, The Resonance Overlap Criterion and the Onset of Stochastic Behavior
    in the Restricted Three-Body Problem, *Astron. Journal*, **85**, 1122–1133

GEOMETRODYNAMICS, CHAOS AND STATISTICAL BEHAVIOUR OF N-BODY SYSTEMS.

P. CIPRIANI
*Physics Dept. - University of Rome "La Sapienza" - ITALY**

G. PUCACCO
*Physics Dept. - University of Rome "Tor Vergata" - ITALY***

D. BOCCALETTI
Mathematics Dept. - University of Rome "La Sapienza" - ITALY

and

M. DI BARI
*Physics Dept. - University of Rome "La Sapienza" - ITALY****

Abstract. We use a geometrical description of the dynamics of Hamiltonian systems to single out the sources of instability (and of *Chaos*, if any). We show that: **A)** the instability is driven by the fluctuations of some geometrical invariants, rather than by their average values; **B)** the most commonly used *invariant* has in general nothing to do with dynamic instability of realistic *many degrees of freedom* systems; **C)** in order to evaluate correctly the relevant quantities entering these geometric invariants, it is necessary the system settles down to a global *virial* equilibrium, and that for this the number of degrees of freedom is crucial; **D)** the gravitating N-body system is peculiar for what it concerns both the dynamical properties and the possibility of a statistical description. So, all the claims that a geometric description of dynamics, in particular for the stellar dynamical problem, gives a *direct* estimate of some *relaxation time* are unjustified. Nevertheless, we point out that the geometrical transcription of hamiltonian systems, if carefully employed, *can give* deep informations about the degree of stochasticity in the dynamics, and very interesting insights on its sources. To overcome some of the limitations of the approach, for systems with few degrees of freedom and/or with time-dependent lagrangians, we introduce an extension, discussed in detail in the companion contribution[3].

Key words: Dynamical systems – Chaos – Statistical Mechanics

1. Introduction.

In this paper we will address some conceptual points, all of them related, in one way or another, with the sources of instability of *trajectories* in hamiltonian (or lagrangian) dynamical systems, and with the long-standing problem of the search for a *dynamical* justification of the statistical description of many degrees of freedom systems. We will discuss some evidences emerged from the work done by some other authors[18] and by ourselves[5, 6, 7]. In particular, we will focus on the chances that the geometrodynamical description of hamiltonian systems could have in gaining deeper insights on both

* e-mail: CIPRIANI@VXRM64.ROMA1.INFN.IT - CIPRIANI@VXRMG9.ICRA.IT
** e-mail: PUCACCO@VAXTOV.ROMA2.INFN.IT - PUCACCO@ATOSCA.ROMA2.INFN.IT
*** e-mail: DIBARI@VXRMG9.ICRA.IT - DIBARI@ATOSCA.ROMA2.INFN.IT

J. C. Muzzio et al. (eds.), Chaos in Gravitational N-Body Systems, 167–172.
© 1996 *Kluwer Academic Publishers. Printed in the Netherlands.*

the aspects mentioned. After recalling the reasons for looking at the occurrence of *Chaos* in the dynamics, we will address to the advantages offered by the geometrical description, and also to the possible shortcomings hidden in a too simplified application. In relation to this last point, we indicate how some of the difficulties implied in the geometrodynamical approach, can in principle be overcome by a suitable generalization, discussed in detail in the companion paper,[3]. We then summarize some of the results (from [5, 7]) obtained using in this framework, the conclusions that can be drawn, and the conceptual points still waiting for a better theoretical assessment.

2. Chaos, Ergodic Theory and Statistical Mechanics.

In the field of Stellar Dynamics it is often recalled the apparent discrepancy between the state of *metastable equilibrium* observed among most elliptical galaxies, and that the usually quoted *relaxation time*, namely the *Chandrasekhar binary* one, τ_b, is longer by several orders of magnitude of the *Hubble time*, H_o^{-1}*. Shortly, it is necessary to look for different mechanisms of *relaxation* to these equilibria. And *Chaos* has been selected as one of the primary candidates in order to justify the *Statistical Approach in Mechanics*. Namely, the presence of *Chaos* guarantees that *conservation laws* do not hold for most of conserved quantities, and it makes favourable an *ergodic-like* behaviour of dynamical trajectories, with all the known consequences, as the replacement of time-averages with integrals of dynamical quantities over the constant energy surface in the *phase-space*. The reasons for which we adopted the geometrical description of the dynamics lie in the following arguments:

- For some (though very abstract) dynamical systems, the geometrical features of the ambient manifold appear to be directly connected with their statistical properties. Namely, it exists, in those cases, a simple relationship between *mixing*, *relaxation* and *instability* time-scales.
- The Krylov criterium, taking into account the warnings added by Sinai, [16], seems to be something more than heuristic; although it has been at the origin of some too simplified treatments, [14, 15].
- When carefully applied it has been shown to work effectively, and to have a wider applicability than usual approaches, when applied to many degrees of freedom hamiltonian systems,[18, 5]. In particular, it offers the possibility, not shared with the usual approach, of quantifying *Chaos* even without the explicit integration of the equations of motion.
- It offers the opportunity of giving a *gauge-independent* characterization of the degree of chaos. And this is important in general, *but it is essential*

---

* This by orders of magnitude difference is largely independent from the *exact* evaluation of the upper cut-off in the Coulomb logarithm, which is however interesting in itself, see e.g. Fukushige contribution to these proceedings.

in the description of *Chaos* in *General Relativistic dynamical systems*. This point is also connected with a deep reconsideration of the *universal* meaning attributed to the LCN's as indicators of *Chaos* (see, e.g. [10]), and also of the usual methods to compute them,[8].

— This approach make it possible to look globally at the ambient manifold, and this yields some interesting insights into the problem of connecting the dynamical behaviour of trajectories, the geometrical features of the ambient manifold and the *statistical* properties, ([5], ch.6).

3. Geometrodynamics of .

In the companion contribution,[3], we present an extension of the approach to more general, or *peculiar* DS, whereas here *briefly* recall the procedure which leads to the geometrical transcription of the dynamics, (see e.g. [5, 18, 19]). The *Maupertuis principle* is used to reduce the f second order eqs. of motion for a (conservative) DS to the eqs. for the geodesic flow $u^i \equiv dq^i/ds$:

$$\frac{d^2 x^i}{dt^2} = -\frac{\partial V}{\partial x^i}, \qquad \longrightarrow \qquad \frac{\nabla u^i}{ds} = 0, \quad (i = 1, \ldots, f); \tag{1}$$

on the (conformal) Riemannian manifold $\mathcal{M}$: $\{[q^i] : V(q^1, \ldots, q^f) \leq \mathcal{E}\}$, equipped with the *Jacobi metric*, $g_{ab} \overset{\text{def}}{=} [\mathcal{E} - V(\mathbf{q})]\eta_{ab} \equiv W(\mathbf{q})\eta_{ab}$, where $\mathcal{E}$ and $V(\mathbf{q})$ are the total and the potential energies, respectively, of the system. In order to investigate the *stability* of the flow on $\mathcal{M}$, we introduced,[5, 6] the so-called *stability tensor*, and its (invariant) trace

$$\mathcal{R}_c^a \overset{\text{def}}{=} R^a_{bcd} u^b u^d, \qquad \text{tr}\mathcal{R} \equiv \mathcal{R}_a^a \equiv R_{bd} u^b u^d \equiv \text{Ric}(u). \tag{2}$$

which is the *Ricci Curvature along the flow*. Indeed, the growth of a perturbation δq^a to a given geodesic, is described by the *Jacobi-Levi-Civita* (JLC) *equation of geodesic deviation*

$$\frac{\nabla}{ds}\left(\frac{\nabla \delta q^a}{ds}\right) = -\mathcal{R}_c^a \delta q^c, \tag{3}$$

and (see[5], ch.4) the evolution of a perturbed geodesic depends on the $(f-1)$ non-vanishing eigenvalues of the *stability tensor*, which are nothing else than the *sectional curvatures* in the $(f - 1)$ *layers* containing the tangent vector to the actual flow.

3.1. LYAPOUNOV EXPONENT AND NEGATIVITY OF CURVATURE.

All the controversies mentioned at the beginning arose from a too naïve extension of some results derived for some particular classes of *very abstract*

DS, [1]. For those DS the *trajectories* are equivalent to the geodesic flow on surfaces of *constant* and *negative* curvature. Consequently, they are *K-systems*, and in that case the instability time-scale and a properly defined *relaxation* time are directly related. What we claimed, however, is that the kind of instability of those DS do not share any resemblance with that of realistic ones.

Paradoxically, we are tempted to say that for those *most unstable* DS the

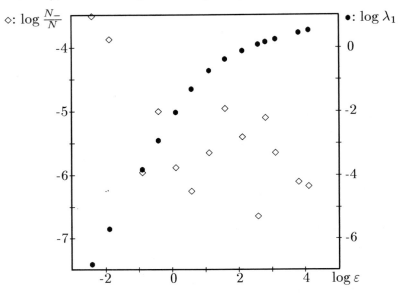

Fig. 1.   Maximal Lyapounov exponent λ_1 (full circles, scale on the right), and frequency of negative values of curvature, over at least $N = 10^9$ time-steps for each ε, (diamonds, scale on the left) for a FPU-β chain at different specific energies ε (f=450).

situation is as clear as for integrable ones! In fact, for surfaces of constant curvature, *Schur's theorem* (see, **e.g.** [12]) assures that *all* the sectional curvatures are constant and have to be equal each other, and then it is a trivial exercise to show ([5], ch.4) that the *constant scalar curvature* $\Re$ gives all the informations about the stability of the flow. If $\Re = -f(f-1)\alpha^2 < 0$, then the perturbations grow exponentially $\delta q \cong \delta q_o\, e^{\alpha s}$, correlations decay on the same time-scale, etc. . . .

But, for realistic DS, it has been shown [5, 6, 7, 18] that:

- Sectional curvatures are not each other equal; they are just the opposite of constant quantities, *i.e.* they are rapidly fluctuating functions of the point on $\mathcal{M}$, and so of the *time* "s"; and they are not all negative!
- The *scalar curvature*, $\Re$, does not enter at all in the evolution of perturbations*. It is also easy to show, [5, 7], that the scalar curvature $\Re$

---

* With the obvious exception of DS with two degrees of freedom!

is *regularly negative* for a relevant fraction of time even for completely integrable systems with a large enough number of degrees of freedom.

- Even adopting an averaging procedure over the sectional curvatures (*i.e.* over the eigenvalues of the stability tensor $\mathcal{R}_c^a$), it is evident that *only the Ricci curvature along the flow*, Ric(u), enters in the evolution of perturbations. And Ric(u) is *usually* positive for most DS, and its average over $\mathcal{M}$ (or $\mathcal{TM}$) is positive for all *physical* systems (see [5, 7]).

- The expression for Ric(u) shows clearly a transition when the system settles down to a global virial equilibrium state, and *all* mdf DS approach this state, self gravitating systems too. When this happens, the dynamics becomes *less chaotic*, the fluctuations of curvature are damped, and it approaches more and more its *ergodic average*, which, as remarked before, is *positive*.
  For the gravitational N-body system, the fluctuations are damped too, still remaining comparatively large, and the *positive* mean value of Ric(u) is only attained on considerably longer time-scales.

- As is shown in fig.1, taken from [5], there is no correlation between the degree of *Chaos*, and the frequency of the negative values of $\mathfrak{R}$ (or even Ric(u)). There, the maximal LCN is calculated from the geometrodynamical approach, and agree with great accuracy with the usually quoted one, computed by the standard algorithm,[2]. It is a striking evidence the *lack of any correlation* between the frequency of the occurrence of a negative value of the curvature and the degree of *Chaos* in the system.

It can be shown, [5, 7, 18], that the JLC equation, when averaged over the orientation of the perturbation can be written as

$$\frac{d^2 X}{dt^2} + Q(t)z \cong 0 \tag{4}$$

where X is the *perturbation*, and

$$Q = 2W^2 k_R + \mathcal{D}(W), \tag{5}$$

in which k_R is the Ricci curvature per degree of freedom *along the flow*; $\mathcal{D}(W)$ is an *almost zero average* fluctuating term, *whose amplitude decrease when the DS goes to the virial equilibrium*. An inspection to the explicit expressions,[5, 7], of these quantities shows that the *stationary* parts are almost everywhere positive and are substantially independent on f. On the contrary, the fluctuating terms have different amplitudes, and the effectiveness of the *virial damping* on them is increasing with the number of degrees of freedom.

4. Conclusions.

The work in [5, 7], yields the following results: for both the *generalized curvatures* the chance of being negative decrease rapidly with the number of degrees of freedom, and with the approach to the virial equilibrium, without regard with the degree of stochasticity. For DS for which the LCN have a well defined behaviour, there is no correlation with the occurrence of negative values of the curvatures. The geometrodynamical approach, for $f \gg 1$, jointly with the use of the theory of *Stochastic Differential Equations*,[20, 4], allows for an analytic evaluation of LCN, which agree very well with the usually quoted (numerical) one.

In particular, for the N-body gravitating system, the peculiarity of the interaction potential, causes the fluctuations to play a major role in driving instability, which really proceed on a dynamical time-scale: the *strong Chaos* there observed is just the direct consequence of the big fluctuations,[7]. The singularity in the newtonian potential, is the source also of a peculiar behaviour with respect to the approach to *ergodicity*, which take place only asymptotically,[7]. Even the lack of any scaling-law is a direct consequence of the nature of the gravitational interaction. So, for general DS, and for self-gravitating N-body too, the (trajectory) instability time-scale $\tau_{t.i.} \approx \lambda_1^{-1}$ has no direct relationship with any relaxation time.

References

1. Arnold,V.I.: 1989, "Mathematical Methods of Classical Mechanics", Springer.
2. Benettin,G. Galgani,L. Strelcyn,J.M.: 1976, *Phys.Rev. A*, **vol.14 - n.6**, 2338.
3. Boccaletti,D. Di Bari,M. Cipriani,P. Pucacco,G.: 1995, *These Proceedings*.
4. Casetti,L. Livi,R. & Pettini,M.: 1994, *preprint* (submitted to *Phys. Rev. Lett.*).
5. Cipriani,P.: 1993, *Ph.D.Thesis*, (in italian), *Univ. of Rome*.
6. Cipriani,P. Pucacco,G.: 1994, *Nuovo Cimento B*, **vol.109, n.3**, 325.
    Cipriani,P. Pucacco,G.: 1994, in "Ergodic Concepts in Stellar Dynamics", (V.G. Gurzadyan - D.Pfenniger eds.) Lect.Notes in Physics: vol.430, Springer, 163.
7. Cipriani,P. Pucacco,G.: 1995a, submitted.
    Cipriani,P. Pucacco,G.: 1995b, submitted.
8. Cipriani,P. Di Bari,M. Pucacco,G.: 1995, submitted.
9. Contopoulos,G.: 1994, in "Galactic dynamics and N-body Simulations", Lecture Notes in Physics, vol.433, Springer, 33.
10. Contopoulos,G.: 1995, *These Proceedings*.
11. Di Bari,M. Boccaletti,D. Cipriani,P. Pucacco,G.: 1995, submitted.
12. Ferrarese, G.: 1994, "Introduzione alla Meccanica Relativistica dei continui." (in italian), Pitagora Editrice, §2.12-2.13.
13. Gallavotti,G.: 1981, "Aspetti della teoria ergodica, qualitativa e statistica del moto" (in italian), Pitagora Editrice, Chapters 1 and 2.
14. Gurzadyan,V.G. Savvidy,G.K.: 1986, *Astron.Astroph.*, **vol.160**, 203.
15. Kandrup,H.E.: 1990, *Ap.J.*, **vol.364**, 420.
16. Krylov,N.S.: 1979, "Works on Foundations on Statistical Physics", Princeton Univ.
17. Miller, R.H.: 1971, *J. Comp. Phys.*, **vol.8**, 449.
18. Pettini, M.: 1993, *Phys.Rev. E*, **vol.47**, 828.
19. Synge, J.L.: 1926, *Phil. Trans. A*, **vol.226**, 31.
20. Van Kampen,N.G.: 1976, *Phys. Rep.*, **vol.24**, 171.

GEOMETRODYNAMICS ON FINSLER SPACES

D. BOCCALETTI
Mathematics Dept. - University of Rome "La Sapienza" - ITALY

M. DI BARI
Physics Dept. - University of Rome "La Sapienza" - ITALY

P. CIPRIANI
Physics Dept. - University of Rome "La Sapienza" - ITALY

and

G. PUCACCO
Physics Dept. - University of Rome "Tor Vergata" - ITALY

Abstract. We have developed a gauge invariant approach to study the dynamical behaviour of Lagrangian systems whose potential depends on both coordinates and velocities (possibly, on time), using a geometrical description. The manifold in which the dynamical systems live is a Finslerian space in which the conformal factor is a positively homogeneous function of first degree in the velocities (the homogeneous Lagrangian of the system).

This method is a generalization of the *standard* geometrodynamical ones which use a Riemannian manifold (see also [11]), as it permits to study a wider class of dynamical systems. Moreover, it is well suited to treat conservative systems with few degrees of freedom and peculiar dynamical systems whose Lagrangian is not "standard", such as the one describing the so-called *Mixmaster Universe*.

We present the method and apply it to some cases of interest: 1) systems with N degrees of freedom described by conservative potentials and 2) Bianchi IX Cosmological Model (*Mixmaster Universe*). This latter example is particularly enlightening, as the introduction of Finsler Geometry overcomes the critical problems which raised some criticisms against the use of the Jacobi (Riemannian) metric to study the chaotic properties of this dynamical system.

Key words: Dynamical systems – Finsler Geometry – Bianchi IX Cosmology

1. Introduction.

In the last years there has been a deep reconsideration of the very meaning of chaos, and the research has been developed towards methods (e.g. spectra of *stretching numbers* [12], geometrical methods [8, 20, 22], see also the companion paper [11]), which generalize the usual tools used to characterize the instability of trajectories, i.e. the Lyapounov exponents. However, for some classes of dynamical systems, other kinds of problems have been raised. For example, the usual geometrical approaches so far used (e.g., Jacobi and Eisenhart metrics) are not able to describe all the possible dynamical systems (e.g. non conservative ones, systems with velocity (and/or time) dependent potentials, Lagrangians with unusual kinetic part).

So, the aim of this work is to present an extension of the geometrical methods capable to overcome these limitations. In order to obtain such a generalization, it is necessary to work not in a Riemannian geometry but in a

173

J. C. Muzzio et al. (eds.), Chaos in Gravitational N-Body Systems, 173–178.
© 1996 *Kluwer Academic Publishers. Printed in the Netherlands.*

manifold whose line element depends on both coordinates and velocities, i.e. in a Finsler space [21] (see also [13]).

The geometrodynamical approach answers also in giving a time gauge invariant characterization of chaos, which is always important, but it becomes essential for General Relativistic dynamical systems.

In these last years many works have been devoted to understand the dynamical nature of the Bianchi IX Cosmological Model [2, 18], but the results are controversial (see e.g.[1, 3, 12, 16, 17]), due to the need of a gauge invariant description and to the peculiar nature of its Lagrangian. In order to solve the first problem mentioned above, some authors ([23] and references therein), adopted the *usual* geometrical description (i.e. Jacobi metric), but even this approach has not been able to give a definitive answer [7], because it has not been taken into account the second problem. Indeed, this dynamical system fully profits of the *Finsler geometrodynamics* as we will discuss briefly in the following sections, in which the proposed solution is also discussed, while the results are presented and interpreted in detail elsewhere [13, 15].

2. Finsler manifold for a Lagrangian system.

In this section we describe the main properties of the Finslerian manifold [13, 21]. The line element of the space, depending on both coordinates and velocities, is

$$ds_F^2 = \Lambda^2(x^i, x^i) = g_{ij}(x^k, x'^k)dx^i dx^j \; ; \qquad g_{ij} = \frac{1}{2}\frac{\partial^2\Lambda^2(x^i, x'^i)}{\partial x'^i \partial x'^j} \; . \quad (1)$$

which reduces to the Riemannian metric if $\Lambda^2(x^i, dx^i) = g_{ij}(x^k)dx^i dx^j$. The $\Lambda(x^i, x'^i)$ function must satisfy the following conditions:

A) $\Lambda(x^i, kx'^i) = k\Lambda(x^i, x'^i) \; ; \qquad k > 0 \; ,$ $\qquad\qquad\qquad (2)$

i.e. $\Lambda(x^i, x'^i)$ is positively homogeneous of degree 1 in the x'^i;

B) $\Lambda(x^i, x'^i) \neq 0 \qquad$ if all $x'^i \neq 0 \; ,$ $\qquad\qquad\qquad\qquad (3)$

C) $\dfrac{\partial^2\Lambda^2(x^i, x'^i)}{\partial x'^i \partial x'^j}\xi^i\xi^j > 0 \qquad \forall \; \xi^i \neq \lambda x'^i \; .$ $\qquad\qquad (4)$

When studying a dynamical system [4], $\Lambda(x^i, x'^i)$ is assumed to be the homogeneous Lagrangian associated to the standard one, $L(x^0, x^\alpha, dx^\alpha/dt)$ $(x^0 = t \; ; \alpha = 1, \ldots, n)$:

$$\Lambda = L\left(x^i, \frac{x'^\alpha}{x'^0}\right)\cdot x'^0 \; ; \qquad x'^0 = \frac{dt}{dw} \; ; \qquad i = 0, \ldots, n \qquad (5)$$

where the prime means $\dfrac{d}{dw}$, w being an additional parameter.

The description of a dynamical system is now "gauge invariant" because Λ is invariant for reparametrization of the "time" w. The equations of motion reduce to the equations of the geodesic flow on Finsler manifold and they are formally analogous to the Riemannian ones.

The information about the possible *chaotic* behaviour of the dynamical system is fully contained in the equation of the geodesic deviation, only formally analogous to the Jacobi-Levi-Civita [8, 9, 20] one. It is given by

$$\frac{\delta^2 z^i}{\delta s^2} + K^i_{jhk}(x^i, x'^i)x'^j x'^h z^k = 0 \tag{6}$$

where $\dfrac{\delta}{\delta s}$ is the so called δ-*differentation* and $K^i_{jhk}(x^i, x'^i)$ is "one" of the curvature tensors which can be defined in this manifold. Its expression contains some terms involving the derivatives in both the coordinates and the velocities.

So, as in Riemannian geometry, we define a *stability tensor* [8, 9, 14]:

$$H^i_k \overset{\text{def}}{=} K^i_{jhk} x'^j x'^h \tag{7}$$

which contains all the informations about the dynamical behaviour of the system.

In the following we derive the main equations for two cases: 1) conservative potentials for N degrees of freedom; 2) the Lagrangian for the Cosmological Bianchi type-IX model. In particular, in the latter and in the former when N is small (here *small* means 2 or 3), we discuss the advantages that the Finsler metric has with respect to the Jacobi one.

2.1. CONSERVATIVE SYSTEMS

As it has been shown in [8, 9, 10, 20], the evolution of a perturbation to a geodesic in the Jacobi metric, for a system with N degrees of freedom, is governed by the Ricci curvature *along the flow*, which is the trace of the stability tensor, given by

$$\text{tr } H_J = \frac{1}{2W^2}\left\{\Delta U + \frac{(\nabla U)^2}{W} + (N-2)\left[\frac{1}{2}\left(\frac{dU}{ds_J}\right)^2 + W\frac{d^2U}{ds_J^2}\right]\right\} \tag{8}$$

where $W \equiv E - U$, E and U being the total and the potential energies, respectively. In analogy, a synthetic indicator of stability for a Finsler geodesic is the trace of H^i_j:

$$\text{tr } H_F = t'^2 \Delta U + Nt'\frac{d^2U}{ds_F^2} + Nt'^2\left(\frac{dU}{ds_F}\right)^2 \tag{9}$$

where $t' = 1/L$ is the inverse of the Lagrangian. The differences in tr H_F, compared to tr H_J, consist on the absence of the gradient and, more relevant, on the presence of Lagrangian instead of the kinetic energy. This fact has very important implications because it allows to overcome the singularity in the conformal factor, unavoidable in the Jacobi metric. Indeed in the definition of the Lagrangian there is a gauge freedom which permits to make it sign-definite (*i.e.* L never vanishes). This is clearly more relevant in the case of few degrees of freedom, or when the system is near to the integrability, as in both cases the chance the kinetic energy vanishes is not negligible.

2.2. FINSLER'S MANIFOLD FOR BIANCHI IX MODEL.

The wider applicability of the Finsler geometrodynamics is strikingly evident in the case of the Bianchi IX model [13] for which the vanishing Hamiltonian is expressed by [19]

$$\frac{\dot{\beta}_+^2}{2} + \frac{\dot{\beta}_-^2}{2} - \frac{\dot{\alpha}^2}{2} + U(\alpha, \beta_+, \beta_-) = 0 \ , \tag{10}$$

where β_+, β_- and α are functions of the scale factors a,b,c of the *Mixmaster model of the Universe*, the dot means differentiation with respect to τ ($d\tau = dt/abc$, t is the cosmological proper time) and U is the potential of the system. We can construct the following homogeneous Lagrangian [5, 13, 15]

$$\Lambda = \frac{1}{2\tau'}(\alpha'^2 - \beta_+'^2 - \beta_-'^2) - V(\alpha, \beta_+, \beta_-)\tau' \ , \tag{11}$$

where $V(\alpha, \beta_+, \beta_-) = -U(\alpha, \beta_+, \beta_-) + c$, in which the gauge is fixed by the choice of the negative constant c, added in order to make the homogeneous Lagrangian positive definite.

For what concerns the geodesic deviation equation, we calculate the trace of the *stability tensor*:

$$H_F = \text{tr } H_j^i = \tau'^2 \Delta U - 3\tau' \frac{d^2 U}{ds_F^2} + 3\tau'^2 \left(\frac{dU}{ds_F}\right)^2 \tag{12}$$

where we have defined, in this case,

$$\Delta U \overset{\text{def}}{=} \frac{\partial^2 U}{\partial \beta_+^2} + \frac{\partial^2 U}{\partial \beta_-^2} - \frac{\partial^2 U}{\partial \alpha^2} \ . \tag{13}$$

In the expression (12) there is a positive term, and two terms whose sign is not clearly definite. This seems to suggest that also in this particular dynamical system, the origin of the dynamical instability is not (or not only) related to the negativity of (some) curvature, but rather to the fluctuations of the geometrical quantities (see e.g. [8, 10, 20]). If this is the case, no rigorous statement can be made about the relationships between the instability time of trajectories and relaxation properties of the system. Let's now consider eq. (10), and introduce it in the expression of the trace of the stability tensor (12). We see that

$$\tau' = \frac{1}{2U - c} \, , \tag{14}$$

with $c < 0$ chosen so that $\tau' > 0$. In order to compare the Finsler metric with the Jacobi one, we calculated also the trace of the *stability tensor* in the Jacobi metric for this dynamical system [7, 8, 20]:

$$H_J = \frac{1}{2W^2} \left[\Delta U + \frac{(\nabla U)^2}{W} + \frac{1}{2} \left(\frac{dU}{ds_J} \right)^2 + W \frac{d^2U}{ds_J^2} \right] \tag{15}$$

where $W = \frac{\dot{\beta}_+^2}{2} + \frac{\dot{\beta}_-^2}{2} - \frac{\dot{\alpha}^2}{2}$. Now it is present also the term

$$(\nabla U)^2 \stackrel{\text{def}}{=} (\frac{\partial U}{\partial \beta_+})^2 + (\frac{\partial U}{\partial \beta_-})^2 - (\frac{\partial U}{\partial \alpha})^2 \tag{16}$$

It is clear why, in this case, the Jacobi metric cannot work: the trace diverges when the "kinetic energy" (or the potential U) vanishes and, along each trajectory, this happens infinitely many times, [7](passing from one *Kasner's epoch* to another) going towards the singularity. On the contrary, this doesn't occur in the Finsler manifold we introduced. We can better understand this by considering the line element of the two metrics, the Jacobi and the Finsler ones:

$$ds_J = -\sqrt{2U} \, d\tau \qquad ds_F = (2U - c)d\tau \tag{17}$$

So, while $ds_J = 0$ if $W = U = 0$ (due to the hamiltonian constraint), and this cannot be avoided adding a constant, we can always have $ds_F \neq 0$, because in this case the conformal factor is the Lagrangian of the system to which we can add a constant without changing the equations of motion. Numerical simulation are running, in order to estimate averages and fluctuations of geometrical quantities related to the Finslerian transcription of BIX dynamics. The results will be presented elsewhere,[15].

3. Conclusions

We have shown how a dynamical system is described in a Finsler manifold, giving some examples. The introduction of a Finsler manifold has been crucial in overcoming the singularities which occurs in the Jacobi metric when the kinetic energy vanishes. The frequency of such an occurence is not negligible when these singularities are inside the manifold, or when the number of degrees of freedom is small or the system is nearly integrable.

The most relevant result we reach is that, by means of this kind of approch, it is possible to give a gauge invariant description of those systems whose potential depends also on velocities (e.g. the restricted three bodies problem [6]) and, moreover, whose Lagrangian is not a positively-definite quadratic form in the velocities. This will be presented elsewhere [6, 14].

References

1. Barrow,J.D.: 1982, *Phys. Rep.*, **vol.85**, 1.
2. Belinskii,V.A. Khalatnikov,I.M. Lifshitz,E.M.: 1970, *Adv.Phys.*, **vol.19** , 525.
    Belinskii,V.A. Khalatnikov,I.M. Lifshitz,E.M.: 1982, *Adv.Phys.*, **vol.31** , 639.
    Belinskii,V.A.: 1991, *Lectures for Ph.D. courses*, (in Italian), *Univ. of Rome.*
3. Berger,B.K.: 1990, *Class. Quantum Grav.*, **vol.7**, 203.
    Berger,B.K.: 1991, *Gen. Rel. Grav.*, **vol.23**, 1385.
4. Boccaletti,D. Di Bari,M. Cipriani,P. Pucacco,G.: 1996, *Nuovo Cim. B*, in press.
5. Boccaletti,D. Di Bari,M. Cipriani,P. Pucacco,G.: 1995, *"Finslerian Geometry for a Gauge-invariant Approach to Chaos in Bianchi IX Cosmological Model"*, Proc. 14th Int. Conf. Gen. Rel. Grav. (Florence, Italy, August 6-12 1995), Abstracts p. B.66.
6. Boccaletti,D. Di Bari,M. Cipriani,P. Pucacco,G.: 1995, *"Restricted three bodies problem on Finsler Manifold"*, submitted.
7. Burd,A. Tavakol,R.: 1993, *Phys. Rev. D*, **vol.47**, 5336.
8. Cipriani,P.: 1993, *Ph.D.Thesis*, (in Italian), *Univ. of Rome.*
9. Cipriani,P. Pucacco,G.: 1994, *Nuovo Cimento B*, **vol.109, n.3**, 325.
    Cipriani,P. Pucacco,G.: 1994, in *"Ergodic Concepts in Stellar Dynamics"*, (V.G. Gurzadyan - D.Pfenniger eds.) **Lecture Notes in Physics: vol.430**, Springer-Verlag, 163.
10. Cipriani,P. Pucacco,G.: 1995, *"Geometrodynamics of Chaos in N-body hamiltonian systems."*, submitted.
11. Cipriani,P. Pucacco,G. Boccaletti,D. Di Bari,M. : 1995, *These Proceedings.*
12. Contopoulos,G.: 1995, *These procedings.*
    Contopoulos,G. Grammaticos,B. Ramani,A.: 1994, proc. Workshop *Deterministic Chaos in General Relativity*, ed. D.Hobill et al. Plenum Press, New York.
    Contopoulos,G. Grammaticos,B. Ramani,A.: 1995, *J. Phys. A: Math. Gen.*, **vol.28**, 5313.
13. Di Bari,M.: 1995, *Ph.D.Thesis*, (in Italian), *Univ. of Rome.*
14. Di Bari,M. Boccaletti,D. Cipriani,P. Pucacco,G.: 1995, *"Dynamical Behaviour of Lagrangian Systems on Finsler Manifolds"*, submitted.
15. Di Bari,M. Boccaletti,D. Cipriani,P. Pucacco,G.: 1995, *"Cosmological Bianchi IX Model on Finsler Manifold"*, submitted.
16. Francisco,G. Matsas,G.E.A.: 1988, *Gen. Rel. Grav.*, **vol.20**, 1047.
17. Hobill,D. Bernstein,D. Welge,M. Simkins,D.: 1991, *Class. Quantum Grav.*, **vol.8**, 1155.
18. Misner,C.W.: 1969, *Phys.Lett.*, **vol.22**, 1071.
19. Misner,C.W. Thorne,K. Wheeler,J.A.: 1977, *Gravitation*, ed. Freeman.
20. Pettini, M.: 1993, *Phys.Rev. E*, **vol.47**, 828.
21. Rund,H.: 1959, *"The differential Geometry of Finsler Spaces"*, Springer-Verlag.
22. Synge, J.L.: 1926, *Phil. Trans. A*, **vol.226**, 31.
23. Szydlowski,M. Lapeta,A.: 1990, *Phys.Lett. A*, **vol.148**, 239.
    Szydlowski,M. Krawiec,A.: 1993, *Phys. Rev. D*, **vol.47**, 5323.

STABILITY OF MOTION OF N-BODY SYSTEMS

A.A. EL-ZANT

Astronomy centre, University of Sussex

Abstract. Complications arising from the non-compact nature of the phase space of N-body systems prevent any asymptotic characterization of chaotic behaviour. This leads us to revisit some of the old results concerning geodesic stability on Lagrangian manifolds. These cannot be applied directly but may be useful when properly reinterpreted. A method for doing that is described along with some tests of its validity and possible future applications.

1. Introduction

If there's a complete lack of particle-particle correlations, an N-body gravitational system is fully describable by a one particle distribution function on the 6 dimensional phase-space. If moreover almost all orbits are regular quasiperiodic then this space divides into a set of 3-D "ergodic components" in the form of KAM tori. A probability density on a 3-torus quickly becomes time independent so that this state of affairs can be fully characterized by stationary solutions of the Collisionless Boltzmann Equation (CBE). If the system is sufficiently far from any chaotic region then neighbouring states cannot be too dissimilar (KAM theory). Thus sufficient near integrability of a N-body system is a sufficient condition for both the existence and stability of solutions of the CBE. These properties however are far from understood in the general case.

It can be shown under fairly general assumptions that the dynamics of gravitational N-body systems is described by the CBE in the limit of $N \rightarrow \infty$ (Braun & Hepp 1977). The question therefore is that of the existence and stability of steady state solutions when a significant amount of chaos is present. Of course chaos in principle does not rule out quasistationary solutions of the CBE. However, chaotic orbits are much more liable to be affected by discreteness noise, thus although solutions may exist in the infinite N limit they would be unstable to perturbations due to discreteness (i.e., for finite N). It is to be noted here that the standard two body relaxation estimates do not guaranty stability against chaotic behaviour since they assume *a priori* that the system being dealt with is integrable (in some sense even linear) and that it remains so under the perturbations due to discreteness. The contradictions arising from this type of analysis have been known for a long time (e.g., Kurth 1957) but are still widely ignored.

A characterization of chaos in the N-Body problem is an essential first step towards understanding its effect. In section 2 we describe some of the

179

J. C. Muzzio et al. (eds.), Chaos in Gravitational N-Body Systems, 179–184.
© 1996 *Kluwer Academic Publishers. Printed in the Netherlands.*

peculiarities of N-body systems which demand the use of a local approach, while in section 3 we describe one such method which makes use of local geometric quantities, sections 4 and 5 are devoted to a derivation and practical application of a criterion that appears to be suitable for N-body systems.

2. What's wrong with this N-body problem?

If a system is mixing (e.g., Sagdeev *et al.* 1988) in an area of its phase-space, it will explore that area and reside most of the time in a region corresponding to the largest number of microstates compatible with some macroscopic state. If moreover the phase space is compact and the mixing takes place for all initial conditions, then Gibb's ergodic hypothesis is satisfied and a Gibb's measure is defined (Pesin 1989). It is usually sufficient for a system to have positive Liapunov exponents on a connected area of the phase space for it to have positive Kolmogorov entropy on this set. This in turn is a sufficient condition for it to be mixing and to approach statistical equilibrium on this set. Thus the Liapunov exponents, which have the appealing property of not requiring any special structures being imposed on the phase space, are ideal for testing whether this state is ever reached.

A number of problems arise however if one tries to apply this simple picture to gravitational systems. First, because of the singularities when the distance between any two particles vanishes, phase space cannot be compact. This problem can only be made worse if the system is unbounded as are realistic astronomical systems. Then there's the problem with the long range of the gravitational forces which prevent most probable steady states from being defined in the conventional sense (Padmanabhan 1990). This property is also responsible for collective instabilities that tend to drive gravitational systems towards more asymmetric and clustered states (such as bar instabilities) as opposed to more conventional thermodynamic behaviour where systems evolve towards more symmetric states which equalize their macroscopic parameters (e.g., temperature). For closed spherical systems this is illustrated by the well known conclusions of Lynden-Bell & Wood (1968) that for a given volume and mass such configurations evolve towards isothermal states only if the energy is high enough to quench the tendency to cluster caused by the long range forces. Otherwise the system clusters and separates into components with heat effectively flowing from the colder regions to the hotter ones. Such is also believed to be the fate of open systems, except that in this case stars gain enough energy to escape and the system eventually dissolves. A system of escaping point masses (with possibly a few binaries left over) has zero Liapunov exponents, thus strictly speaking a gravitational N-body system cannot be chaotic in the asymptotic regime no matter what the transient dynamics had been.

3. Towards a local characterization of chaos

The existence of the Liapunov exponents is guaranteed by virtue of the Osledec theorem which bypasses the need for a particular metric description. The problem of course is that they cannot describe the transient dynamics well and are therefore unsuitable when this is what is required. One might try to redefine the Liapunov exponents as local averages as is done in their calculation using the standard algorithm (Pettini 1993). However, in this case, the averages are over temporal states and not trajectories. Such a procedure would not distinguish between non-trivial mixing that leads to evolution and trivial effects such as phase mixing.

To make further progress one can define a metric structure with the help of which the local deviation of trajectories can be quantified. Being a quadratic form, the kinetic energy generates such a metric on the configuration manifold M (the one spanned by the generalized coordinates) of a Hamiltonian system with complete information on the dynamics (positions *and* velocities) contained in the associated tangent space TM (e.g., Arnold 1989). Then from the Jacoby equation which describes the divergence of infinitely close geodesics on M, one can derive the equation for the stability of trajectories to perturbations normal to the geodesic velocity $\mathbf{u}$. This reads (Anosov 1967; Gurzadyan & Savvidy 1986)

$$\frac{d^2 \parallel \mathbf{n} \parallel^2}{ds^2} = -k_{\mathbf{u},\mathbf{n}} \parallel \mathbf{n} \parallel^2 + 2 \parallel \nabla_u \mathbf{n} \parallel, \tag{1}$$

where $k_{\mathbf{u},\mathbf{n}}$ is the two dimensional curvature in a plane defined by $\mathbf{u} \times \mathbf{n}$. If k is negative everywhere on a compact manifold for all $\mathbf{n}$ normal to $\mathbf{u}$ then (1) describes the foliations of expanding and contracting spaces characteristic of Anosov C-systems. Again, one does not expect gravitational systems to possess these characteristics. However there are two properties that make this type of analysis helpful: first we are considering deviations only normal to the motion thus avoiding artifacts arising from differences in phases. Also, the second term in the right hand side of (1) vanishes if one uses Fermi coordinates following the motion (Pesin 1989). Thus estimates of the divergence using the two dimensional curvatures are *local* and compare divergences of trajectories and not states. Therefore, one may wonder if it is possible to find some averaged criterion that takes advantage of these properties without imposing such strict conditions as required for C-systems. Before investigating such possibilities however I would like to clarify two points that seem to cause misunderstanding. First, it should be clear that *the negativity of the two dimensional curvature on compact manifolds is a sufficient but not necessary condition* for mixing and the approach to equilibrium, these properties belong to all systems with positive Kolmogorov entropy of which C-systems are but one (extreme) subclass. Second, the negativity of the two dimensional curvature on a subset m of M is a *sufficient condition for chaos on m only if this is the only region visited by the system.*

4. Ricci curvature and applications

During its motions a system is continually subjected to perturbations. These will change direction very quickly compared with the evolution time so as to span most directions $\mathbf{n}$ in the space of directions normal to $\mathbf{u}$. To represent this, let S be that space and $\mathbf{Z}$ be a random vector field uniformly distributed on S with constant magnitude (alternatively one may visualize it as a single vector rotating in S). The time evolution of its average magnitude is then given by

$$-\frac{d^2 \int Z^2 d\mathbf{A}}{ds^2} = \int k_{\mathbf{A}} Z^2 d\mathbf{A}$$

in Fermi coordinates and where the (Lebesgue) integrals are taken over all directions $\mathbf{A}$ in S. It is well known (e.g., Eisenhart 1926) that the quantity

$$r_{\mathbf{u}} = \sum_{\mu=1}^{3N-1} k_{\mathbf{n}_\mu, \mathbf{u}}(s) = R_{ij} \frac{u^i u^j}{\| \mathbf{u} \|^2}$$

(where R_{ij} is the Ricci tensor) does not depend on the particular set of normal directions $\mathbf{n}$ chosen so that $r_{\mathbf{u}}/(3N-1)$ can be seen as the average value of k over S. In addition, $r_{\mathbf{u}}$ does not directly depend on the Riemann tensor, therefore it provides a particularly convenient framework for characterizing instabilities of dynamical systems. Moreover, since Z does not depend on $\mathbf{A}$ we have

$$-\frac{d^2 Z^2}{ds^2} = r_{\mathbf{u}}/(3N-1)Z^2. \tag{2}$$

Negativeness of the Ricci (or mean) curvature $r_{\mathbf{u}}$ will therefore imply that there is average instability under random perturbations. The more negative the curvature the more unstable the system is likely to be (precise definitions are given in Gurzadyan & Kocharyan 1988 and El-Zant 1995). This is a far cry from any rigorous proof of the existence of chaos or the like but is useful for a local quantification of the stability of motion. One can also derive a rough exponentiation time-scale for regions of negative curvature

$$\tau_e \sim (\frac{3N}{-2\bar{r}_{\mathbf{u}}})^{1/2} \times \frac{1}{\bar{T}} \tag{3}$$

where the bars represent averages over trajectories and T is the kinetic energy.

5. What's wrong with these two body theories?

In El-Zant (1995) the Ricci curvature was calculated for systems of 231 particles started from sheet-like configurations and integrated with very high precision (energy conservation of at least ten digits) for a few dynamical times.
Once contributions from close encounters are removed the Ricci curvature
was found to be almost always negative and became more so when macroscopic instabilities were present (e.g., violent relaxation and the two types
of evolutionary instabilities present in gravitational systems were discussed
in section 2 — namely clustering instabilities and those driving the system
towards more symmetric states exhibiting more isotropic velocities).

It is important to check that predictions using the Ricci curvature give
the correct description of the approach towards statistical equilibrium when
such a state exists, especially since the Ricci curvature is only an averaged
quantity. Some investigations were made of softened systems enclosed in
"elastic" boxes and which had sufficient energy for stable isothermal spheres
to exist. They were started from homogeneous density states in virial equilibrium with (anisotropic) velocities decreasing exponentially with radius.
The spheres were divided into ten concentric shells and the trace of the
velocity dispersion tensor σ_t was calculated in each of them. The relative
dispersion of σ_t

$$\sigma_d = \frac{\sqrt{\frac{1}{N}\Sigma(\sigma_t - \bar{\sigma})^2}}{\bar{\sigma}} \tag{4}$$

(where the bars denote the average over shells) was calculated as a function
of time for systems ranging from $N = 250$ to $N = 2500$. Preliminary results
were not found to depend sensitively on N, although data for systems with
fewer particles was noisier and harder to interpret of course. It was found
that these configurations relax towards isothermality in a strikingly short
period of a few dynamical times (Fig. 1). Moreover, these state were verified to last for a few hundred dynamical times for systems with $N = 250$
and $N = 500$. Clearly, these results contradict two body relaxation theory.
And while there is significant departure from virial equilibrium during the
evolution, all evolution cannot be attributed to departure from dynamical
equilibrium (violent relaxation) as was verified by considering systems where
particles had different masses (Fig. 2). Calculations of the Ricci curvature
predict an exponentiation time-scale of the order of a dynamical time. This
is in good agreement with the results described above for it means that
complete relaxation should occur within a few dynamical times. It must
be stressed however that in some other cases (e.g., for most systems starting with isotropic velocities which decrease with radius) the relaxation time
was considerably longer. It was still however usually short compared to the
binary relaxation time. A full account of these results is in preparation.

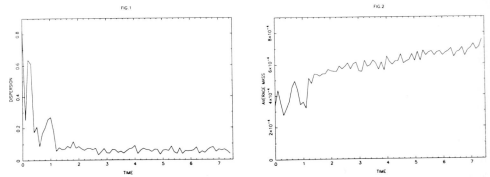

Fig. 1.   Relative dispersion of the trace of the velocity dispersion tensor as a function of the crossing time for a system of 2500 equal mass particles as given by eq. (4). Fig. 2. Average mass in innermost cell for system of 2500 particles with salpeter mass function with cutoff mass 10 times greater than the minimum mass and total mass=1.

6. Concluding remarks

It is difficult to define chaos on non-compact phase spaces. However, what we hope to obtain using the Ricci curvature method is a *local characterization of the stability of motion*. We have described some tests of the method for some idealized systems, the next step will be to see how it could be applied in realistic situations. Since this is obviously not a trivial problem one may start by examining the stability of N-body realizations of systems starting near steady state solutions of the CBE—that is mapping out this region of the phase-space. This type of analysis should be a powerful aid in the interpretation of the results of short time integrations of such systems which might help in the understanding of their evolutionary behaviour.

References

Anosov D.V., 1967, Geodesic flows on closed Riemanian manifolds with negative curvature. Proceeding of the Steklov institute number 90
Arnold V.I., 1989, Mathematical methods of classical mechanics. Springer, New York
Braun W., Hepp K., 1977, Commun. Math. Phys.. 56,101
Eisenhart L.P., 1926, Riemanian geometry. Princeton Univ. press, Princeton
El-Zant, 1995, A & A Submitted
Gurzadyan V.G., Savvidy G.K., 1986, A & A 160,203
Gurzadyan V.G., Kocharyan A.A., 1987 A & SS 135,307
Kurth R., 1957, Introduction to the mechanics of stellar systems. Pergamon press, London
Lynden-Bell D., Wood R., 1968, MNRAS 138,495
Padmanabhan T., 1990, Phys. Rep. 188,285
Pesin Ya.B., 1989, General theory of smooth hyperbolic dynamical systems. In: Sinai Ya.G. (ed) Encyclopedia of mathematical sciences: Dynamical Systems II, Egodic theory
Pettini M., 1993, Phys Rev. E47,828
Sagdeev R.Z., Usikov D.A., Zaslavsky G.M., 1988, Nonlinear physics: from the pendulum to turbulence and chaos. Harwood academic, chur.

MICROSCOPIC DYNAMICS OF ONE-DIMENSIONAL
SELF-GRAVITATIONAL MANY-BODY SYSTEMS

T. TSUCHIYA

National Astronomical Observatory, Mitaka, 181, Japan

N. GOUDA

Department of Earth and Space Science, Osaka University, Toyonaka, 560, Japan

and

T. KONISHI

Department of Physics, Nagoya University, Nagoya, 464-01, Japan

Abstract. Very long time numerical simulations and analyses of the microscopic dynamics reveal the nature of the phase space dynamics of one-dimensional gravitating systems. There is a barrier which interferes with diffusion of an orbit out to the rest of the phase space. It takes a very long time to go through the barrier: e.g., for the case of the water-bag distribution it takes $4 \times 10^4 N$ crossing times.

1. Introduction

One-dimensional self-gravitating many-body systems consist of N identical parallel sheets which have uniform mass density m and infinite in extent in the (y, z) plane. We call the sheets *particles* in this paper. The particles are freely free to move along the x axis and accelerate as a result of their mutual gravitational attraction. The Hamiltonian of this system has a form of

$$H = \frac{m}{2} \sum_{i=1}^{N} v_i^2 + (2\pi G m^2) \sum_{i<j} |x_j - x_i|, \qquad (1)$$

where m, v_i, and x_i are the mass (surface density), velocity, and position of ith particle respectively.

This system is originally introduced as a model for the motion of stars in a direction normal to the disk of a highly flattened galaxy(Oort, 1932; Camm, 1950). In 1970s and 1980s this system was used to study mechanism of relaxation and evolution of galaxies. This system was not realistic model of galaxies but the reason why they used it was because the gravitational force is uniform in one-dimensional systems, thus the equations of motion are reduced to algebraic equations. This fact allows us to compute the evolution numerically in a very high accuracy and for a long time.

These works yielded some inconsistent results of relaxation time. Hohl (Hohl and Broaddus, 1967; Hohl and Feix, 1967) insisted that the system relaxes in a time scale of $N^2 t_c$. Later, one of initial conditions, the water-bag, was found not to relax after $t \sim 2N^2 t_c$ (Wright et al., 1982). On the

185

J. C. Muzzio et al. (eds.), Chaos in Gravitational N-Body Systems, 185–190.
© *1996 Kluwer Academic Publishers. Printed in the Netherlands.*

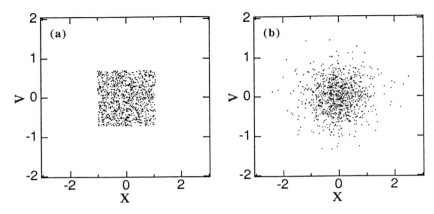

Fig. 1.   The water-bag(a) and the isothermal(b) distributions in one-body phase space.

other hand, another initial conditions, such as counter streamed, relax in a
time scale of $N\,t_c$ (Luwel et al., 1984).

Our aim is to answer this problem. What allow us to get successful results
are (1) progress of computers, (2) new algorithm, which is similar to the
HEAP SORT when updating the particles' positions, and (3) new analyses,
in which we pay attention to fluctuations of individual particles' energy. In
the earlier works the authors concerned mainly with relaxation of smoothed
distribution, which we refer to as *macroscopic* properties, where we analyze
the motions of individuals, which are *microscopic* dynamics.

2.   Numerical simulations

In our numerical simulations, we use the system size, the total mass, and
crossing time t_c, as the units of length, mass, and time. In these units, N is
the only parameter which discriminates the Hamiltonian.

Our initial condition, the water-bag distribution (Fig. 1a), has a homogeneous phase density, and is in virial equilibrium. Putting particles in a
rectangle region is just for convenience. In a few crossing times, the corners disappear and become round but still remains to be uniform. In such
a short time scale, the system behaves like *collisionless* system, and in that
case the water-bag distribution is one of the stationary solutions. One of
the other stationary solutions is the *isothermal* distribution (Fig. 1b). It
is true thermal equilibrium, the maximum entropy state, and then all the
other distributions should transform into the isothermal distribution. However, the stationary water-bag has been believed not to relax to the thermal
equilibrium.

We have studied the very long time evolution of the stationary water-bag distribution, and found it eventually transformed into the isothermal distribution(Tsuchiya et al., 1994). At $t \sim 4 \times 10^6 t_c$ the macroscopic distribution changed, but in a much shorter time scale we found another kind of relaxation. It can be seen in the microscopic dynamics. To see that, we concern with the individual (specific) energy fluctuation $\varepsilon_i(t)$, where i is index of the particles. If the system is ergodic, which is expected in usual thermal equilibrium, the infinite time average of $\varepsilon_i(t)$ gives unique value for the equipartition:

$$\lim_{T \to \infty} \frac{1}{T} \int_0^T \varepsilon_i(t)dt = \overline{\varepsilon}_i \equiv \varepsilon_0. \tag{2}$$

Even in the thermal equilibrium, there exist the thermal fluctuations thus a finite time average,

$$\overline{\varepsilon}_i(t) = \frac{1}{t} \int_0^t \varepsilon_i(t')dt', \tag{3}$$

shows small deviation from the equipartition. Here we introduce a averaged deviation from equipartition until time t,

$$\Delta(t) \equiv \varepsilon_0^{-1} \sqrt{\frac{1}{N} \sum_{i=1}^N (\overline{\varepsilon}_i(t) - \varepsilon_0)^2}. \tag{4}$$

The statistical theory tells us that if $\varepsilon_i(t)$ behaves like thermal noise, $\Delta(t) \propto t^{-1/2}$. Therefore we can make use of $\Delta(t)$ for test of thermalization.

Figure 2 shows the time variation of $\Delta(t)$. In the figure we can find two distinct time scales. The plateau at the beginning represents the collisionless phase, because in the collisionless phase the individual energies are conserved. After $t \sim 100$, $\Delta(t)$ begins to decrease as $t^{-1/2}$, which means that the fluctuation behaves as the same manner as the thermal noise. The transition from constant of $\Delta(t)$ to the power law, $\Delta(t) \propto t^{-1/2}$ determines the *microscopic relaxation*. If the water-bag distribution is the thermal-equilibrium, then no more change is expected and $\Delta(t)$ goes to zero as t increases. However, it is found that $\Delta(t)$ increases at some $10^6 t_c$. At that time the macroscopic distribution transforms from the water-bag into the isothermal distribution. It is the *macroscopic relaxation*.

3. Mechanism of two relaxations

Dependence of relaxation time on the number of particles gives us much insight about its mechanism. Figure 3 shows dependence of the microscopic relaxation time on the number of particles. The doted line stands for

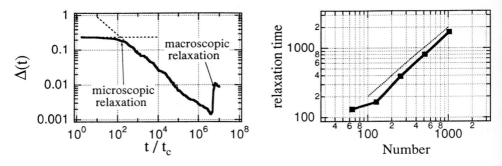

Fig. 2.   (left) Time variation of $\Delta(t)$.

Fig. 3.   (right) Dependence of the microscopic relaxation time on N.

the linear dependence on N. Thus the microscopic relaxation time $\sim N\,t_c$. It can be explained that the microscopic relaxation is a *diffusion process* caused by random force created from thermal fluctuation of mean field distribution(Miller, 1991). Even though the microscopic dynamics shows the same property as the thermal equilibrium, the macroscopic distribution still remains the water-bag one.

The mechanism of the macroscopic relaxation time is not so clear. The macroscopic relaxation has some different properties to the microscopic relaxation. In the case of the microscopic relaxation of the water-bag distributions, different microscopic distributions (created by different random seeds) yield the definite relaxation time scale. For the macroscopic relaxation, however, we found a distribution of relaxation time which has a range over an oder of magnitude. Therefore the mechanism of the macroscopic relaxation is surmised to be different from that of the microscopic relaxation. In this case, the distribution of the relaxation time is a big clue to find the mechanism.

Figure 4 shows the probability distribution of the relaxation time of the system with $N = 64$. We took 100 different initial states, statistical ensemble, of the macroscopically same water-bag distribution, which are created by different seeds of the random number generator The relaxation time of each run, T_M, is determined by the figure of $\delta(t)$, but there are wide variations among the ensemble. The relaxation times are divided into bins with the interval of $2 \times 10^6\,t_c$. The solid diamonds represent the results of the numerical simulations. The vertical axis is scaled such that $P(T_M)\,dT_M$ gives the fraction of the number of runs, which fall into the interval dT_M. It is clear from the figure, that the distribution has a exponential distribution,

$$P(T_M) = \frac{1}{\langle T_M \rangle} e^{-T_M/\langle T_M \rangle}, \quad \text{where } \langle T_M \rangle = 2.8 \times 10^6 t_c, \quad \text{for } N = 64, (5)$$

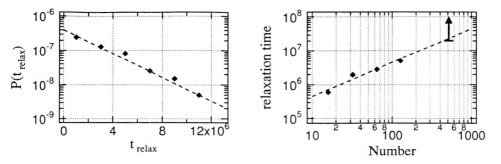

Fig. 4.   (left) Probability distribution of macroscopic relaxation time

Fig. 5.   (right) Dependence of the macroscopic relaxation time on N.

and this gives the expectation value of the relaxation time.

Before we proceed to speculate the mechanism it is useful to remind us the phase space dynamics of the system.

Any state of the system with N particles can be described by a certain point in the $2N$ dimensional phase space (Γ space). Each point yields some macroscopic distribution, such as the water-bag or the isothermal. In Γ space, there is a region where all phase points yield the water-bag distribution. At the beginning the phase point is located in the region and then it moves out of the region as the system evolves. The macroscopic distribution is the water-bag during the phase point stays in the region. When the point escapes from the region, the macroscopic distribution is transformed from the water-bag distribution into the isothermal distribution because the isothermal distribution is defined as the maximum entropy state, which means that it occupies the largest, and usually most region in the Γ space.

Now we should explain the following facts:

1. The water-bag exhibits several properties of thermal equilibrium, such as convergence of Lyapunov exponents, and equipartition of energy, which suggest ergodicity.

2. The probability distribution of the macroscopic relaxation time is a exponential function.

One possibility to explain these facts is that there exists a barrier which is an obstacle for a phase point to get away from the water-bag region. Due to this barrier a phase point is restricted in the region for a long time, then it travels all over the region as if the water-bag region is ergodic. However, the barrier is not perfect and there is a small gate from which a phase point in the water-bag region can escape. A phase point travels in the region in a very complicate way and almost ergodic, thus the point finds the gate and escape at random. Now we show a simple model which might explain the numerical results well as follows: as the simplest case, we assume that the escape probability is uniform in the region. Suppose that we have an ensemble of the phase points in the region. At the beginning the ensemble contains $n(0)$ points, but they escape from the region with constant rate $1/\langle T_M \rangle$, then the number of points which stay in the region at t decreases as $n(t)$. It is well known that $n(t)$ has the same form as eq.(5). Therefore this simplest model can explain the facts obtained by the simulations. This kind of stochastic escape is basically same as decay of unstable nuclear, which also results the exponential distribution of life time.

Next, in order to investigate the dependence of the time scale on the number of particle, the same procedure was applied to the system with different N; $N = 16, 32, 128$, and 512. Figure 5 shows the results. Especially for $N = 512$, we observed that 50 runs of the maximum integration until the time $T = 10^6 t_c$ did not relax. Thus we can not determine the time of the relaxation for $N = 512$, but by assuming the exponential probability distribution, we can restrict the region that the *true* relaxation time lies probably. The arrow indicated the region of 90% confident level. These data is approximated by a linear relaxation,

$$\langle T_M \rangle = 4 \times 10^4 \, N \, t_c, \tag{6}$$

which is shown by the dashed line.

4. Discussions

From our previous work, we found that this system shows ergodicity, eventually. However, in present work we show that the diffusion of the phase point is not uniform but there are some regions which are enclosed by a sort of barriers. They are the cause of slow relaxation to the thermal equilibrium. Existence of these structure in such a large degree system seems very important for the basic of statistical physics. We will discuss these problems further in another paper(Tsuchiya et al., 1996) and future works.

References

Camm, G. (1950).:*Mon. Not. Roy. Astron. Soc*, **110**, 305.
Hohl, F. and Broaddus, D. T. (1967).:*Phys. Lett. A*, **25**, 713.
Hohl, F. and Feix, M. R. (1967).: *Astrophys. J.*, **147**, 1164.
Luwel, M., Severne, G., and Rousseeuw, P. J. (1984).: *Astrophys. Space Sci.*, **100**, 261.
Miller, B. N. (1991).: *J. Stat. Phys.*, **63**, 291.
Oort, J. H. (1932).: *Bull. Astr. Inst. Netherlands*, **6**, 289.
Tsuchiya, T., Konishi, T., and Gouda, N. (1994).: *Phys. Rev. E*, **50**, 2706.
Wright, H. L., Miller, B. N., and Stein, W. E. (1982).: *Astrophys. Space Sci.*, **84**, 421.
Tsuchiya, T., Gouda, N., and Konishi, T. (1996).: *Phys. Rev. E*, in press.

NUMERICAL EXPLORATION OF THE DYNAMICS OF SELF-ADJOINT S-TYPE RIEMANN ELLIPSOIDS

A. BRUNINI, C.M. GIORDANO and A.R. PLASTINO
*PROFOEG - Facultad de Ciencias Astronómicas y Geofísicas, Paseo del Bosque (1900)
La Plata, Argentina.*

Abstract. In this paper we present a numerical exploration of the phase-space structure for the Self-Adjoint S-Type Riemann ellipsoids via Poincaré Surfaces of Section, which reveal a rich and complex dynamical behaviour.

Both the occurrence of chaos for certain values of the parameters of the system as well as the existence of periodic orbits are observed.

We also considered ellipsoids embedded in rigid, homogeneous, spherical halos, obtaining evidence of the stabilizing effect of halos even in the case of finite-amplitude oscillations.

1. Introduction

Dirichlet (1860) studied the oscillations of self-gravitating, incompressible, homogeneous, ellipsoidal configurations with velocity fields whose components are linear functions of the coordinates. The theory was elucidated by Chandrasekhar and others (see Chandrasekhar 1987 - hereafter C87 - and references therein). Recently, Rosensteel and Tran (1991) applied group theoretical methods to this problem obtaining new interesting results, proving that the concomitant equations of motion constitute a Hamiltonian dynamical system.

It was Rossner (1967) (hereafter R67) who first numerically integrated some orbits of a particular case of Dirichlet's problem, namely, the finite amplitude oscillations of Maclaurin's spheroids, corresponding to the S-type Riemann ellipsoids (C87). As far as we know, no further numerical integrations of this problem have been carried out, although Rossner's system represents one of the few cases where the dynamical evolution of a self-gravitating fluid can be followed in detail. There lies our motivation to numerically explore its phase-space structure, which we have done by the method of Poincaré Surface of Section.

2. Dynamical Equations and Numerical Methods

We refer to Chandrasekhar (1987) for a thorough description of the problem. Within Dirichlet's formalism, the evolution of a self-gravitating fluid ellipsoid is described by a set of ten ordinary differential equations allowing the integrals of energy, angular momentum and circulation. In the particular

J. C. Muzzio et al. (eds.), Chaos in Gravitational N-Body Systems, 191–196.
© *1996 Kluwer Academic Publishers. Printed in the Netherlands.*

case of the S-type Riemann ellipsoids (R67) - for which the angular velocity, angular momentum, vortex velocity and circulation are aligned with the principal axes - the system reduces to a set of second order ordinary differential equations in the semiaxes a_1, a_2 and a_3 (C87). The relevant equations are:

$$\ddot{a}_i = \frac{2K^2}{(a_1 + a_2)^3} - 2A_i a_i + \frac{2p_c}{\rho a_i} \quad (i = 1, 2), \tag{1}$$

where K is proportional to the integral of *angular momentum*, the A_i are known functions of the semiaxes (given in terms of incomplete elliptic integrals of the second kind), ρ denotes the constant density and the central pressure p_c is a given function of the semiaxes and their first time-derivatives (see C87). Let us notice that the equation for a_3 is unnecesary because of the incompressibility condition $\frac{d}{dt}(a_1 a_2 a_3) = 0$ (R67).

In our experiments, the elliptical integrals in A_i have been computed by means of *Numerical Recipes's* subroutines (Press *et al.* 1989). Equations (1) have been numerically integrated by Bulirsh and Stoer extrapolation method (Gear 1971), the truncation error being controlled so as to keep the *energy integral* constant up to its twelfth significative figure.

In computing our Surfaces of Section, we have represented each trajectory in the subspace $(a_1 - a_2,\ a_1 + a_2,\ \dot{a}_1 + \dot{a}_2)$ and considered its successive intersections (positive crossings only) with the "Surface of Section" defined as the bidimensional manifold $a_1 - a_2 = 0$ (spheroidal state).

3. Results

3.1. DYNAMICAL BEHAVIOUR OF THE SELF-ADJOINT S-TYPE RIEMANN ELLIPSOIDS

Computation of trajectories for several initial conditions has been performed for various assigned values of the total energy E. Following R67 we have assigned to K the numerical value corresponding to a given Maclaurin spheroid of semiaxes $a_1 = a_2 = a_r$ (in the sequel *semiaxis of reference*) and binding energy W.

In our numerical experiments we adopted $a_r = 1$ and prescribed values for the total energy E which are in excess of the binding energy W by different fractional amounts $f = (E - W)/W$.

Figure 1a reveals no chaotic motion for $f = 0.05$. Regular curves cover the whole available area and there is no irregular orbit. For $f = 0.09$ (Figure 1b) the sections of the invariant manifold break into islands and become complicated curves. When the energy is further increased (see Figure 1c) the chaotic region tends to fill the whole available space.

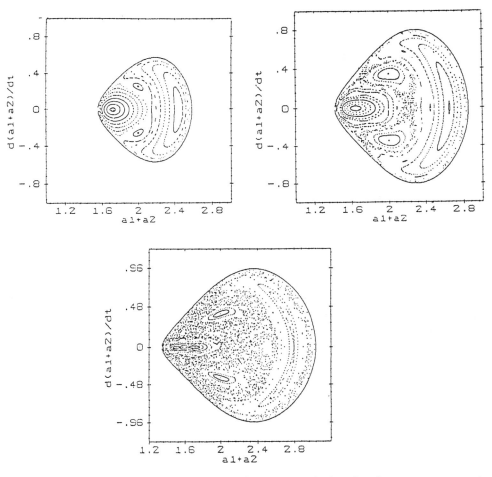

Fig. 1.   Surfaces of Section corresponding to semiaxis of reference $a_r = 1$; 1a : $f = 0.05$, 1b : $f = 0.09$, 1c : $f = 0.12$.

The readers should be reminded that a real self-gravitating fluid body with initial conditions corresponding to an ellipsoidal figure within Dirichlet's formalism may, under a small disturbance, depart from that kind of configuration. This may happen especially at large amplitude. In order to explore this possibility one should carry out stability studies that are beyond the scope of the present work.

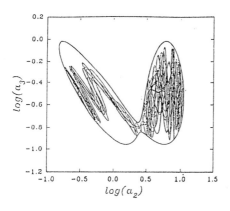 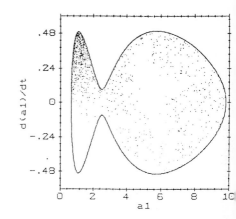

Fig. 2. An initially unstable Maclaurin spheroid with semiaxis of reference $a_r = 2.5$ and $f = 0.003$; 2a : a trajectory; 2b : a Surface of Section corresponding to the same orbit.

3.2. THE CASE OF UNSTABLE MACLAURIN REFERENCE ELLIPSOIDS

Figure 2a displays a zero-velocity curve in the plane $(\log a_2, \log a_3)$, for the unstable Maclaurin spheroid with $a_r = 2.5$ and an orbit already investigated by Rossner (1967). The trajectory hovers permanently whithin one of the lobes during a rather long interval of time, which encompasses Rossner's short-term numerical integration. Thus, Rossner's result suggests that the trajectory is *"effectively trapped"* whithin one lobe (C87). However, our long-term numerical integration reveals that the orbit does pass through the neck of the zero-velocity curve many times.

It is instructive to see how this situation shows up in an suitable Poincaré section. In Figure 2b we plot, for the same orbit, a_1 vs. $\dot{a}_1$ each time the ellipsoid crosses the surface $\log a_3 = -0.8$. We observe that this one is a chaotic orbit that wanders in the stochastic web.

3.3. ELLIPSOIDS IN HALOS

The stabilizing effect of a halo gravitational field on the linear oscillations of the Maclaurin spheroids has been studied by Durisen (1978) and Durisen and Bacon (1981).

The relevant dynamical equations for a uniform density spherical halo are (Durisen 1978)

$$\ddot{a}_i = \frac{2K^2}{(a_1 + a_2)^3} - 2A_i a_i + \frac{2p_c}{\rho a_i} - \frac{4}{3} a_i m \quad (i = 1, 2), \tag{2}$$

where m is the ratio between the halo and spheroid densities.

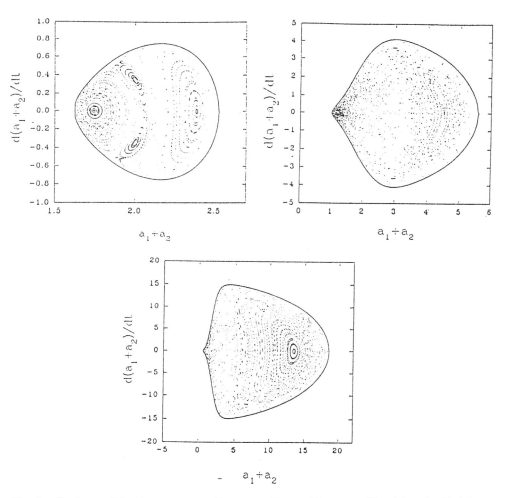

Fig. 3. Surfaces of Section corresponding to an S-type Riemann ellipsoid embedded in a spherical halo. $a_r = 1; m = 0.5; 3a : f = 0.12, 3b : f = 2, 3c : f = 20$.

Surfaces of Section shown in Figure 3 correspond to ellipsoids undergoing toroidal oscillations for three different values of the fractional energy f, when the value $m = 0.5$ is adopted. The comparison of these surfaces with those of Figure 1 gives evidence of the stabilizing effect of halos even for finite amplitude oscillations. For example, for $f = 0.12$ while in the no-halo case the accessible region of motion is almost filled with chaotic orbits, no evidence of chaos is found at the resolution of the figure when the halo is included.

The stabilizing effect of halos admits of a clear physical interpretation. For higher energies and corresponding high amplitude oscillations, the self-gravitating effects become less important, and the harmonic terms in the potential, due to the presence of the halo, dominate the dynamical system which behaves approximately as a linear non-chaotic one.

4. Conclusions

Our numerical experiments show that the oscillations of self-adjoint S-type Riemann ellipsoids display a remarkable structure when viewed in a Poincaré map. The Surfaces of Section reveal the existence of several periodic orbits and the occurrence of chaos for moderate values of the energy. Moreover, the regular behaviour for low energies suggests that a second quasi-integral of motion may be obtained by means of an analytic method (Contopoulos 1963, Gustavson 1966).

We have also considered ellipsoids embedded in rigid, uniform-density, spherical halos. Our experiments show that the stabilizing effect of halos, well-known in case of small oscillations, is also present for finite amplitude oscillations.

Acknowledgements

We wish to thank Prof. J.C. Muzzio for his constant encouragement through-out this research. It is also a pleasure to thank useful discussions with Prof. G. Contopoulos, Prof. S. Ferraz-Mello and Prof. J. Nuñez and the valuable comments of the referee Prof. R. H. Miller.

We acknowledge the financial support from Fundation Antorchas, and one of us, A. Brunini, that from CONICET.

References

Chandrasekhar, S., 1987, "Ellipsoidal Figures of Equilibrium", Dover Publications, Inc., New York.
Contopoulos, G., 1963, AJ, 68, 1.
Dirichlet, G. L., 1860, J. Reine Angew. Math., 58, 217.
Durisen, R. H., 1978, ApJ. 224, 826.
Durisen, R. H., Bacon, D. A., 1981, ApJ. 245, 829.
Gear, C. W., 1971, "Numerical Initial Value Problems in Ordinary Differential Equations", Englewood Cliffs, N. J.: Prentice-Hall.
Gustavson, F. G., 1966, AJ, 71, 670.
Press, W. H., Flannery, B. P., Teukolsky, S. A., et al., 1989, "Numerical Recipes, The Art of Scientific Computing", Cambridge University Press.
Rosensteel, G., Tran, H. Q., 1991, ApJ, 366, 30.
Rossner, L. F., 1967, ApJ, 149, 145.

ON THE POSSIBILITY OF 2D DYNAMICAL CHAOS
IN ASTROPHYSICAL DISKS

A.M. FRIDMAN and O.V. KHORUZHII

Institute of Astronomy, Russian Academy of Sciences
48 Pyatnitskaya St., Moscow, 109017, Russia

March 6, 1996

Abstract. It is shown, that, unlike the small-scale turbulence, the large-scale turbulence in astrophysical disks can be described by 2-D nonlinear dynamical equations. Derived necessary and sufficient conditions of 2-D description are fulfilled rather seldom in astrophysical disks. Under this conditions we derived the 2-D equation with scalar and vector nonlinearities describing the dynamics of gaseous disk rotating in solid-body manner. From our study it follows that in area of solid-body rotation different vortices exist: solitary and double ones. The interaction of the latters leads to the appearance of vortex turbulence. This circumstance lets to clarify the observed correlation of the enhanced star formation in the paired Seyfert galaxies with the presence of abnormally large areas of solid-body rotation (Keel 1993, 1995).

1. Astrophysical Disks Are One of the Most Interesting Dynamical Systems in the Universe. Some Unsolved Problems.

Among different astrophysical objects disks have the most various dynamical structures and different kind of turbulence. So far the origin of many observed structures in disks is puzzle as well as turbulence mechanisms of different kinds of disks. The problem of the galactic spiral structure is waiting for own solution more than one half century. The origin of narrow Uranian rings and Cassini division with its complex inner structure, the cause of the turbulent viscosity many orders greater than the molecular one in the accretion disks, non-Kolmogorov turbulence spectrum of the Milky Way and at last recently discovered enhanced star formation in large areas of solid-body rotation of interacting Seyfert galaxies are still unsolved problems.

2. Traditional Conditions for 2D Description of the Dynamics of Astrophysical Disks

The seldom cases of the use of the 3D description of astrophysical disk dynamics are restricted by the study of the processes of bending and warping of the disk and by the formation of bars and bulges.

Due to the fact that the semithickness H of the majority of astrophysical disks is much less than their radius R, $H \ll R$, the dynamical processes and

J. C. Muzzio et al. (eds.), Chaos in Gravitational N-Body Systems, 197–206.
© 1996 *Kluwer Academic Publishers. Printed in the Netherlands.*

structures in the disks were as a rule studied in the frame of 2D approximation. In doing so two conditions assumed to be fulfilled.

First: the structures and processes are symmetric about the disk plane.

Second: the typical scales of the processes and structures L are much
greater than the semithickness of the disk, $L \gg H$.

But these conditions were never obtained in a rigorous manner as a complete set of sufficient conditions. Consequently our first question is:

**"What is the complete set of sufficient conditions to describe the
dynamics of astrophysical disks by 2D dynamical equations?"**

3.   On the Possibility to Study the Disk Dynamics in the Frame of 2D Approximation

3.1.   What are the initial equations we have to write?

The 2D dynamical equations for astrophysical disks should be derived from
general 3D equations. Analysis of this problem results in the following conclusions (Fridman, Khoruzhii and Libin, 1994).

If the equation of state of the gaseous disk has

1) the general form, $P = P(\rho, S)$ (where P, ρ and S are pressure, density
and entropy, respectively), then the derivation of 2D equations is problem-
atic;

2) the barotropic form, $P = P(\rho)$, then we come to the closed 2D set of
integro-differential equations, the solution of which is complicated;

3) the polytropic form, $P = A \cdot \rho^\gamma$ (where A and γ are constants, γ is
polytropic index), then we come to the system of 2D partial differential
equations, but with additional terms, which were not written before.

3.2.   On the sufficient conditions of the correctness of 2D approximation

These conditions can be written in the following simple form (Fridman,
Khoruzhii and Libin, 1994; Fridman and Khoruzhii, 1994):

$$\frac{H^2}{L^2} \ll 1, \qquad \frac{H^2 R}{\zeta^2 L} \ll 1, \qquad \omega^2 \ll \frac{c^2}{H^2}. \qquad (1)$$

Here H and R are semithickness and radius of the disk, respectively; L and
ζ are typical radial and azimuthal scales of perturbations; ω is the typical
frequency of the process.

First two strong inequalities are well-known conditions of large-scale
approximation.

The third condition was not used before.

In most cases this condition is reduced to

$$\omega^2 \ll \Omega^2, \tag{2}$$

where Ω is the angular velocity of a disk. As a rule this condition was not fulfilled in previous 2D studies of the disk dynamics.

Particularly, as it is easy to see, the majority of 2D theories of spiral structure do not fit this condition!

4. Typical Mistakes in Works Devoted to 2D Dynamics of Astrophysical Disks

4.1. THE CASE OF HIGH-FREQUENCY PERTURBATIONS, $\omega \sim \Omega$ (GRAVI-SOUND BRANCH: GRAVITATING AND HYDRODYNAMICAL INSTABILITIES, THEORIES OF SPIRAL STRUCTURE OF GALAXIES, THEORY OF JEANS TURBULENCE, ETC.)

Schematically the reason of a mistake can be illustrated by the simplest example. Let us consider the 3D equations of motion

$$\frac{dv_\perp}{dt} = -\nabla_\perp \chi, \tag{3}$$

$$\frac{dv_z}{dt} = -\frac{\partial}{\partial z}\chi, \tag{4}$$

where

$$\chi \equiv \mathcal{P} + \Psi, \tag{5}$$

$\mathcal{P}$ is the function of a "volume" pressure P_v,

$$\mathcal{P} \equiv \int dP_v/\rho; \tag{6}$$

Ψ is gravitational potential; ρ is the volume density.

For low-frequency perturbations (2) the term dv_z/dt in equation (4) may be omitted, and the χ does not depend on z. Hence the right-hand side of equation (3) does not depend on z also. As a result the left-hand side does not depend on z and the equation of motion (3) can be reduced to the 2D form.

The latter argumentation is not true for the high-frequency perturbations, $\omega \sim \Omega$. In this case 2D approximation can be correct only in some special cases (Fridman and Khoruzhii, 1994):

1) isothermal disk in a strong outer gravitational field, and
2) self-gravitating disk with polytropic index $\gamma = 2$.

4.2. The case of low-frequency perturbations, $\omega \ll \Omega$ (vortex and dissipative branches)

In this case in principal 2D approximation is correct if the 2D equations are derived from 3D ones.

As a rule the incorrect 2D dynamical equations are used where all functions are integrated over z. Let us illustrate it by the simplest example of 2D equation of motion.

4.2.1. A polytropic disk in outer gravitational field
a) Traditional form

$$\frac{d\mathbf{v}}{dt} = - \frac{1}{\sigma}\nabla P_s - \nabla\Psi_c = - A_s\gamma_s\sigma^{\gamma_s-2}\nabla\sigma - \nabla\Psi_c, \tag{7}$$

where $P_s = A_s\sigma^{\gamma_s}$ is "flat" pressure, Ψ_c is gravitational potential in the plane $z = 0$, A_s and γ_s are constants, γ_s is a "flat" polytropic index.

b) Correct form (Fridman and Khoruzhii, 1994)

Substituting in (6) the expression for the 3D polytrope $P_v = A_v\rho^{\gamma_v}$, where A_v and γ_v are constants, γ_v is a "volume" polytropic index, we obtain $\mathcal{P} = A_v\rho^{\gamma_v-1}\gamma_v/(\gamma_v - 1)$. Finding hence ρ and integrating it over z from $-\infty$ to $+\infty$, we obtain the expression for the surface density

$$\sigma(r,\varphi,t) = \left(\frac{\gamma_v - 1}{A_v\gamma_v}\right)^{\frac{1}{\gamma_v-1}} \int\limits_{-\infty}^{+\infty} [\chi(r,\varphi,t) - \Psi(r,\varphi,z,t)]^{\frac{1}{\gamma_v-1}} \, dz. \tag{8}$$

The condition of the infinitesimal thickness of a disk lets to write a relationship between σ and χ for an arbitrary function Ψ, using its expansion in the neighborhood of the plane $z = 0$

$$\Psi(r,\varphi,z,t) = \Psi_c(r,\varphi,t) + \Psi_c'(r,\varphi,t)z + \frac{1}{2}\Psi_c''(r,\varphi,t)z^2. \tag{9}$$

Substituting (9) to (8) and integrating over z we obtain

$$\sigma(r,\varphi,t) = \sqrt{\pi} \left(\frac{\Psi_c''}{2}\frac{\gamma_v - 1}{A_v\gamma_v}\right)^{\frac{1}{\gamma_v-1}} \frac{\Gamma\left(\frac{\gamma_v}{\gamma_v-1}\right)}{\Gamma\left(\frac{\gamma_v}{\gamma_v-1} + \frac{1}{2}\right)}.$$

$$\cdot \left[\left(\frac{\Psi_c'}{\Psi_c''}\right)^2 + \frac{2(\chi - \Psi_c)}{\Psi_c''}\right]^{\frac{\gamma_v+1}{2(\gamma_v-1)}}. \tag{10}$$

Whence

$$\chi = C\sigma^{2\lambda} + \Psi_c - \frac{(\Psi_c')^2}{2\Psi_c''}, \tag{11}$$

where $\lambda \equiv (\gamma_v - 1)/(\gamma_v + 1)$, Γ is gamma–function,

$$C \equiv \left[\frac{\Psi'' }{2\pi} \frac{\Gamma^2 \left(\frac{\gamma_v}{\gamma_v - 1} + \frac{1}{2} \right)}{\Gamma^2 \left(\frac{\gamma_v}{\gamma_v - 1} \right)} \right]^{\lambda} \left(\frac{A_v \gamma_v}{\gamma_v - 1} \right)^{2/(\gamma_v + 1)}, \qquad (12)$$

Eq.(3) after the substitution in it (11) has a desired form of 2D equation of motion for 2D functions

$$\frac{d\mathbf{v}}{dt} = -\,2\lambda C \sigma^{2\lambda - 1} \nabla \sigma \, - \, \nabla \Psi_c \, - \, \underline{\sigma^{2\lambda} \nabla C} \, - \, \nabla \left[\frac{(\Psi')^2}{\Psi''} \right]_c , \qquad (13)$$

The terms in (13) different from (7) are underlined.

4.2.2. A self-gravitating disk
In this case 2D dynamics can be described by the equation similar to (7). The difference lies in the form of expression for A_s.
  a) Traditional form (Hunter 1972)

$$(A_s)_H = \frac{\pi^{(3/2 - 1/\gamma_v)} \Gamma \left(2 - 1/\gamma_v \right)}{2^{(2 - 1/\gamma_v)} \Gamma \left(5/2 - 1/\gamma_v \right)} A_v^{1/\gamma_v} G^{1 - 1/\gamma_v}, \qquad (14)$$

  b) Correct form (Fridman and Khoruzhii, 1994)

$$(A_s)_{corr} = \frac{2^{1/\gamma_v} \pi^{(1 - 1/\gamma_v)}}{3 - 2/\gamma_v} A_v^{1/\gamma_v} G^{1 - 1/\gamma_v}, \qquad (15)$$

Their ratio is not equal to unity:

$$\frac{(A_s)_H}{(A_s)_{corr}} = \frac{\pi^{1/2}}{2} \frac{\Gamma(\gamma_s/2 + 1/2)}{\Gamma(\gamma_s/2)}, \qquad \gamma_s \equiv 3 - \frac{2}{\gamma_v}. \qquad (16)$$

The reason of the demonstrated difference is very simple: in right-hand side of the correct 2D equation of motion stands the gradient of the function of the volume pressure, $\nabla P = \nabla \int dP_v/\rho$, but not the gradient of a "flat" pressure, $\sigma^{-1} \nabla P_s \equiv \sigma^{-1} \nabla \int P_v dz$, as in equation (7). The latter term has not the physical sense of the force.

Why?
Let us consider a tube of a conic section. Let be $P_v = $ const. Then a gas in the tube must be $unmoved$. But P_s in different sections is different. If we compare two positions with different cross-sections $z_1 > z_2$ we obtain

$$P_{S_1}(x_1) \equiv 2 \int_0^{z_1} P_v(x_1)dz \, > \, P_{S_2}(x_2) \equiv 2 \int_0^{z_2} P_v(x_2)dz. \qquad (17)$$

Hence, according to equation (7) the gas must move from the section S_1 to the section S_2!

We come to the nonsense.

5. The Basic Dynamical Equation in 2D Approximation for Astrophysical Disks in Outer Gravitational Field

Using above-written vector equation of motion (13) for disk in outer gravitational field for slow–frequency perturbations (2) we derive (Fridman and Khoruzhii, 1996) the following nonlinear dynamical equation for a solid–body rotating part of a disk, $\Omega = \text{const}$,

$$\frac{\partial}{\partial t}\left(\tilde{\chi} - a_R^2 \Delta \tilde{\chi}\right) + U_R \frac{\partial \tilde{\chi}}{\partial y} - \frac{c_s^2}{8\Omega^3} J(\tilde{\chi}, \Delta \tilde{\chi}) + \frac{(\ln C)'_x}{4\Omega} \frac{\partial \tilde{\chi}^2}{\partial y} = 0. \quad (18)$$

Here $\tilde{\chi}$ is perturbation of χ (see (11)); $a_R \equiv c_s/2\Omega$ is Rossby radius; $U_R \equiv -2a_R^2\Omega \cdot (\ln \sigma_0)'_x$ is Rossby velocity; c_s is the sound speed determined by the "flat" functions

$$c_s^2 \equiv \sigma \left(\frac{\partial \chi}{\partial \sigma}\right)_0 ; \qquad (19)$$

$$J(A, B) \equiv \frac{\partial A}{\partial x}\frac{\partial B}{\partial y} - \frac{\partial A}{\partial y}\frac{\partial B}{\partial x} \qquad (20)$$

is Jacobian; x, y are Cortesian coordinates with x along radius and y along azimuth; c_s is the sound speed (or velocity dispersion for stars); the function C is determined above in (12).

Equation (18) contains the vector and scalar nonlinearities and is the first 2D nonlinear dynamical equation for astrophysical disks in an outer gravitational field derived from 3D initial equations.

6. Two Types of Vortices

Derived for astrophysical disks nonlinear dynamical equation (18) is similar to well-known in hydrodynamics the Charney-Obukhov equation. In plasma physics the similar equation was derived by Hasegava and Mima (1978). The use of well worked out theory and laboratory modelling of these equations in hydrodynamics and plasma physics leads to the following result (Fridman and Khoruzhii, 1996)

Equation (18) has two kinds of stationary solutions, which describes, respectively, two types of solitary vortices: single and double vortices.

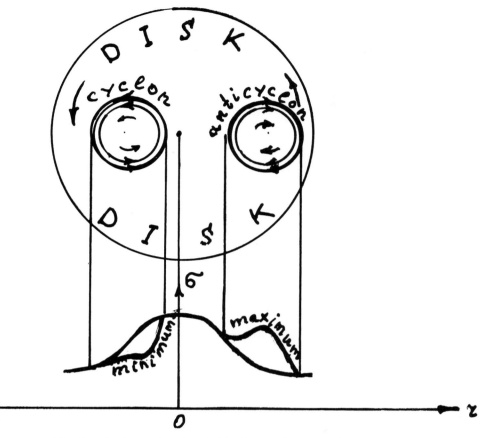

Fig. 1.   Single vortices in solid body rotating part of galactic disk.

Sizes a of these structures are restricted:

$$1 \; < \; a/H \; < \; (R/H)^{1/3}, \tag{21}$$

where H is the disk thickness, R is the typical scale of the density inhomogeneity.

6.1. SINGLE VORTICES: CYCLONES AND ANTYCYCLONS

On the certain distance from the center of the disk it is possible to form only one kind of solitary vortices: cyclones or anticyclones (Fig.1). The kind depends on the profile of the initial surface density of the disk. Cyclone is characterized by a minimum of the surface density. Anticyclone is characterized by a maximum of the surface density.

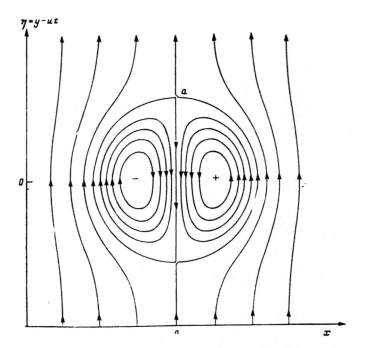

Fig. 2.   Velocity field (streamlines) of the double vortex.

6.2. DOUBLE VORTICES: MODONS

Double vortex – modon represents the cyclone–anticyclone pair and has the form shown in the Figure 2.

6.3. MODON SEA

Generation of several modons results in their interaction and in the formation of vortex turbulence. In the Fig.3 the modon sea on the surface of the shallow water is shown (Nezlin and Snezhkin, 1993).

6.4. ON THE STRONG VORTEX TURBULENCE

The vortex turbulence is distinct from the wave one in principle. The wave turbulence is strong if the amplitude of perturbed value $\tilde{A}$ is closed to the stationary one A_0 or more. The vortex turbulence can be strong under the condition

$$\tilde{A} \ll A_0, \tag{22}$$

as the time of vortex–vortex interaction is much longer than that for wave–wave interaction.

Fig. 3.   Modon sea on the surface of the shallow water.

7. On the Connection of Enhanced SFR with Large Areas of Solid-Body Rotation

Enhanced star formation is found for interacting galaxies with large areas of solid-body rotation (Keel 1993).

The paired Seyfert galaxies show specially a striking number of solid-body rotation curves, 80% of the 39 observed (Keel 1995).

One of the conclusions of Keel's work (1993) is: "The frequent presence of large areas of solid-body rotation, extending to a medial radius of 2.1 kpc, is not closely linked to the presence of optical bars."

We try to clarify the reason of the dynamical activity of a gaseous disk in the absence of a bar, but with a large solid-body rotation area as the development of strong vortex turbulence.

References

Fridman A.M., Khoruzhii O.V., Libin A.S., 1994, Appendix I In N.N. Gor'kavyi, A.M. Fridman "Physics of planetary rings". M.: Nauka, 259. (In Russian).

Fridman, A.M., Khoruzhii, O.V., 1994, Appendix II *ibid*.

Fridman, A.M., Khoruzhii, O.V., 1996, "Lectures on Nonlinear Graviphysics: Solitons, Vortices and Turbulence", Scuola Normale Superiore, Pisa.

Hunter, C. , 1972, Ann. Rev. Fluid Mech., **4**, 219.

Keel, W.C., 1993, Astronomical J., **106**, 1771.

Keel, W.C., 1995, "Kinematic Instabilities, Interactions, and Fuelling of Seyfert Nuclei", (to be published).

Nezlin, M.N., Snezhkin, E.N., 1993, "Rossby Vortices, Spiral Structures, Solitons", Berlin: Springer–Verlag.

THE OBSERVED TURBULENT SPECTRUM OF CLOUDY POPULATION OF THE MILKY WAY AS AN EVIDENCE OF A WEAK ROSSBY WAVE TURBULENCE

A.M. FRIDMAN and O.V. KHORUZHII

Institute of Astronomy, Russian Academy of Sciences
48 Pyatnitskaya St., Moscow, 109017, Russia

Abstract. The most general nonlinear equation has been derived, describing the dynamics of perturbations in a rotating gravitating system of nonuniform density. For small–scale perturbations the equation can be reduced to the well known nonlinear equation for Rossby waves – Charney-Obukhov equation. The spectrum of weak turbulence obtained from this equation leads to the observed by Mayers and Solomon et al. relations between the fundamental parameters of the gas cloudy population of the Milky Way.

1. Observed Turbulent Spectrum of Cloudy Population.

Up to now there are numerous investigations devoted to measurements of the turbulent spectrum in both the every gaseous cloud and of the ensemble of clouds. Kaplan in 1955 found the correlation function $B_{rr} \sim r^{0.71}$ what is very close to the Kolmogorov spectrum (see in Kaplan & Pikel'ner, 1963). Larson (1981) result was also close to that: $\Delta v \sim l^{0.38}$. But later more accurate investigations resulted in more steeper spectrum. Mayers (1983), Henriksen & Turner (1984), and Vereschagin & Solov'ev (1990) obtained the spectrum $\Delta v \sim l^{0.5}$. Sanders, Scovill and Solomon (1985) obtained $\Delta v \sim l^{0.62}$. These spectrums are different from the Kolmogorov one and we should take into account the anisotropy to explain them.

As the basis observed correlations we adopt: 1) correlation for the velocity fluctuations $\Delta v \sim l^{0.5}$; 2) correlation for the density fluctuations (Mayers 1983) $\rho \sim l^{-1}$; and 3) correlation for mass spectrum (Masevich & Tutukov 1988) $N(m)/N(m_0) \sim (m/m_0)^{-1.5}$. These correlations mutually agree if we adopt the assumption of virial equilibrium on all scales.

2. General Nonlinear Equation for Rossby Waves and Correspondent Turbulent Spectrum.

The following general nonlinear equation can be derived in the low-frequency approximation ($\omega \ll \Omega_0$) for the solid body rotating selfgravitating cloud:

$$\left(\frac{\partial}{\partial t} + \frac{1}{2\Omega_0} [\nabla_\perp \tilde{\chi}, \nabla_\perp]_z \right) \left(\Delta_\perp \tilde{\Psi} - \frac{\omega_0^2}{4\Omega_0^2} \Delta_\perp \tilde{\chi} \right) - \frac{\omega_0^{2'}}{2\Omega_0} \frac{1}{r} \frac{\partial \tilde{\chi}}{\partial \varphi} = 0. \qquad (1)$$

J. C. Muzzio et al. (eds.), Chaos in Gravitational N-Body Systems, 207–208.

Here $\omega_0^2 \equiv 4\pi G\rho_0$, $\omega_0^{2\prime} \equiv d(\omega_0^2)/dr$; other notations coincide with that from the previous paper of authors in this volume.

Small-scale perturbations correspond to the case $\tilde{\Psi} \ll \tilde{\mathcal{P}}$ when Eq.(1) can be reduced to

$$\left(\frac{\partial}{\partial t} + \frac{1}{2\Omega_0}\left[\nabla_\perp\tilde{\mathcal{P}}, \nabla_\perp\right]_z\right)\Delta_\perp\tilde{\mathcal{P}} - 2\Omega_0\frac{\rho_0'}{\rho_0}\frac{1}{r}\frac{\partial\tilde{\mathcal{P}}}{\partial\varphi} = 0. \tag{2}$$

The latter equation is similar to well known in hydrodynamics the Charney-Obukhov equation. This analogy allows to use results from the well elaborated fields to describe the dynamics of the perturbations in gravitating molecular clouds. Particularly in accordance with Sazontov (1981) and Mikhailovskii *et al.*(1988) the nonstationary solution of Eq.(2) describes Rossby waves turbulence with energy spectrums $w_k^{(1)} \sim k_y^{-3/2}k_x^{-2}$, and $w_k^{(2)} \sim k_y^{-3/2}k_x^{-3}$.

On the limit of the theory application $k_x \simeq k_y \simeq k_\perp$ and taking into account characteristic property of Rossby waves $k_\perp \simeq k$ we have $w_k^{(1)} \sim k^{-3.5}$, and $w_k^{(2)} \sim k^{-4.5}$. According to Hasegawa *et al.*(1979) $w_k \sim k^{-4}$. The last relation gives (Landau & Lifshitz, 1984) $v_k^2 \sim \int_k^\infty E_k dk = \int_k^\infty w_k k^2 dk \sim k^{-1} \sim \lambda$, that is 1) $v_\lambda \sim \lambda^{0.5}$. On the other hand $P_\lambda \simeq n_\lambda m_\lambda v_\lambda^2 \simeq \rho_\lambda v_\lambda^2 = \text{const}$ that is the density spectrum 2) $\rho_\lambda \sim \lambda^{-1}$. Finally as $m_\lambda \simeq \rho_\lambda \lambda^3 \simeq \lambda^2$ one obtains the mass spectrum 3) $n_\lambda \sim m_\lambda^{-1}v_\lambda^{-2} \sim m_\lambda^{-1.5}$.

We see that the obtained turbulent spectrums corresponds to the observed spectrums that is the evidence of a week Rossby waves turbulence in a cloudy population of the Milky Way.

References

Henriksen, L.N., Turner, I.R., 1984, Ap. J., **287**, 200.

Hasegawa, A., McLennan, C.G., Kodama, J., 1976, Phys. Fluids **22**, 2122.

Kaplan, S.A., Pikel'ner, S.B., 1963, Interstellar Medium, M.: Fizmatgiz. (in Russian).

Landau, L.D., Lifshitz, E.M., 1984, Fluid Mechanics, Pergamon Press.

Larson, R.B., 1981, Month. Not. RAS, **194**, 809.

Masevich, A.G., Tutukov, A.V., 1988, Stellar Evolution: Theory and Observations, M.: Nauka. (in Russian).

Mayers, P.C., 1983, Ap. J., **270**, 105.

Mikhailovskii , A.B., Novakovskii, S.V., Lakhin, V.P., Makurin, S.V., Novakovskaya, E.A., Onishchenko, O.G., 1988, Preprint Inst. of Space Res., N. 1356, Moscow. (in Russian).

Sanders, D.B., Scovill, N.Z., Solomon, P.M., 1985, Ap. J., **289**, 373.

Sazontov, A.G., 1981, Preprint Inst. of Appl. Phys., N. 3, Gor'kii. (in Russian).

Vereschagin, S.V., Solov'ev, A.V., 1990, Astronom. Zh., **67**, 188. (in Russian).

ON THE ACCURACY IN THE NUMERICAL TREATMENT
OF SOME CHAOTIC PROBLEMS

P.E.ZADUNAISKY and C.FILICI

Comisión Nacional de Actividades Espaciales, Buenos Aires, Argentina.

e-mail: pez@conae.gov.ar, Fax: 54-1-3310189

Abstract. The object of this article is to analyse and show in some examples the application of two methods to estimate global errors in the numerical integration of initial problems of the form $\mathbf{y}' = \mathbf{f}(x, \mathbf{y})$, $\mathbf{y}(x_0) = \mathbf{y}_0$ that model some chaotic problems. These methods are a) The reverse test and b) The method of the neighbouring problem. For illustration three examples are considered: 1) The heliocentric unperturbed motion of a Planet, 2) The Lorenz Problem and 3) The Pyt hagorean Problem of Three Bodies.

1. Introduction

It is known that chaotic systems are quite sensitive to initial data and they tend to amplify truncation and round off errors in their numerical treatment. Although it is possible to establish upper bounds for global errors they are often too large, then a good estimation of errors has capital importance.

The validity of the "Reverse Test" is not established rigorously and its results may not be validated in certain cases. However the method is frequently used without any further explanation or justification. By exception in a recent paper it is still used although *"it has its shortcomings"*.

In Section 2 we give some results of the asymptotic theory of error propagation that will be used to analyse the behaviour of the methods of error estimation. In Section 3 we study the reverse test (**RT**) and explain how the estimation of global errors can be either justified or rejected. In Section 4 we present the method of the "Neighbouring Problem" (**NP**) for estimation of errors developed earlier by one of us; about it exists at present an extensive literature of successful applications.

For illustration we will consider the following three examples: 1)The heliocentric unperturbed motion of a planet, 2)The Lorenz problem and 3)The Pythagorean problem. The last two examples are chaotic problems requiring for our purposes the application of some special methods.

2. Some basic results of the theory of error propagation

For simplicity we shall consider here a system of first order equations

$$\mathbf{y}' = \mathbf{f}(x, \mathbf{y}) \quad \mathbf{y}(x_0) = \mathbf{y}_0 \tag{1}$$

J. C. Muzzio et al. (eds.), Chaos in Gravitational N-Body Systems, 209–214.
© *1996 Kluwer Academic Publishers. Printed in the Netherlands.*

to be solved numerically through a **one-step method**

$$\mathbf{y}_{n+1} = \mathbf{y}_n + h\boldsymbol{\Phi}(x_n, \mathbf{y}_n) \tag{2}$$

All our results can be extended also to multistep methods.

Concerning the estimation of **global errors** let us describe some theorems presented by P.Henrici (1962) and G.W.Gear (1971).

2.1. Truncation errors in one step methods

THEOREM 2.1. *If the local truncation error can be expressed as*

$$\mathbf{T}_n = h^{p+1}\boldsymbol{\Psi}(x_n, \mathbf{y}_n) + O(h^{p+2})$$

*where $\boldsymbol{\Psi}$ is called the **principal error function**, then the global truncation error after n steps $\mathbf{e}_n = \mathbf{y}_n - \mathbf{y}(x_n)$ can be estimated by*

$$\mathbf{e}_n = h^p \mathbf{e}(x_n) + O(h^{p+1})$$

where $\mathbf{e}'(x) = \mathbf{G}(x, \mathbf{y}(x))\mathbf{e}(x) + \boldsymbol{\Psi}(x, \mathbf{y}(x))$
with $\mathbf{e}(0) = 0$ and $\mathbf{G}(x, \mathbf{y}(x))$ is the Jacobian matrix $\mathbf{G}_{ij} = \frac{\partial f_i}{\partial y_j}$.

THEOREM 2.2. *Let y_n be a sequence of vectors satisfying*

$$\mathbf{y}_{n+1} = \mathbf{y}_n + h[\boldsymbol{\Phi}(x_n, \mathbf{y}_n) + h^q \boldsymbol{\Theta}_n K] \tag{3}$$

$n = 0, 1, 2, \ldots$, $x_n \in [a, b]$ where $K \geq 0$ and $q \geq 0$ are constants and where $\boldsymbol{\Theta}_n$ are vectors such that $\|\boldsymbol{\Theta}_n\| \leq 1$. Then for $x \in [a, b]$ and $h \leq h_0$

$$\|\mathbf{y}_n - \mathbf{y}(x_n)\| \leq h^r[Nh_0^{p-r} + Kh_0^{q-r}]E_L(x_n - a) \tag{4}$$

where $r = \min(p, q)$, E_L is the "Lipschitz function" and $L \simeq norm(\frac{\partial f_i}{\partial y_j})$.

3. Analysis of the Reverse Test

Let us assume that a numerical integration of the system (1) has been performed on an interval (x_i, x_f), arriving at x_f with an error as $O(h^q)$. To estimate such error by the reverse test let us assume that an integration is accomplished backwards starting at x_f with initial values affected by errors a s $O(h^q)$. Let us also assume that such integration is performed through a one-step algorithm of order p and as $\mathbf{y}_n = \mathbf{y}(x_n) + O(h^q)$ we may write

$$\mathbf{y}_{n-1} = \mathbf{y}(x_n) + h\boldsymbol{\Phi}(x_n, \mathbf{y}(x_n)) + h^q(1 + \boldsymbol{\Theta}_n)K$$

where $\|\boldsymbol{\Theta}_n\| \leq 1$, $K > 0$ is a constant and $h < 0$. This expression is similar to (3) of Theorem 2.2 and, according to (4), if $p > q$, the starting errors $O(h^q)$

will prevail upon the discretization errors $O(h^p)$. Consequently in the interval $x_f \to x_i$ the global errors of the reverse test will remain approximately equal to the starting errors at x_f. They will be determined by subtracting from the final results of the reverse process the initial values at x_i.

4. The "Neighbouring Problem" method

Let us consider the "original problem" (1) to be solved numerically. In order to estimate global errors a neighbouring problem may be constructed: after applying the numerical process through N steps one obtains the numerical solutions $\tilde{\mathbf{y}}_n$. Then it is possible to find a vector of polynomials $\mathbf{P}(x)$ that represent in the best way possible each component of $\tilde{\mathbf{y}}_n$. Let us put

$$\mathbf{D}(x) = \mathbf{P}'(x) - \mathbf{f}(x, \mathbf{P}(x))$$

then the neighbouring problem takes the form

$$\mathbf{z}' = \mathbf{f}(x, \mathbf{z}(x)) + \mathbf{D}(x) \quad \mathbf{z}(x_0) = \mathbf{P}(x_0) \tag{5}$$

which exact solution is evidently $\mathbf{z}(x) = \mathbf{P}(x)$. Furthermore if the empirical functions $\mathbf{P}(x)$ are properly determined they can differ as little as we wish from $\tilde{\mathbf{y}}_n$. Then the formula for estimating errors becomes

$$\mathbf{w}_n = \tilde{\mathbf{z}}_n - \mathbf{z}(x_n) = \tilde{\mathbf{z}}_n - \mathbf{P}(x_n).$$

The function $\mathbf{D}(x)$ is called the "Defect". An heuristic criterion of validity is that the estimated errors should have at least the same order of magnitude as the actual errors provided that in (x_i, x_f) the order of magnitude of $\|\mathbf{D}(x)\|$ is no larger than either the local truncation or round-off error. Now, by applying some results of the asymptotic theory of error propagation with a method of order p and error constant c it is possible to prove that if

$$\frac{1}{c} \left| \frac{d^p \mathbf{D}_i(x)}{dx^p} \right| \le O(h^\eta), \ \eta \ge 1 \tag{6}$$

then $\|\mathbf{e}_n - \mathbf{w}(x_n)\| \le O(h^{p+\eta}) E_L(x_n - x_i)$.

Concerning the round-off error, a similar conclusion can be derived.

Another possibility is to improve the results of a numerical integration by subtracting from them the estimated errors, which is called "Defect Correction" (Böhmer et al(1984), Brunini et al(1993), Zadunaisky(1976)(1979)).

4.1. EFFICIENCY OF THE METHOD

If the analytical solution of (1) is known, or if a "reference solution" (**RS**) is calculated, we can find the "exact errors" $\mathbf{e}_{in}$ of the numerical integration and compare them with the estimation furnished by the neighbouring problem method. This is achieved by defining the 'Efficiency'

$$\mathbf{EFF}_i(x_n) = -\log_{10}(|\mathbf{e}_{in} - \mathbf{w}_{in}|/10^{-r}) \tag{7}$$

TABLE I

Unperturbed Orbit of Jupiter

Interval (a,b)=(0,10000)	Errors at b:	Pos=1E-9	Vel=3E-12	
RT(AM10,h=40)	Differences:	Pos=3E-10	Vel=4E-13	
NP(RKF78,h≤100,τ=E-5)	Efficiencies:	Max=5.6	Min=2.9	Mean 3.8
RS(Keplerian Ellipse)				

where $\mathbf{w}_{in}$ is the estimated error and r is the larger, in absolute value, of the two exponents of the floating forms of $\mathbf{e}_{in}$ and $\mathbf{w}_{in}$. If $\mathbf{EFF}_i(x_n) > 0$ then its integer part is equal to the number of figures, to the rth decimal order of correctly estimated error. And $|\mathbf{EFF}_i(x_n)| \leq 1$ indicates that at least the order of magnitude of the error has been correctly attained.

5. Examples

Example 1: Unperturbed heliocentric motion of Jupiter.

This is a non chaotic problem and due to its smoothness and the exact solution being known it helps to check our results and procedures.

The numerical integration from $x_i = 0$ to $x_f = 10000$ days, was performed with the Runge-Kutta-Felbherg method of embedded orders 7/8, with a tolerance $\tau = 10^{-5}$. The maximum stepsize was 100 days. The reverse test was performed with RKF7(8) at the initial stage and then with Adams-Moulton method of order 10 and a constant step size $h = 40$ days.(See Table I).

Example 2: The Lorenz Problem.

The Lorenz system is given by

$$\dot{y}_1 = \sigma(y_2 - y_1) \quad \dot{y}_2 = \rho(y_1 - y_2 - y_1 y_3) \quad \dot{y}_3 = -\beta y_3 + y_1 y_2$$

with $y_1(0) = 0$, $y_2(0) = 1$, $y_3(0) = 0$, where $\sigma = 10$, $\rho = 28$ and $\beta = \frac{8}{3}$.

This problem has not a known exact solution and a numerical one will be affected by global errors that we want to check. Thus, a reference solution was obtained by a 32-term Taylor series expansion (Coomes et al.(1995)).

For the numerical experiment we performed an integration applying the Grag-Bulirsch-Stoer method (ODEX version, Hairer et al.(1987)). Comparing with the reference solution we found that the "global errors" at $x_f = 1.3$ were of order $O(10^{-9})$. To c heck these errors the reverse test was performed and the results compared against to the reference solution showed a clear deviation leading to a very pessimistic estimation of the global errors. The

TABLE II

Lorenz Problem

Interval $(\mathbf{a},\mathbf{b})=(0,1.3)$	Errors at **b**:	rms=3E-10
RT(RKF78,h$\leq$E-3,τ=E-5)	Differences:	rms=1E-6
NP(ODEX,h$\leq$E-3,τ=E-12)	Efficiences:	Max=5.4 Min=2.2 · Mean 3.7
RS(TAYLOR(32),h=E-4)		

explanation may be found through formula (4) wh ere the first term within brackets is negligible and the upper bound for the global error becomes

$$\|\mathbf{y}_n - \mathbf{y}(x_n)\| \le Kh_0^q E_L(|x_n - x_f|) = Kh_0^q(e^{L|x_n - x_f|} - 1)/L.$$

The Lipschitz constant L (see Theorem 2.2) experiences a great increase near $x_i = 0$ and the global error correspondingly increases too.

On the other hand, the neigbouring problem performed using polynomials of 10th degree, worked satisfactorily estimating correctly 3 to 5 significant figures of the global errors along the whole interval.(See Table II).

Example 3: The Pythagorean Three Body Problem.

In a recent paper "an attempt is made to establish regions of ... chaotic motions. Time-reversed solutions are employed to ensure reliable numerical results". In other words the crux of the whole process is the reliability of the numerical results and in some cases "This introduces fundamental difficulties in establis hing chaotic regions" (Inverted commas enclose quotations from S.J.Aarseth et al.(1994)). In this example we compare the estimation of numerical errors obtained through the reverse test and the neighbouring problem method. We arrive at the conclusion that the results of both methods are acceptable although the second one is more accurate.

Our computing process was performed as follows. Direct integration from $x_i = 0$ to $x_f = 7$ was performed with ODEX and in order to avoid singularities during close approaches that would make the neighbouring problem impracticable, the Kustaanheimo-Sti efel regularization method was used.

The neighbouring problem was solved using polynomials of 10th degree. And due to the non-smoothness of the original variables, the regularized ones were those which we interpolated, hence all the error estimations and efficiencies are referred to the se variables.

Finally, the reverse test was performed and its results were compared against a reference solution. The test succeeds in obtaining an estimation of the error committed at $x_f = 7$. On the other hand, the neighbouring problem achieves positive efficiencie s in nearly the whole interval of integration, reaching the correct estimation of one significant figure of the errors during close approaches.(See Table III).

TABLE III

Pythagorean Problem

Interval (a,b)=(0,7.04)	Errors at b:	Pos=7E-10	Vel=5E-10	
RT(RKF78,h≤E-2,τ=E-10)	Differences:	Pos=3E-10	Vel=3E-10	
NP(ODEX,h≤E-2,τ=E-8)	Efficiencies:	Max=1.4	Min=-.2	Mean=.3
RS(RKF78,h≤5E-4,τ=E-13)				

6. Conclusions

The reverse test may give accurate estimations of the global errors propagated in the "direct integration" from x_i to x_f. To make it sure it is necessary to perform the reverse integration with a method several orders higher in precision than the method used in the direct sense. In this way, according to formula (4), the global errors at x_f are transported to x_i multiplied by the Lipschtiz function $E_L(|x - x_f|)$. In the Lorenz Problem example the method failed because this function grew too much.

The method of the neighbouring problem may give good or acceptable results provided that the sufficient, but not necessary condition of validity (6) were fulfilled. If the "defect function" were small along the whole interval of integration it w ould be possible to apply the "defect correction method".

The computational effort in both methods is approximately equivalent.

The reverse test may be used at each step instead of at the end of the interval of integration; that would require modifications and justifications that will be given in a more extended forthcoming article.

References

Aarseth, S.J., et.al. (1994) Cel. Mech. and Dyn. Astr., **58**, p.1-16, Kluwer.

Böhmer, K. and Stetter, H.J.(1984) Computing Suplementum, **5**, Springer Verlag, Berlin.

Brunini, A. and Zadunaisky, P.E. (1993) Computers in Physics, **7**, No 1, pp. 81-86.

Coomes, B.A., et.al. (1995) Numer. Math., **69**, pp. 401-421.

Dormand, M. et.al.(1989) Comput. Math. Appl., **18**, p.835.

Gear, G.W. (1971)*Num. IVP in ODEs*. Prentice Hall, Inc., Englewood Cliffs, New Jersey.

Hairer, E. ,et.al.(1987)*Solving ODEs I: Nonstiff Problems*. Springer Verlag, Berlin.

Henrici, P. (1962)*Discrete Variable Methods in ODEs*. John Wiley and Sons, Inc., N. York.

Zadunaisky, P.E. (1976) Numerische Mathematik, **27**, p.21-39, Springer Verlag, Berlin.

Zadunaisky, P.E. (1979) Celestial Mechanics, **20**, p.209, Reidel.

THE LIAPUNOV EXPONENT AS A TOOL
FOR EXPLORING PHASE SPACE

O. C. WINTER

Grupo de Dinâmica Orbital e Planetologia - UNESP,
CP 205 - CEP 12500-000, Guaratinguetá, SP, Brazil.

and

C. D. MURRAY

Astronomy Unit, Queen Mary and Westfield College, London, U.K.

February 12, 1996

Abstract. We have used the Liapunov exponent to explore the phase space of a dynamical system. Considering the planar, circular restricted three-body problem for a mass ratio $\mu = 10^{-3}$ (close to the Jupiter/Sun case), we have integrated $\sim$16,000 starting conditions for orbits started interior to that of the perturber and we have estimated the maximum Liapunov characteristic exponent for each starting condition. Despite the fact that the integrations, in general, are for only a few thousand orbital periods of the secondary, a comparative analysis of the Liapunov exponents for various values of the 'cut-off' gives a good overview of the structure of the phase space. It provides information about the diffusion rates of the various chaotic regions, the location of the regular regions associated with primary resonances and even details such as the location of secondary resonances that produce chaotic regions inside the regular regions of primary resonances.

Key words: Chaos – Liapunov exponents – Orbits

1. Introduction

The aim of the present work is to show that it is possible to get a reasonable idea of the structure of the phase space of a dynamical system through the use of Liapunov characteristic exponents even when their precise values are not known due to insufficient integration time. The structure of the phase space is mainly determined by resonances between the main frequencies of the system; these define the location and size of the regular and chaotic regions.

A chaotic orbit is sensitively dependent on the initial conditions such that small changes in position or velocity produce a very different final state. As a result two initially nearby trajectories diverge exponentialy from each other in a chaotic region. A measurement of local divergence of nearby trajectories is the Liapunov characteristic exponent. Our computation of the Liapunov characteristic exponent was based on the definition of its maximum value, γ, given by Benettin *et al.* (1980). The value of γ is used as an indicator of chaotic or regular behaviour. The maximum Liapunov characteristic exponent has to be computed by numerical integration, ideally for infinite time,

215

J. C. Muzzio et al. (eds.), Chaos in Gravitational N-Body Systems, 215–219.
© *1996 Kluwer Academic Publishers. Printed in the Netherlands.*

with its 'final' value determining the nature of the trajectory ($\gamma = 0$ and $\gamma > 0$ correspond to a regular and a chaotic orbit respectively.

One crucial parameter in the computation of the Liapunov characteristic exponent is the length of the integration. For how long should the system be integrated in order to determine the character of the trajectory using the Liapunov characteristic exponent method? After some integration time this method may indicate that the trajectory gives the appearance of being regular while an extended integration may eventually reveal a chaotic nature (see, for example, Winter & Murray 1995).

The dynamical system considered here is the planar, circular restricted three-body problem, which consists of primary and secondary bodies moving in circular orbits about their common centre of mass, and a test particle which moves under the gravitational effect of both masses without affecting their motion. The system conserves neither angular momentum nor energy and the only integral of the motion is the Jacobi constant.

In our computations intermediate values of γ were computed every 10 orbital periods of the secondary mass after which the separation vector of the nearby trajectory in phase space was rescaled in order to maintain measurement of the local divergence of the two orbits.

2. Determination of the Chaotic Regions

Approximately 16,000 starting conditions were considered and all were chosen to be initially at pericentre on the line joining the two main bodies, and the majority of them was selected to be equally spaced. In general, the orbits were integrated for three to four thousand orbital periods of the perturbing mass. The only exceptions were for the starting conditions whose integration was stopped due to an approach within 0.01 of the secondary, in units in which the distance between the two masses is unity. The starting conditions that resulted in close approaches were assumed to be chaotic; although it is possible to find regular, periodic orbits that have such close approaches, the probability that one of these orbits could be found in our integrations is neglegible.

Here we make a statistical analysis of the data on the Liapunov characteristic exponents that we have computed. Our task is to determine the chaotic regions based on the last value of $\log \gamma$ for each orbit. We denote this value as $\log \gamma^*$ to distinguish it from the logarithm of the *actual* characteristic exponent which can only be determined by integrations over infinite time. A histogram showing the distribution of $\log \gamma^*$ is given in Fig.1. From the bimodal nature of this plot we can estimate that all orbits with $\log \gamma^* \geq -2.2$ are chaotic. A plot showing the starting conditions for all these chaotic orbits in the a–e plane with $\varpi = 0$ (where a, e, ϖ are the orbital elements known

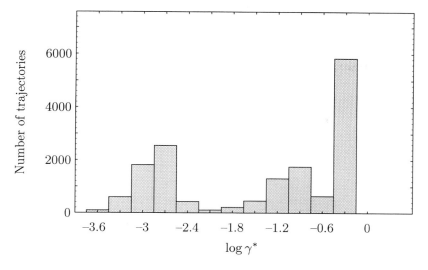

Fig. 1.   Histogram of the number of trajectories within given ranges of $\log \gamma^*$, the last value of $\log \gamma$ in our integrations.

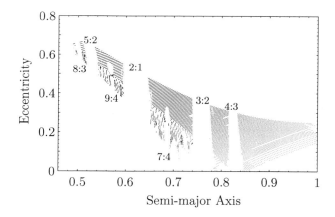

Fig. 2.   Plot showing the starting conditions in the $a - e$ plane with $\varpi = 0$ which have $\log \gamma^* \geq -2.2$. The numbers indicate the mean motion resonances associated with the regular (empty) regions.

as semimajor axis, eccentricity and longitude of pericentre, respectively) is shown in Fig.2. The two different patterns for the starting points shown in Fig.2 (and also in Figs 3 and 4 below) are simply a consequence of the way in which we chose our starting conditions in these regions; they have no physical meaning.

The resonances which are associated with the empty regions are indicated in the plots. Although they have a similar appearance, these gaps should not

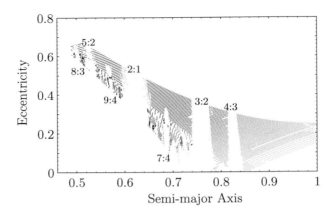

Fig. 3. The same as in Fig.2, but for the starting conditions which have $\log \gamma^* \geq -2.5$.

be confused with the Kirkwood gaps in the asteroid belt. A close inspection of the 9:4 gap shows that there is a chaotic region inside the gap. This is due to the existence of a secondary resonance in that region; this can easily be verified by looking at the Poincaré surfaces of section for values of Jacobi constant between 3.038 and 3.054 given in Winter & Murray (1994). It is important to note that there can still be chaotic regions in the regular zones associated with resonances.

When we consider the cut-off value to be lower, $\log \gamma^* \geq -2.5$ (see Fig.3) and $\log \gamma^* \geq -2.8$ (see Fig.4), new sets of starting conditions that can give rise to chaotic trajectories appear. The empty regions get narrower due to starting conditions that give rise to chaotic orbits that have a low value for $\log \gamma^*$ because they belong to the separatrix of a secondary resonance about the main primary resonances or they are trajectories that are suffering the effect of the 'stickiness' phenomenon (Winter & Murray 1995). Note also that there are additional points in the lower part of the diagrams corresponding to large values of Jacobi constant. These correspond to chaotic trajectories that are in regions with a low diffusion rate; for these values of the Jacobi constant the 'area' of the phase space in the surface of section where particle motion is permitted is smaller than for lower values of the Jacobi constant (see fig.2 of Winter & Murray 1995).

3. Conclusions

The main conclusion of this work is that it is possible to get information about the nature of trajectories by monitoring the behaviour of γ as the integration progresses, without necessarily obtaining its final value.

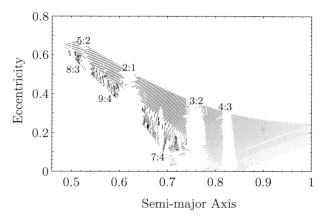

Fig. 4.   The same as in Fig.2, but for the starting conditions which have $\log \gamma^* \geq -2.8$.

We have shown that even for relatively short integration times, the value of γ^* can provide information about the diffusion rates of the various chaotic regions, the location of the regular regions associated with primary resonances and even details such as the location of secondary resonances that produce chaotic regions inside the regular regions of primary resonances.

Acknowledgements

OCW thanks CAPES-Brazil for the support given through Proc. 445/90-1. CDM is grateful to the UK PPARC for the award of an Advanced Fellowship.

References

Benettin, G., Galgani, L., Giorgilli, A. and J.-M. Strelcyn (1980). Lyapunov characteristic exponents for smooth dynamical systems and for Hamiltonian systems; a method for computing all of them. Part 1: Theory. *Meccanica* **15**, 9–20.

Winter, O.C. and C.D. Murray (1994). *Atlas of the Planar, Circular, Restricted Three–Body Problem. I. Internal Orbits.* QMW Maths Notes No. 16, Queen Mary and Westfield College, Mile End Road, London E1 4NS, U.K.

Winter, O.C. and C.D. Murray (1995). Project CRISS-CROSS: A Preliminary Analysis. In *From Newton to Chaos: Modern Techniques for Understanding and Coping with Chaos in N-Body Dynamical Systems* (A.E. Roy and B.A. Steves, Eds.) NATO ASI Series, Plenum Press, New York, 193–198.

INFORMATION ENTROPY AS A TOOL FOR SEARCHING PERIODICITY

P. M. CINCOTTA, A. HELMI,* M. MÉNDEZ** and J. NÚÑEZ**

Facultad de Ciencias Astronómicas y Geofísicas, Universidad Nacional de La Plata

Abstract. In this paper we introduce the mathematical formalism of an algorithm previously described by Cincotta, Méndez & Núñez. It uses the Information Entropy as an indicator of periodic behavior within a time series. We also include here the analysis of a quasi–periodic signal formed by two nearly equal frequencies, proving some advantages of this method in comparison to classical or modified Fourier analysis.

Key words: Numerical Methods – Data Analysis, Algorithms for

1. Introduction

Both in theoretical and experimental physics, periodic and quasi–periodic phenomena are usually the key to understand the basic dynamics of a complex system. While experimental data often hide some recurrence associated to an underlying physical process, periodic or quasi–periodic motions play a significant role in the topology of nonlinear systems. Although it is possible to study the behavior of a dynamical system at a fixed time step (through the numerical solution to the equations of motion), this is not always the case for experimental measurements, where data are often unevenly sampled. This circumstance usually discourages the use of traditional (i.e., discrete Fourier methods) period analysis techniques (see Scargle 1989). Other methods (developed to deal with astronomical time series) have been proposed to handle these situations (String Length Methods and Phase Dispersion Minimization methods; see Cuypers 1987 for a review). Most of them lack in a mathematical proof, or require a large amount of computing time. Several authors have developed modified versions of Fourier transforms to deal with unevenly sampled data (Barning 1963; Vaníček 1971; Lomb 1976; Ferraz–Melo 1981; Scargle 1989). Although intrinsically robust and mathematically correct, these algorithms are complex and time consuming too (see for instance Press and Rybicki 1989).

In a previous paper (Cincotta, Méndez & Núñez 1995, hereafter CMN95) we have proposed a method that is suitable for both situations. Data need not to be evenly sampled in time, thus being particularly useful in the experimental situation. However, being conceptually so simple and so fast, it might be preferred even when uniformly sampled data is available.

* Fellow of CONICET, Argentina
** Member of the Carrera del Investigador Científico y Tecnológico, CONICET, Argentina

221

J. C. Muzzio et al. (eds.), Chaos in Gravitational N-Body Systems, 221–226.
© 1996 *Kluwer Academic Publishers. Printed in the Netherlands.*

In Section 2 we define the appropriate tools needed to formalize the algorithm within the framework of the Information Theory, leaving the full mathematical justification for a subsequent paper (Cincotta, Helmi, Méndez & Núñez, in preparation). In Section 3 we apply the method to a quasi–periodic signal formed by two independent frequencies. This is a very powerful test of the internal accuracy and the "resolving power" of the algorithm. For the sake of brevity we do not show here further testing, but refer the reader to CMN95.

2. The Mathematical Formalism

In what follows we will briefly summarize the mathematics of this method. Due to the lack of space, the full description is postponed to another paper (Cincotta, Helmi, Méndez & Núñez, in preparation).

Let us consider a time series $u(t)$ with the following properties:

i) $u\colon I \to F, \qquad I, F \subset \mathbf{R}$,

ii) $\forall t \in I, \exists k \in \mathbf{R}^+ \colon \|u(t)\| < k \qquad$ (boundness),

iii) $\forall t \in I, u(t + T) = u(t) \qquad$ (periodicity).

We construct a one–to–one mapping φ_{pn}, such that:

iv) $\varphi_{\mathrm{pn}}\colon Q_{\mathrm{pn}} \to [0, 1)$,

v) $\varphi_{\mathrm{pn}}(t) = \frac{t - t_0}{p} - n, \qquad t_0 \in I, \qquad p \in \mathbf{R}^+ - \{0\}$,

vi) $n = \left[\frac{t - t_0}{p}\right], \qquad Q_{\mathrm{pn}} = (np, (n + 1)p] \subset I, \qquad \bigcup_n Q_{\mathrm{pn}} = I$,

where $[\ ]$ is the integer function and p is a parameter. As defined, φ_{pn} (often called the phase) is the time modulo the trial period p. Now, consider the map g_{pn} which is defined by:

vii) $g_{\mathrm{pn}} \equiv u \circ \varphi_{\mathrm{pn}}^{-1}$,

viii) $g_{\mathrm{pn}}\colon [0, 1) \to F_{\mathrm{pn}}$,

ix) $F_{\mathrm{pn}} = u(Q_{\mathrm{pn}})$,

x) $F_{\mathrm{pn}} \subset F, \qquad \bigcup_n F_{\mathrm{pn}} = F$.

As u is bounded, g_{pn} has an upper and a lower limit:

$$\alpha_{\mathrm{pn}} \leq g_{\mathrm{pn}}(\xi) \leq \beta_{\mathrm{pn}} \qquad \forall \xi \in [0, 1); \qquad \alpha_{\mathrm{pn}}, \beta_{\mathrm{pn}} \in \mathbf{R}.$$

Finally, we define the non–invertible map $g_{\mathrm{p}}\colon [0, 1) \to F$ as

$$g_{\mathrm{p}} = \bigcup_n g_{\mathrm{pn}} = \bigcup_n u \circ \varphi_{\mathrm{pn}}^{-1} = u \circ \varphi_{\mathrm{p}}^{-1}.$$

It is always possible to define a normalized map $G_p: [0, 1) \to [0, 1]$:

$$G_p = \frac{g_p - \alpha_p}{\beta_p - \alpha_p}, \quad \text{where} \quad \alpha_p = \sup_n\{\alpha_{pn}\}, \quad \beta_p = \inf_n\{\beta_{pn}\}.$$

Thus, the point $x_p = (\varphi_p(\eta), G_p(\xi))$ belongs to a cylinder (2–dimensional manifold) C, where we can define a measure μ.

Assuming that the time series u is a discrete signal $\mathcal{F} = \{(t_i, u_i), i = 1, ..., N\}$, by the transformation described above we obtain the set $\mathcal{G}_p = \{x_{pi}, i = 1, ..., N\} \subset C$. Then, if T is the period of u, given an arbitrary small real number ϵ, of all the sets $\{\mathcal{G}_p\}_{p=p_i,...,p_f}$, those $\mathcal{G}_p$ such that $|p - T| < \epsilon$ will be more ordered than the others. This degree of order can be measured by means of the *Information Entropy* (see CMN95 for more details):

Let us take the function $Z: [0, 1] \to A \subset \mathbf{R}$:

$$Z(x) = -x \ln x \qquad 0 < x \le 1, \qquad Z(0) = 0,$$

(Z is continuous, $Z \ge 0$, and $\frac{d^2 Z}{dx^2} < 0$), and let $\alpha = \{a_i\}_{i=1,...,q}$ be a measurable partition on C such that

$$\mu\left(\bigcup_{i=1}^{q} a_i - C\right) = 0, \qquad \mu(a_i \cap a_j) = 0 \qquad i \neq j.$$

Let us define on C a probability density $\rho(\mathbf{x})$ as

$$\rho(\mathbf{x}) = \frac{1}{N}\sum_{i=1}^{N} \delta(\mathbf{x} - \mathbf{x}_{pi}),$$

where $\mathbf{x}_{pi} \in \mathcal{G}_p$, and δ is the *delta* function, from which follows that $\int_C \rho(\mathbf{x})d^2x = 1$. So, now we can define the measure $\mu(a_i)$ as:

$$\mu(a_i) = \int_{a_i} \rho(\mathbf{x})d^2x,$$

and therefore, by definition, the entropy of the partition α is

$$S(\alpha) = \sum_{i=1}^{q} Z[\mu(a_i)].$$

The entropy measures the amount of disorder in the system. For example, if the a_i's are such that $\mu(a_i) = \mu(a_j), \forall(i, j)$, i.e. a completely disordered signal (an ergodic system), then the entropy reduces to $S = \ln q$, and $S \to \infty$ as $q \to \infty$. Conversely, if $\mu(a_i) = 1$ for one i and $\mu(a_j) = 0, \forall j \neq i$, i.e. a completely ordered signal (a non–ergodic system), $S = 0$. Thus we should expect low values of S for nearly ordered system and high values for disordered ones.

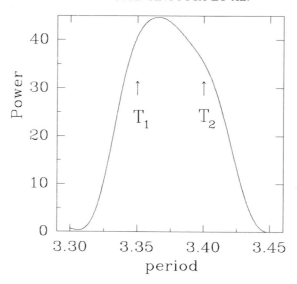

Fig. 1.   The periodogram, using modified FFT, of a time series formed by 2 periodic signals of almost the same frequency. Although not symmetric, this periodogram only shows one maximum, as the 2 periods do not satisfy Loumos & Deeming criterion (see text).

3. Application to a Quasi–periodic Signal

In order to test the method we constructed a time series and examined it using the modified Fast Fourier Transform (FFT) for non–evenly sampled data, and the Information Entropy algorithm previously described. The data set consisted of the sum of two periodic functions of the form:

$$u_j(t) = A_{0j} + \sum_{n=1}^{3} A_{nj} \sin\left(\frac{2n\pi t}{T_j}\right), \ j = 1, 2 \tag{1}$$

(both signals having the same amplitude) plus uniformly distributed noise (r.m.s of 20% of the peak–to–peak of the signal). The first signal had a period of 3.35, while the second one was tuned up to the resolution limit of the Entropy method. We generated a sample of 220 random (thus avoiding aliasing or pseudo–aliasing) arguments t_i spanning over ~ 150 units of time, and calculated the ordinates according to (1). This time series may be represented by the set $\{(t_i, u_i), i = 1, ..., N\}$, with $N = 220$, $t_1 = 0.824$ and $t_N = 149.746$. These data were studied using Press & Rybicki (1989) modified FFT algorithm, with an oversampling factor in frequency of 16. Loumos & Deeming (1978) and Kovacs (1981) showed that Fourier techniques are able to separate two frequencies only if $\Delta f > 1/\tau$, ($f = 1/T$, and $\tau = t_N - t_1$). For the data set described here this means that FFT

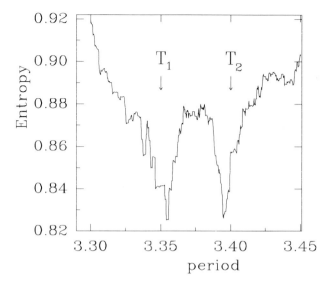

Fig. 2.   Same as Figure 1, but using the Information Entropy. In this case both periods may be identified separately, indicating that the internal precision is better in this method than in Fourier analysis.

analysis can resolve the second period only if $T_2 \geq 3.43$. In Figure 1 we show the periodogram for a data set consisting of 2 periods $T_1 = 3.35$ and $T_2 = 3.40$, calculated using modified FFT. Of course these 2 periods do not satisfy the above mentioned criterion, so only one peak may be resolved in the periodogram.

On the other hand, Figure 2 shows the periodogram calculated with the Information Entropy method, using a partition of 3×3 elements (remember that right periods were those that minimized the Entropy). The period range for the analysis and the number of periods tested were similar as for the FFT. It is worth noticing that this method is capable of resolving the two independent periods, even in this case where FFT could not. We have to add, however, that the periods detected using this method are not exactly the same as those in the original signal (the difference is of 0.12 % for T_1, and 0.14 % for T_2). Anyhow, the same effect occurs in the FFT method, even for periods not as close as in this case (as was also shown by Loumos & Deeming 1978). This experiment revealed that the Information Entropy is a powerful tool for studying the periodic behavior of a given (not necessarily evenly sampled) signal, and that the underlying period may be determined with very good (intrinsic) precision.

4. Conclusions

We have developed and tested a new algorithm that uses the Information Entropy as an indicator of periodic behavior within a time series. As String Length Methods or Phase Dispersion Minimization techniques, this one is well suited for the analysis of unevenly sampled data. It is however much faster than those ones, it is computationally straightforward, and just requires counting points within a binning scheme.

We tested this algorithm on a quasi–periodic signal consisting of 2 periodic functions of almost the same frequency. We have shown that this method is capable of resolving these 2 frequencies even in cases where Fourier techniques cannot.

Finally, one important point that still remains to be done is to give the statistical significance of the detected periods (minima in the entropy periodogram). We are now working on that point, and we hope to address this issue in a future paper.

5. Acknowledgements

We are very grateful to Sylvio Ferraz–Mello and Jacques Laskar for valuable comments and questions during the oral presentation of this paper. This work was partly supported by Fundación Antorchas, Consejo Nacional de Investigaciones Científicas y Técnicas de la República Argentina, and Facultad de Ciencias Astronómicas y Geofísicas de la Universidad Nacional de La Plata.

References

Barning, F. J. M.: 1963, *Bull. Astron. Inst. Neth.* **17**, 22.
Cincotta, P. M., Méndez, M. & Núñez, J.A.: 1995, *ApJ* **449**, 231 (CMN95)
Cuypers, J.: 1987, *Preprint of the Astronomisch Instituut van de Katholieke Universiteit Leuven* **92**.
Ferraz-Melo, S.: 1981, *AJ* **86**, 619.
Kovacs, G.: 1981, *Ap&SS* **69**, 485.
Lomb, N. R.: 1976, *Ap&SS* **39**, 447.
Loumos, G.L., Deeming, T.J.: 1978, *Ap&SS* **56**, 285.
Press, W. H., Rybicki, G.B.: 1989, *ApJ* **338**, 277.
Scargle, J. D.: 1989, *ApJ* **343**, 874.
Vaníček, P.: 1971, *Ap&SS* **12**, 10.

PAINLEVE ANALYSIS OF SIMPLE KALUZA-KLEIN MODELS

AMINA HELMI * and HÉCTOR VUCETICH **

Facultad de Ciencias Astronómicas y Geofísicas, Paseo del Bosque s/n, 1900 La Plata, Argentina

Abstract. We study the integrability of isotropic and homogeneous Kaluza-Klein models, using the Painlevé property of differential equations. We can conclude that the system is generically nonintegrable due to the presence of movable algebraic branch points.

Key words: Kaluza-Klein theories – Integrability – Painlevé property

1. Introduction

Kaluza-Klein theories (Kaluza 1921; Klein 1926) provide a way to the unification of all the interactions in Nature. They constitute a generalization of General Relativity, in which extra dimensions are introduced, thus giving new degrees of freedom to the metric, which now represents a $4 + N$ dimensional space-time. These new degrees of freedom are associated to scalar and tensor fields representing the rest of the interactions in Nature (Chodos & Detweiler 1980).

The extra dimensions should have been relevant during the early epochs of evolution of the Universe (Segré 1989, Weinberg 1984), near the Planck era, when they should have had the same physical size as that of the usual 3-space dimensions. Due to the fact that the internal manifold has not yet been detected, the present size of the extra dimensions should be very small, that is, they should be compactified (Gillies 1988, Sisterna & Vucetich 1995).

In this article we will study very simple Kaluza-Klein models. Although quantum effects should have been important and may have determined their evolution, and for example account for compactification, we will only deal with classical models (Rubin 1984). This is mainly due to the fact that, even in this case, very interesting behaviour can be found because of the highly non-linear character of the resulting equations of motion. And we might expect that, when one complicates the system through quantum corrections, this behaviour will still be present and be of great importance.

As is well-known, Einstein equations conform a highly nonlinear set of differential equations. But they have not yet been thoroughly studied in the search for chaotic behaviour, although this is an increasing field of interest

---

* Fellow of CONICET
** Member of the Carrera del Investigador del CONICET

J. C. Muzzio et al. (eds.), Chaos in Gravitational N-Body Systems, 227–232.

among people working in General Relativity and other theories of gravity
(Calzetta & El Hasi 1993, Rugh 1994, Contopoulos et al. 1993, Latifi et al.
1994).

In the examples we consider in this article, nonintegrability can be called
a generic feature of the model. This result is obtained through an analytical
study of the differential equations involved based on the Painlevé property
(Painlevé 1900, 1902).

The method we use to determine whether the system is integrable or not
is based on the analysis of the singularities of the solution in the complex
plane of time (the independent variable). If its only movable singularities
are poles, then it can be expanded in a Laurent series, and the solution so
obtained will be generic. The algorithm to follow is known as ARS algorithm
(Ablowitz et al. 1978, 1980), and consists of 3 steps:

  − determination of leading order behaviours of the Laurent series around
    the movable singular point,
  − determination of the resonances, that is, the powers at which arbitrary
    constants enter into the Laurent series,
  − verification that a sufficient number of arbitrary constants exists with-
    out the introduction of logarithmic movable critical points.

If the system fails to pass anyone of these steps, then it is possible to
conclude that the system is not integrable. It is important to point out that
the test provides with necessary but not sufficient conditions for a system
to be integrable (for a complete review, see Ramani et al. 1989).

The equations of motion of a $4 + N$ dimensional space-time, which is
isotropic and homogeneous are (Kolb & Turner 1990):

$$3\frac{\ddot{a}}{a} + N\frac{\ddot{b}}{b} = \frac{-8\pi \hat{G}}{N+2}\rho[(N+1) + 3\omega + N\omega'], \tag{1}$$

$$\frac{\ddot{a}}{a} + 2(\frac{\dot{a}}{a})^2 + N\frac{\dot{a}}{a}\frac{\dot{b}}{b} + \frac{2k_a}{a^2} = \frac{8\pi \hat{G}}{N+2}\rho[1 + (N-1)\omega - N\omega'], \tag{2}$$

$$\frac{\ddot{b}}{b} + (N-1)(\frac{\dot{b}}{b})^2 + 3\frac{\dot{a}}{a}\frac{\dot{b}}{b} + \frac{(N-1)k_b}{b^2} = \frac{8\pi \hat{G}}{N+2}\rho[1 - 3\omega + 2\omega'], \tag{3}$$

$$\dot{\rho} + 3(1+\omega)\rho\frac{\dot{a}}{a} + N(1+\omega')\rho\frac{\dot{b}}{b} = 0. \tag{4}$$

where $a(t)$, $b(t)$ are the scale factors corresponding to the usual 4-dim.
space-time $R \times M^3$, and to the internal space M^N respectively, and k_a, k_b
the corresponding curvature constants. These equations are supplied with
equations of state for the density and pressures in the different spaces. In

TABLE I
Resonances for Branch 2, $\omega = 0$.

If $A = \frac{N+1}{2} + \frac{(N-1)(4+5N)^{1/2}}{6\sqrt{3}N^{3/2}}i$
x -1
x $-1 + \frac{1+2N}{3NA^{1/3}} + A^{1/3}$
x $-1 - \frac{1}{2}[\frac{1+2N}{3NA^{1/3}} + A^{1/3}] \pm \frac{\sqrt{3}}{2}[-\frac{1+2N}{3NA^{1/3}} + A^{1/3}]i$

this case we have considered a perfect fluid $p_3 = \omega\rho$ in M^3 and $p_N = \omega'\rho$ for M^N.

2. Analysis of the system

2.1. LEADING ORDER BEHAVIOUR

The leading order terms are found by taking $a(t) \approx a_0(t - t_0)^\alpha$, $b(t) \approx b_0(t-t_0)^\beta$, $\rho(t) \approx \rho_0(t-t_0)^\gamma$ in the equations of motion. The branches obtained are:

1. $\alpha = \frac{2-N(1+\omega')}{3(1+\omega)}$, $\beta = 1$, $\gamma = -2$.

2. $\alpha = 1$, $\beta = \frac{2-3(1+\omega)}{N(1+\omega')}$, $\gamma = -2$.

3. derived from the following quadratic equations:
$$3(1+\omega)\alpha + N(1+\omega')\beta - 2 = 0,$$
$$(1 + (N-1)\omega - N\omega')[\beta(\beta - 1) + (N-1)\beta^2 + 3\alpha\beta] =$$
$$= (1 - 3\omega + 2\omega')[\alpha(\alpha - 1) + 2\alpha^2 + N\alpha\beta].$$

These results show that the first two branches can lead to integer or rational exponents, depending on the value of ω and ω'. When one studies the third branch, the quadratic equation is found to have as coefficients very complex algebraic functions of N, ω and ω'. That is, in general the system has algebraic branch points, even in some very simple cases, like for instance if $\omega = 0$ or $\omega' = 0$, the exponents are found to be complex or irrational numbers.

The case $\omega = \omega'$ is one observed to have rational leading orders. With this in mind we will restrict ourselves to this case in the following.

TABLE II

Resonances for Branch 2, $\omega = \frac{1}{3}$.

	If $A = \frac{1}{4} + \frac{3}{8N} + \frac{(N+3)(24+17N)^{1/2}}{24\sqrt{3}\,N^{3/2}}\,i$
x	-1
x	$-\frac{1}{2} + \frac{6+5N}{12NA^{1/3}} + A^{1/3}$
x	$-\frac{1}{2} - \frac{1}{2}\left(\frac{6+5N}{12NA^{1/3}} + A^{1/3}\right) \pm \frac{\sqrt{3}}{2}\left(-\frac{6+5N}{12NA^{1/3}} + A^{1/3}\right)i$

2.2. RESONANCES

As we have just stated, we will consider what happens if $\omega = \omega'$. For the second branch, one gets a 5th. degree algebraic equation for the resonances, whose solutions are $x = 0$ and those satisfying the following homogeneous equation:

$$a_4 x^4 + a_3 x^3 + a_2 x^2 + a_1 x + a_0 = 0, \tag{5}$$

with
$$a_4 = N(1+\omega)^4,$$
$$a_3 = N(1+\omega)^2(4 + 2\omega - 2\omega^2),$$
$$a_2 = (1-\omega)(1+\omega)^2(4N - 1 - (3N+6)\omega - (3N+9)\omega^2),$$
$$a_1 = (-(N+3) - (11N+17)\omega - (6N+18)\omega^2 + (14N+26)\omega^3$$
$$+ (7N+21)\omega^4 - (3N+9)\omega^5),$$

$$a_0 = -2(1+\omega)(-(N+1) - (2N+4)\omega - (4N+2)\omega^2 - (2N+12)\omega^3$$
$$-(3N+9)\omega^4.$$

The resonances are listed in tables (I, II, III) for different physically meaningful values of ω. As it is immediately seen, one finds complex resonances. There is one exception when $\omega = \frac{1}{N+3}$ and $N = 6$. For that case, we have studied the first branch, and found irrational values for x (see Helmi & Vucetich, 1995 for further details). The presence of this kind of resonances leads to movable algebraic branch points, which in turn makes the system nonintegrable.

3. Discussion and Conclusions

The application of ARS algorithm has lead us to some very interesting conclusions. In the more generic case, that is for ω, ω' and N arbitrary, the

TABLE III

Resonances for Branch 2, $\omega = \frac{1}{N+3}$.

If $A = \frac{N+6}{2N} + \frac{(N-6)(24+5N)^{1/2}}{6\sqrt{3}N^{3/2}}i$
x $\quad -1$
x $\quad \frac{N+2}{N+4}[-1 + \frac{2(N+3)}{3NA^{1/3}} + A^{1/3}]$
x $\quad \frac{N+2}{N+4}[-1 - \frac{1}{2}(\frac{2(N+3)}{3NA^{1/3}} + A^{1/3}) \pm \frac{\sqrt{3}}{2}(-\frac{2(N+3)}{3NA^{1/3}} + A^{1/3})i]$

leading order terms found for the third branch have as exponents irrational or complex numbers, thus showing that the system is not integrable in the most general case, due to the presence of movable algebraic branch points. However, further analysis was needed for some special cases, when $\omega = \omega'$. Depending on the values for ω the leading behaviours were given by integer or rational numbers, which could have been related to the Painlevé, or the weak-Painlevé property respectively.

The values found for the resonances for those cases clearly show that none of them pass Painlevé test. The resonances are either irrational or complex numbers, showing again the presence of movable algebraic branch points. We can assert from these results that, no matter what nonzero value N takes, the system will not be integrable.

Some more information about the behaviour of the system can be obtained by studying the structure of the distribution of singularities in the complex plane of time. Due to the fact that some resonances are complex, the distribution of the singularities in the neighbourhood of a given one t_0 (which is an initial condition) is in the form of a logarithmic spiral. If one changes the initial condition to a neighbouring point, one will find this structure repeated, but now around this new point. This shows the selfsimilar nature of the distribution of the singularities in the complex t-plane.

The parameters of this structure are obviously determined by the position of the principal singularity t_0 and by the value of the resonance x (which depends only on the value of N, for each value of ω analysed). If we denote $t_n = r_n e^{i\phi_n}$ for the different singularities located in those structures, one finds that

$$\log r_{n+1} - \log r_n = \pi x_2/(x_1^2 + x_2^2), \qquad \text{and} \qquad \phi_{n+1} - \phi_n = \pi x_1/(x_1^2 + x_2^2).$$

Due to the fact that $\bar{x}$ is also a solution of the resonance equation (5), there are indeed two logarithmic spirals associated with each t_0, one with increasing and one with decreasing 'radius'. This last fact shows a singularity accumulation in the complex t-plane. This phenomenon has been found in Henon-Heiles system (Weiss 1981), while similar studies revealed such singularity spirals in several other non-integrable Hamiltonian systems (Chang et al. 1982; Bier 1987). It has been suggested that such spiralling patterns resulting from complex resonances can lead to the formation of *natural boundaries* in the complex *t*-plane (Chang et al. 1982).

In summary, we have shown that a large class of simple Kaluza-Klein cosmological models are nonintegrable and, probably, chaotic. The latter property, being a global one, should be addressed through some global method.

4. Acknowledgements

We wish to thank Fundación Antorchas, Facultad de Ciencias Astronómicas y Geofísicas de la Universidad Nacional de La Plata and CONICET for financial support.

References

Ablowitz, M. J., Ramani, A. & Segur, H.: 1978, *Lett. Nuovo Cimento* **23**, 333.
Ablowitz, M. J., Ramani, A. & Segur, H.: 1980, *J. Math. Phys.* **21**, 715.
Bier, M.: 1987, Ph. D. Thesis, Dept. of Mathematics, Clarkson Univ. Postdam, N. Y.
Calzetta, E. & El Hasi, C.: 1993, *Clas. Quantum Grav.* **10**, 1825.
Chang, Y., Tabor, M. & Weiss, J. : 1982, *J. Phys. A* **21**, 33.
Chodos, A. & Detweiler, S.: 1980, *Phys. Rev. D* **28**, 4, 772.
Contopoulos, G., Grammaticos, B. & Ramani, A.: 1993, *J. Phys. A: Math. Gen.* **26**, 5795.
Gillies, G. T.: 1988, eds. de Sabbata, V. & Melnikov, V., Gravitational Measurements, Fundamental Metrology and Constants, Kluwer Academic Publ., 191.
Helmi, A. & Vucetich, H.: 1995, *Integrability and chaos in Kaluza-Klein theories*, Preprint of the Universidad Nacional de La Plata, Argentina.
Kaluza, Th.: 1921,*Sitzungsber. Preuss. Akad. Wiss. Phys. Math. Kl.* **LIV**, 966.
Klein, O.: 1926, *Z.Phys.* **37**, 895.
Kolb, E. & Turner, M.: 1990, The Early Universe, Addison-Wesley, 470.
Latifi, A., Musette, M. & Conte, R.: 1994, *Phys. Lett. A* **194**, 83.
Painlevé, P.: 1900, *Bull. Soc. Math.* **28**, 201.
Painlevé, P.: 1902, *Acta Math.* **25**, 1.
Ramani, A., Grammaticos, B. & Bountis, T.: 1989, *Phys. Rep.* **180**, 3, 159.
Rubin, M.: 1984, *Kaluza-Klein Cosmology: Techniques for Quantum Computations*, Inner Space/Outer Space, 583.
Rugh, S.: 1994, *Chaos in Einstein equations*, Preprint of Niels Bohr Institute, Denmark.
Segré, G.: 1989, *Physics is more than 4 dimensions: Another look at the Kaluza Klein theory*, eds. Alschuler, D. & Nieves, J., Proceedings of the First Winter School of Physics: Cosmology and Elementary Particles, World Scientific, 99.
Sisterna, P. & Vucetich, H.: 1995, , eds. Babour, J. & Pfister, H., Proceedings of the Conference on 'Mach's Principle', Birkhausen, 375.
Weinberg, S.: 1984, *Physics in Higher Dimensions*, Inner Space/Outer Space, 571.
Weiss, J.: 1981, eds. Tabor, M. & Treve, Y., Integrability in Hydrodynamics and Related Dynamical Systems, AIP Conf. Proc. 88., AIP: New York, 243.

NONLINEAR ANALYSIS OF A CLASSICAL COSMOLOGICAL MODEL

S. BLANCO and O. A. ROSSO
Instituto de Cálculo, UBA, CONICET
Pabellón II , Ciudad Universitaria.
1428 Bs. As. , Argentina.

and

A. COSTA
Instituto de Astronomía y Física del Espacio. CONICET
C.C. 67 Suc. 28, 1428 Bs. As. , Argentina

Abstract. We study the dynamics of a Friedmann-Robertson-Walker Universe considered as an autonomous Hamiltonian. The time evolution of this Hamiltonian presents numerical instabilities so we apply a symplectic integration method. We conclude that chaotic behavior is possible in this spatially closed Universe.

1. Introduction

Hamiltonian systems are not structurally stable in mathematical sense and standard numerical integration schemes neglect important features of the dynamics, in particular the fact that the time δt map of phase space is symplectic [1 − 3]. That is, the motion of the phase-space points from time t to time $t + \delta t$ is obtained by a sucession of infinitesimal canonical transformations. If the numerical integration scheme is not symplectic (e.g. Runge-Kutta) the conservation of Poincaré invariants must be tested after each integration step. Moreover, even when this condition is numerically satisfied, it does not necessarily imply that the time evolution obtained is the real dynamical description of the Hamiltonian system under study because an arbitrary displacement in phase space does not necessarity have to be consistent with the Hamiltonian equations [3]. These are the main reasons why analytical solutions, via qualitative theory of ordinary differential equations, and analytical-numerical solutions, via symplectic integration, must both be exhaustively studied for Hamiltonian systems.

In this work, we study the dynamics of the Friedmann-Robertson-Walker Universe conformally coupled to a real, self-interacting, massive scalar field, within the framework of Dynamical System Theory [4 − 7]. The original Hamiltonian model has a logarithmic singularity in the universe radius $a = 0$. We consider a conformal coupling between the field and gravity that allows to parametrize the field so as to simplify the Hamiltonian expression shifting the singularity to infinite. In a previous work [6], the numerical instability found in the integration of this Hamiltonian model near resonant

J. C. Muzzio et al. (eds.), Chaos in Gravitational N-Body Systems, 233–238.

initial conditions with a Runge-Kutta scheme was the main difficulty found, even when the numerical results were those predicted by analytical calculations. In order to avoid the difficulties found in the numerical integration of our cosmological Hamiltonian model and confirm that the chaotic behavior observed was not numerical noise, we implemented a symplectic integration derived from infinitesimal canonical transformations [7].

2. The model

The dynamical system studied has two degrees of freedom, a and ϕ, and Hamiltonian

$$H = \frac{1}{2}\,[-(p_a^2 + k\,a^2) + (p_\phi^2 + k\,\phi^2) - \mu^2 a^2 \phi^2 + \frac{\lambda}{2}\phi^4 + \frac{\Lambda}{2}a^4] \qquad (1)$$

where Λ is the cosmological constant, μ^2 is the mass of the field and $\lambda > 0$ is the self-interacting coefficient. k represents the spatial curvature. Here we restrict our analysis to close universes $k = +1$ [6 − 7].

Bearing in mind that in General Relativity the value of the Hamiltonian is enforced by the theory to be equal zero, we introduce it as a constraint on allowable initial conditions ($H = 0$), which is clearly satisfied by the dynamics.

3. Dynamical Analysis

The fixed points of a Hamiltonian dynamical system play an important role and clearly have a characteristic *local* behavior, and can be thought of as the "organizing centers" of a system's phase space dynamics [8].

The fixed points, $\mathcal{P} \equiv (a, \phi, p_a, p_\phi)$, for our Hamiltonian system are [6 − 7]:

$$\mathcal{P}_1 = (\,0\,,\,0\,,\,0\,,\,0\,)\,, \qquad (2)$$

$$\mathcal{P}_2 = (\,\pm\sqrt{k/\Lambda}\,,\,0\,,\,0\,,\,0\,)\,, \qquad (3)$$

$$\mathcal{P}_3 = (\,\pm\sqrt{k\,\frac{(\lambda - \mu^2)}{(\Lambda\cdot\lambda - \mu^4)}}\,,\,\pm\sqrt{-k\,\frac{(\Lambda - \mu^2)}{(\Lambda\cdot\lambda - \mu^4)}}\,,\,0\,,\,0\,) \qquad (4)$$

with

$$\Lambda\cdot\lambda - \mu^4 \neq 0\,. \qquad (5)$$

The fixed points must have real coordinates because they belong to the phase space of the system. This imposes restrictions on the parameters Λ, λ and μ [6 − 7]. In addition the fixed points must satisfy the Hamiltonian constraint, $H(\mathcal{P}) = 0$. Only $\mathcal{P}_1$ and $\mathcal{P}_3$ satisfy this condition. $H(\mathcal{P}_3) = 0$

introduces a new relation among the three parameters of the model, $\Lambda + \lambda + 2\,\mu^2 = 0$

From the corresponding linear stability analysis we can conclude that, $\mathcal{P}_1$ is an elliptic point and $\mathcal{P}_3$ is an hyperbolic point. The co-existence of elliptic and hyperbolic points is a first evidence of possible chaotic behavior in the system [8]. In order to confirm the existence of chaotic behavior we evaluated the Poincaré section of the dinamical flux.

4. Numerical Analysis

An infinitesimal canonical transformation (written in symplectic notation) can be expressed in terms of the new variables as [2, 7]

$$\xi = \eta + \delta\eta, \tag{6}$$

where η is a column matrix of $2n$ elements with $\eta_i = q_i$, $\eta_{i+n} = p_i$ and $i \leq n$. The infinitesimal change, $\delta\eta$, satisfies

$$\delta\eta = \varepsilon\, \mathbf{J}\, \frac{\partial G}{\partial \eta}, \tag{7}$$

where

$$\mathbf{J} = \begin{pmatrix} \mathbf{0} & \mathbf{1} \\ -\mathbf{1} & \mathbf{0} \end{pmatrix} \tag{8}$$

and $G(\eta)$ is the generating function of the infinitesimal canonical transformation. The above expression written in terms of Poisson brackets is

$$\delta\eta = \varepsilon\,[\,\eta\,,\,G\,]. \tag{9}$$

If we now consider an infinitesimal canonical transformation in which the continuous parameter is t so that $\varepsilon = dt$ and let the generating function G be the Hamiltonian, then the equation of the infinitesimal canonical transformation is

$$\delta\eta = dt\,[\,\eta\,,\,H\,] = \dot{\eta}\,dt = d\eta. \tag{10}$$

This leads to a new interpretation where the canonical transformation that depends on the time parameter can be seen as moving the system point along a continuous trajectory in phase space. Then, a finite transformation $t_0 \rightarrow t$, can be regarded as the sum of an infinite sucession of infinitesimal canonical transformations, each of which corresponds to an infinitesimal transformation along a curve in the phase space. It is formally possible to obtain the finite transformation by integrating the expression of the infinitesimal displacements.

A formal expression for the time evolution of the variable η can be obtained by expanding $\eta(t_0 + \delta t)$ in a Taylor series around the initial conditions $\eta_0 = \eta(t_0)$,

$$\eta (t_0 + \delta t) = \eta_0 + \sum_{n=1}^{\infty} \frac{\delta t^n}{n!} f_n(\eta) \Big|_{t_0} \tag{11}$$

with

$$f_1(\eta) = [\eta , H], \tag{12}$$
$$f_{n+1}(\eta) = [f_n(\eta) , H]. \tag{13}$$

We can use this formal expression to derive an integration method. Choosing a time step δt small enough, we can truncate the series (11) up to order N.

Then in order to minimize the error, we solve in a selfconsistent way the system

$$\eta = \eta_0 + \sum_{n=1}^{N} \frac{\delta t^n}{n!} f_n(\eta) \tag{14}$$

In this way, we have derived an integration method, which is symplectic since it is obtained from an infinitesimal canonical transformation. The series expansion (14), which gives the values of the new coordenates Q_i and impulses P_i at time $t_0 + \delta t$ (up to order N in δt) and the symplectic integration scheme proposed by Scovel's [9] are equivalent in both cases up to $N = 3$.

We implemented the symplectic integration scheme described above, with a time step $\delta t = 0.001$ and an error $\varepsilon_\eta = 10^{-15}$ for the resolution of eq. (14), that, in consequence, gives an error in the numerical evaluation of the Hamiltonian $\varepsilon_H \sim 10^{-12}$. In Fig. 1 we present the Poincaré map (a , p_a) evaluated for 5000 data points for a Poincaré section defined by $\phi = 0$ for the parameter set $\lambda = 0.25$, $\Lambda = 0.001$. The values of the initial conditions are $a = 0.2$, $\phi = 0$. The p_a value was slowly varied between $0.0 - 4.5$ and the corresponding initial values of p_ϕ were obtained from the Hamiltonian constraint. Note the rich pattern displayed in the map. It gives a clear example of what happens when a regular orbit of an integrable system is disturbed by a nonintegrable perturbation. Depending on the different initial conditions regular or completely irregular motion results. There are torii with different degrees of breakdown while the original torus decomposes into smaller and smaller torii. Some of these newly created torii are again stable according to the KAM theorem, but between the stable torii the motion is completely irregular.

5. Conclusions

We have applied standard nonlinear dynamical methods to the Friedmann - Robertson - Walker cosmology. We analyzed the fixed points and confirmed

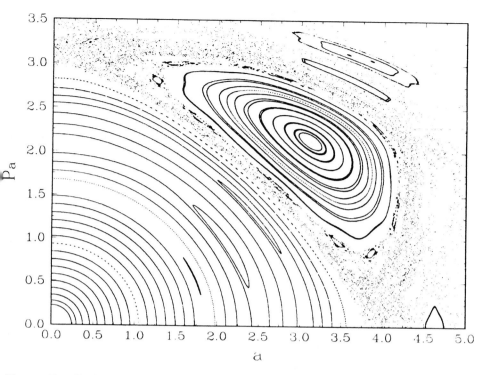

Fig. 1.   Detail of Poincaré map (a, p_a), for the parameter set $\lambda = 0.25$ and $\Lambda = 0.001$ with the different initial conditions described in the text, showing the sucessive torii breakdown.

the existence of chaotic behavior for this closed Universe. This is due to the simultaneous presence of elliptic and hyperbolic points. This chaotic behavior can be observed through the Poincaré maps.

Due to the structural unstable characteristic of this Hamiltonian system, in a mathematical sense, we stress the importance of the symplectic numerical integration method. This algorithm preserves global geometrical structure in phase space and shows long time stability.

Acknowledgement

The authors wish to thank Dr. A. Held (Editor of General Relativity and Gravitation) and Plenum Publishing Corp. for allowing us to use some materials previously published in General Relativity and Gravitation, 27 (1995) 1295.

References

1. Arnold, V.I.: 1989, *Mathematical Methods of Classical Mechanics*, Second Edition. Springer-Verlag, New York.
2. Goldstein H.:1980, *Classical Mechanics*, Second Edition. Addison-Wesley, New York.
3. Miller R. H.: 1993, "A horror story about integration methods", *J. Comp. Phys.* **93** 469.
4. Calzetta E. and El Hasi C.: 1993, "Chaotic Friedmann-Robertson-Walker cosmology", *Class. Quantum Grav.* **10** 1825.
5. Calzetta E. and El Hasi C.: 1995, "Nontrivial dynamics in the early stadge of inflation", *Phys.Rev. E* **51** 2713.
6. Blanco S., Domenech G., El Hasi, Rosso O. A.: 1994, "Chaos in classical cosmology", *Gral. Rel. and Grav.* **26** 1131.
7. Blanco S., A. Costa, Rosso O. A.: 1995, "Chaos in classical cosmology (II)", *Gral. Rel. and Grav.* (in press).
8. Gutzwiller M. C.: 1990, *Chaos in Classical and Quantum Mechanics*, Springer Verlag, Berlin.
9. Channell P. J. and Scovel C.: 1990, "Symplectic integration of Hamiltonian systems", *Nonlinearity* **3** 231.

NONTRIVIAL DYNAMICS AND INFLATION

CLAUDIO D. EL HASI

Instituto de Ciencias, Universidad de General Sarmiento.
Paunero 1721, 1663 San Miguel, Pcia. de Buenos Aires - Argentina.
and
Instituto de Astronomía y Física del Espacio
CC 67 Suc. 28, 1428 Buenos Aires - Argentina.

Abstract. The scope of the present work is to show how the application of some usual tools in Dynamical System Theory can bring ligth on several cosmological problems. We analize several effects in classical cosmology due to the nonlinearity of the equations of evolution that govern the Universe. We will find the existence of irregular or chaotic motion and the formation of some structures, in the phase space, that lead to a nontrivial dynamics of the system.

1. Introduction.

By definition the Universe is the physical system that contains all other systems. Therefore, it could be considered the most complete and complicated physical system. Nevertheless, it can be studied under great simplifications that lead to models obeying some known mathematical equations.

Despite of the clumpy and quiet picture of the sky that we can see any starry night, our Universe is, when considering at large scale, isotropous, homogeneous and expanding. These assertions lead to a new cosmological framework: the Standard Cosmological Model (SCM)[6]. It is supposed that our Universe was born from a cosmic singularity, a state of so high temperature and pressure that nothing, neither energy (or matter) nor even time did exist as we known them today. By some mechanism not yet well know, but probably a quantum mechanical effect, there was a huge explosion of this singularity (the *Big Bang*). At first matter and radiation were presumably in thermal equilibrium at a very high temperature, with radiation driving any physical process (the so called *radiation dominated era*). As the Universe subsequently expanded, the primitiae mixing of both matter and radiation cooled. Eventually, when temperature had dropped enough the free electrons joined atoms, so that the opacity dropped sharply, breaking the thermal contact between radiation and matter. Whatever radiation existed at that epoch has since been enormously red-shifted, but it still fills the space around us.

The SCM rests on three pillars of evidence. First, there is the observation by Edwin P. Hubble in the late 20's that light from galaxies is shifted toward the lower-frequency end of the spectrum in proportion to their distance from us. This is an indication that they are moving away from us, and

J. C. Muzzio et al. (eds.), Chaos in Gravitational N-Body Systems, 239–244.
© 1996 *Kluwer Academic Publishers. Printed in the Netherlands.*

the fact that the recessional velocity is proportional to distance confirms that the Universe is expanding. (It is the same behavior than the spots on the surface of a balloon when it is inflated.) By extrapolating towards the past, from the current rate of expansion, astronomers have inferred that the Big Bang occurred 10.000 or 20.000 million years ago. The imprecision reflects uncertainty over the distances of galaxies, the amount of mass in the Universe (which affects the rate of expansion) and other factors.

The second pillar of evidence is the cosmic microwave background radiation (CMBR), which was discovered in 1965 by Arno A. Penzias and Robert Wilson of Bell Telephone Laboratories. The CMBR is a remnant of the primordial cosmic fire ball still surrounding us. Recent measurements of COBE satellite corroborate that this background is isotropous and homogeneous with great accuracy, with a temperature $T \sim 2.73$ K and a deviation of $\frac{\Delta T}{T}|_Q \sim 5 \times 10^{-6}$. This was a great success of SCM.

Finally, there are the observed abundances of elements in the Universe: hydrogen accounts for roughly three quarters of the total mass and helium one quarter, with heavier elements occurring in traces. Various physicists have calculated that a "hot" big bang would forge elements in just such proportion through conventional nuclear fusion, also known as nucleosynthesis.

Due to the very intensive forces involved, in the SCM the gravitational field is described by the Einstein's General Relativity Theory (GRT). The Universe is so a four dimensional space-time and the distance would be measure by a second rank metric tensor $(g_{\mu\nu})$. Taking into account the proprieties of homogeneity and isotropy of the Universe $g_{\mu\nu}$ can be modeled as a diagonal tensor with just a temporal function: $a(t)$, known as the scale factor (or more loosely: the *radius* of the Universe).

The SCM had some drawbacks. For example, the sometimes called homogeneity problem. Calculations suggested that the primordial fireball expanded too fast for all its parts to exchange radiation and so to reach thermal equilibrium. Why then the CMBR have the same temperature no matter where it is measured?

It is also the so-called flatness problem that stems from GRT, which states that matter, through the influence of gravity, causes space to bend. This effect has fateful consequences in an expanding Universe. If the amount of matter per unit volume in the Universe — that is, mass density — is above a certain "critical" level, space curves in on itself in such a way that parallel lines converge. Even more significantly, gravity eventually halts the expansion of the Universe and causes to collapse in a "big crunch". Such a universe is said to be closed. If the mass density falls bellow the critical level, parallel lines diverge and the Universe expands forever; this is known as an open universe. If the density is precisely at the critical level the Universe keeps expanding but at an ever decreasing rate. This is called a flat universe, because space has no curvature: parallel lines behave just as Euclid said they do and as they seem to in our world. Indeed, present observations strongly suggest that today the Universe is almost flat. Various cosmologists had argued that the apparent flatness of the Universe represents an extraordinary coincidence. If the Universe had had a bit more matter than it does, it would have collapsed long ago; a bit less matter and it would have flown apart too swiftly for stars and galaxies to form. This is a very peculiar situation corresponding to very restricted (or "fine tuning") initial conditions at the very beginning.

To solve these problems it is postulated the Inflationary Universe Model[1]. In the Very Early Universe (VEU) there was a period of exponential expansion where the scale factor grew exponentially fast. In this way it is possible to solve not only the "fine tuning" but most of the other problems of the SCM. The period of inflation is powered by the effect of the energy of a scalar field (the inflaton field). Because its proper dynamics during inflation the effect of this field can be taken just as a cosmological constant Λ causing a time dependence $a(t) \sim e^{\Lambda t}$ of the scale factor.

According to this scenario, the entire observable Universe is just a tiny bubble in a vastly greater cosmos, and most of its matter was created virtually from nothing. All radiation within this small region was able to keep thermal equilibrium and there is no contradiction with present observations.

Again Inflation provides an answer for the flatness problem: just as blowing up a beach ball to 1.000 times its normal size would make its surface appear flatter to a nearby observer (to be more precise to someone who *lives on* the surface), so would inflation flatten out the region of the Universe that we can observe. In such a flat Universe, the density of matter would be set at critical level.

Inflation even provides an explanation for the formation of galaxies. According to quantum physics, any energy field constantly fluctuates in intensity at the subatomic level. The peaks created by these quantum fluctuations in the VEU would have became large enough, after inflation, to serve as seeds of stars and galaxies. Gravity would have done the rest of the work.

The Inflationary Universe Model has to match with the SCM, then at some time inflation has to stop and the Universe must evolve in a radiation dominated way. In the most simplified model this can by done by mimic radiation with a single scalar field ϕ. In principle this field could be considered massless and nontrivially coupled to gravity, but by coupling with the inflaton field it is possible that ϕ develops a mass (due to a spontaneous symmetry breaking process) and so we will explore such possibility.

2. Dynamical Evolution.

The model we will use can be considered within the context of the Chaotic Inflationary Scenario[4], where inflation is powered by the vacuum energy of the inflaton field when slow rolling down to the minimum of its potential. As it is expected, once inflation begins the Universe undergoes a DeSitter like (or exponential) expansion[3]. Nonetheless it is possible to find extremely complex behavior in prior inflation stages.

In the massless case (technically: under conformal symmetry), the two degrees of freedom system (a the scale factor, and ϕ the radiation field) is exactly integrable, a and ϕ are decoupled, the system poses two hyperbolics fixed points joined by two heteroclinics orbits in phase space: *the separatrix.* Within the separatrix motion is quasiperiodic and confined to KAM tori. Inflationary orbits are unbounded and approach asintotically the unstable manifolds of the static solutions. In other words, under conformal symmetry there is a clear cut distinction between inflationary and recolapsing universes as can be seen (indicated by solid lines) in Fig. 1. Once the system is perturbed (due to the mass term), internal resonances are induced and as a consequence the separatrix and KAM tori are destroyed and replaced by a new kind of structure: *the stochastic layer*[5]. The structure of the layer can be well analyzed by means of the Poincaré Space of Section. The separation between inflating and recolapsing initial conditions becomes less clear cut, orbits below the original separatrix are able to scape and become inflationary through a Hamiltonian diffusion process (see Fig. 1). It can be seen that the reason for this behavior is that the action variable associated to the scalar field is nonconserved[2].

However the localization of the stochastic layer in phase space is proper of a two degrees of freedom system. Such a localization does not occur in higher dimensional systems, for example: considering inhomogeneities in the field or the geometry. A second model to be considered, adding the first inhomogeneous mode to the scalar radiation field, represents a first step in this way.

There is a fundamental property that distinguishes systems with more than two degrees of freedom: *Arnold's Diffusion*[7]. In two degrees of freedom systems, chaotic regions are always divided by invariant tori, they act as insurmountable barriers in the Poincaré Sections. But with three or more degrees of freedom there are extra dimensions in phase space that trajectories can use to avoid such invariant tori and, in principle, to explore the whole phase space. Unfortunately this behavior is very difficult to visualize numerically and can only be seen in the boundary of the chaotic region.

Fig. 2 corresponds to a map in the initial condition space (fixing all initial conditions but varying only two of them). Giving more than 2000 initial conditions we let the system evolve. Plus symbols corresponds to unstable

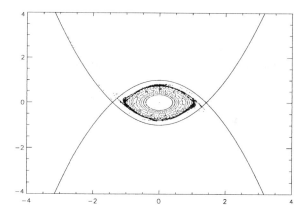

Fig. 1. Solid lines corresponds to DeSitter like (inflationary) solutions. Dots are perturbed solutions, notice as they approach inflationary behavior

trajectories, that is to say, orbits that become inflationary. The dots indicate stable trajectories that iterate at least 1000 times on the map. These numerical simulations confirm previous results[2]. It is clearly seen the absence of a net division between stable and unstable trajectories. Inflationary orbits can also be seen well within the regular region so, any perturbation can send them to an inflationary stage. The picture suggests the existence of a web (Arnold's Web), connecting (almost) the whole phase space. Let's observe that we explore just a very little region in the initial conditions space, the process is very sensible to them. We can conclude that even in case the system evolves to an inflationary expansion, this process is in no way trivial. There is a very rich structure behind this apparently simple model.

3. Final Comments.

In recent years, two branches of Physics have deserved special interest: Dynamical Systems and Cosmology. The former has even exceeded the physics's concerns and it is applied in any situation involving nonlinear interactions. Meanwhile, cosmological surveys stems on the hope to built a model that allows us to understand the origin and development of our Universe.

As the equations of motion describing the VEU are non linear, it seems reasonable to link both branches of knowledge and search for further consequences. So, we apply several methods and techniques of DST to the study of cosmological problems.

We show that dynamical system effects, such as: internal resonances, breakdown of separatrices and KAM tori, and Arnold's diffusion appear

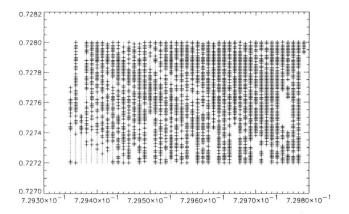

Fig. 2. A map in initial condition space of mommenta associated to the scale factor and homogeneous mode of the radiation field. Each + denotes an inflationary trajectory

very frequently in Cosmology. This effects lead to nontrivial behavior in the VEU, that also match with present accepted believes.

Of course, to analyze this subjects further will require much more sophisticated analytical and numerical techniques. We are given a first step in this direction, because we think that chaos can be considered as a major key to understand the origin and evolution of our Universe.

Acknowledgements

This work has been partially supported by Fundación Antorchas, CONICET, UBA, and the Directorate General of Science, Research and Development of the Commission of the European Communities Nr. CI 1* - CJ 94-0004. I am grateful to Esteban Calzetta and Graciela Domenech for their useful comments and encouragement.

References

[1] Börner, G.: 1988, 'The Early Universe', (New York. Springer-Verlag) (Second Ed. 1992).
[2] Calzetta E. and El Hasi C., 1995 'Nontrivial Dynamics in the Early Stages of Inflation', *Phys. Rev.* **D51**, 2713.
[3] Gibbons, G., and Hawking, S. W.: 1977, *Phys. Rev.* **D15**, 2738.
[4] Linde, A. D.: 1983, Phys. Lett. **129B**, 177.
[5] Reichl L. E. and Zheng W. M.: 1987, 'Non Linear Resonance and Chaos in Conservative Systems', in Directions in Chaos ed. Hao Bai-Lin (Singapur, World Scientific).
[6] Weinberg, S.: 1972, 'Gravitation and Cosmology', (New York, John Wiley).
[7] Zaslavsky, G. M., Sagdeev, R. Z., Usikov, D. A., and Chernikov, A. A.: 1991, 'Weak Chaos and Quasi Regular Patterns', (Cambridge, Cambridge University Press).

A MODEL DISTRIBUTION FUNCTION FOR VIOLENTLY RELAXED N-BODY SYSTEMS

N. VOGLIS and C. EFTHYMIOPOULOS

Department of Astronomy, University of Athens
Panepistimiopolis, GR 157 84 Athens, Greece

1. Introduction

A system of N particles collapsing under their collective self-gravity serves as a model to study galaxy formation from density inhomogeneities in the expanding universe. Such a system is being subject to a 'violent relaxation' towards a quasi-equilibrium state reached after a few dynamical times $\tau_d = (G\bar{\rho})^{-1}$, $\bar{\rho}$ is the mean spatial density of particles). The violent relaxation is driven mainly by the rapid oscillations of the 'smooth' self-gravitational potential $\psi(\mathbf{r}, t)$ derived by Poisson equation $\nabla^2 \psi(\mathbf{r}, t) = 4\pi G\rho(\mathbf{r}, t)$. During the potential oscillations the particles' individual energy changes with time. A fraction of particles may acquire positive energies and escape. The rest form a bound configuration made of a tight 'core' and a loose 'halo'.

By applying statistical mechanics similar to a Fermi gas, under the assumption of total mass and energy conservation, Lynden-Bell (1967) derived the coarse-grained distribution function of the 'maximum entropy' equilibrium state of a violenty relaxed system:

$$f(\mathcal{E}) = \eta \frac{e^{\beta(\mathcal{E}-\mu)}}{e^{\beta(\mathcal{E}-\mu)} + 1} \tag{1}$$

with $\mathcal{E} = -E > 0$, where E is the binding energy of individual particles, μ is the so-called "chemical potential", β is is a constant measuring the "inverse temperature" and η is the constant initial density of phase elements.

All initially 'cold' systems collapsed violently enough should reach the distribution function (1). However, N-Body simulations have repeatedly shown (e.g. Hénon 1964, Lecar & Cohen 1972, White 1979, May & van Albada 1984, Voglis et al. 1991, Voglis et al. 1995) that cold-collapsed systems relax only partially, i.e. some memory of initial conditions survives after violent relaxation. It is thus of interest to find modifications of the distribution function (1) in order to obtain agreement with the results of N-Body simulations.

Such a model (Voglis 1994a) was shown to nicely fit the coarse grained distribution of a nearly-spherical anisotropic cold collapsed N-body system. This model has a 'core' and a 'halo' component, describing two populations having a different level of mixing and obeying different statistics.

J. C. Muzzio et al. (eds.), Chaos in Gravitational N-Body Systems, 245–252.
© 1996 *Kluwer Academic Publishers. Printed in the Netherlands.*

Here we show that the Voglis 1994a model is applicable to systems relaxed from a range of cold spherical initial conditions created by imposing a spherical density perturbation of the form:

$$s(r) = \frac{\delta\rho}{\rho}(r) \propto \frac{1}{r^{\frac{n+3}{2}}} \tag{2}$$

to a homogeneous sphere resolved into 5616 particles of equal mass which initially mimics expansion in an Einstein-de Sitter model universe, i.e. interparticle distances grow as $t^{\frac{2}{3}}$. We choose for n five different values, $n = -3$, $-2.5, -2, -1$, and 1, consistent with an rms profile of $s(r)$ in the hierarchical CDM scenario (e.g. Peebles 1980, Davies et al. 1985, Voglis 1994b).

The particles' positions and velocities during the linear phase of expansion ($\frac{\delta\rho}{\rho} \ll 1$), are evolved analytically (Palmer and Voglis, 1983). The non-linear collapse phase is simulated by the tree-code (Hernquist, 1987) and the slow phase mixing towards equilibrium with a smooth potential code (Allen, Palmer and Papaloizou, 1990, see Voglis 1994a for details).

2. The Fitting Model

For spherical systems, the DF is a composite function of the integrals of the motion i.e. the energy $\mathcal{E} = \Phi(r) - \frac{v^2}{2} = -E$ ($\Phi(r) = -\Psi(r)$) and the angular momentum L^2, and can be written as (Spitzer & Shapiro, 1972):

$$f(\mathcal{E}, L^2) = \frac{N(\mathcal{E}, L^2)}{4\pi^2 T_r(\mathcal{E}, L^2)}. \tag{3}$$

$N(\mathcal{E}, L^2)$ is the 'number density' i.e. the number of stars per unit energy and square angular momentum, and $T_r(\mathcal{E}, L^2)$ is the radial period, i.e. twice the time of motion from pericenter to apocenter:

$$T_r(\mathcal{E}, L^2) = 2\int_{r_p}^{r_a} \frac{dr}{\sqrt{2[\Phi_o(r) - \mathcal{E}] - L^2/r^2}}, \tag{4}$$

In our N-Body experiments $\Phi_o(r)$ was taken equal to the monopole term in the potential expansion of each system.

The model for the distribution function was obtained by combining the separate fits of the number density $N(\mathcal{E}, L^2)$ and of the radial period $T_r(\mathcal{E}, L^2)$.

The inverse of $T_r(\mathcal{E}, L^2)$ measures the density of states of given $\mathcal{E}$ and L^2. For the isochrone (Hénon, 1959) and Keplerian potentials, $T_r \propto \mathcal{E}^{-3/2}$ i.e. independent on L^2. It is found that, in the realistic N-Body potential, $T_r(\mathcal{E}, L^2)$ is independent of L^2 to within an error of only a few percent. T_r is $\propto \mathcal{E}^{-m}$ with $m = 3/2$ for $\mathcal{E}$ small and $2 > m > 1$ for $\mathcal{E}$ large. We used a continuous polynomial interpolation of $\log T_r$ vs. $\log \mathcal{E}$.

The model for the number density $N(\mathcal{E}, L^2)$ is given by an appropriate modification of Lynden-Bell's formula (Eq.1), namely:

$$N(\mathcal{E}, L^2) = \frac{N_h(\mathcal{E}, L^2) + N_c(\mathcal{E}, L^2)}{\exp[\beta(\mathcal{E} - \mathcal{E}_c)] + 1}. \tag{5}$$

The two terms $N_c(\mathcal{E}, L^2)$ and $N_h(\mathcal{E}, L^2)$ in Eq. (5) correspond to a 'core' and a 'halo' population respectively. The core particles are trapped in nearly circular orbits while the halo particles move in elongated orbits.

The denominator of Eq.(5) is the denominator of Eq. 1, if μ is replaced by the circular energy function $\mathcal{E}_c(L^2)$:

$$\mathcal{E}_c(L^2) = \Phi_o(r_c) - \frac{L^2}{2r_c^2}, \tag{6}$$

where $r_c(\mathcal{E}, L^2)$ is the radius of the circular orbit of given $\mathcal{E}, L^2$ in the potential $\Phi_o(r)$. Thus the chemical potential $\mu = \mathcal{E}_c(L^2)$ becomes a single-valued function of the angular momentum (Shu, 1969). The energy of the circular orbit gives a natural chemical potential, i.e. it gives the least work needed in order to make unbound the motion of a particle in circular orbit with angular momentum L^2.

The denominator of equation (5) works as a cut-off of the distribution of particles with $\mathcal{E} > \mathcal{E}_c$. With analogy to the Fermi-Dirac distribution, β is the 'inverse temperature' $\beta = \frac{1}{kT}$. In perfectly spherical systems, no particle is allowed to have $\mathcal{E} > \mathcal{E}_c$, so in Eq. (5) $\beta \to \infty$ and $T \to 0$. Our realistic potential, however, has a small multipole contribution. Thus, β takes a finite value, allowing for some particles to exceed the energy of the circular orbit of the monopole term of the potential.

The 'core' term $N_c(\mathcal{E}, L^2)$ in the numerator of equation (5) is given by:

$$N_c(\mathcal{E}, L^2) = A(\mathcal{E}_c)\mathcal{E}^p. \tag{7}$$

The function $A(\mathcal{E}_c) = e^{b\mathcal{E}_c}/\mathcal{E}_c^p$ can be used (Voglis, 1994a) but we found that $A(\mathcal{E}_c) = const$ is an even better choice. Eq. 7 describes a popytropic core, meaning that the core has not reached a state of complete mixing. An 'isothermal' core corresponds to $p \to \infty$.

The 'halo' term $N_h(\mathcal{E}, L^2)$ in the numerator of (5) was found by fitting the contours of the number density $N(\mathcal{E}, L^2)$, in the region of energies where the halo population dominates, i.e. not very close to the energy of the circular orbit. The contours of any given N_h are nicely fitted by a function of the Gram-Charlier type (Voglis 1994a), of the form:

$$L^2 = L_m^2(N_h)(x^2 - 1/4)\exp(-x^2 + 5/4), \tag{8a}$$

$$x = \frac{\log(\mathcal{E}/\mathcal{E}_m(N_h))}{\sigma(N_h)} + \sqrt{5}/2, \qquad x > 0.5. \tag{8b}$$

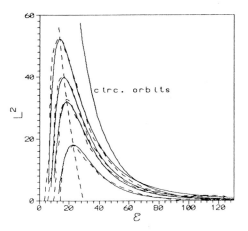

Fig. 1.   Contours of the histogram of $N(\mathcal{E}, L^2)$ at equilibrium after the collapse of the $n = 1$ N-Body experiment. The halo dominates. The dashed curves are the numerical contours while the solid curves are the fittings by Eq. 8a and 8b. The circular energy function $\mathcal{E}_c(L^2)$ is plotted. The dashed line that crosses the points $(\mathcal{E}_m, L_m^2)$ is given by eliminating $y(N_h)$ between the two first equations of (8c).

The contour diagram of $N(\mathcal{E}, L^2)$ for an experiment where the halo population dominates is shown in Fig.1 (dashed lines) while the solid lines represent the family of contours, parametric in N_h, given by Eq. 8a,b. For given N_h, the quantities $\mathcal{E}_m(N_h)$ and $L_m^2(N_h)$ correspond to a characteristic point of the contour at which L^2 takes a maximum value $L^2 = L_m^2$ at the value of energy $\mathcal{E} = \mathcal{E}_m$. The function $\sigma(N_h)$ acts as a dispersion function. Looking for a correlation between N_h and the constants $L_m^2, \mathcal{E}_m^2$, and σ, it was found that the latter are suprisingly well fitted as linear functions of $log(log(N_h))$, noted $y(N_h)$, namely:

$$L_m^2 = L_o^2(P - y(N_h)), \quad log\mathcal{E}_m = log\mathcal{E}_0 + \nu y(N_h), \quad \sigma = \sigma_o - \sigma_1 y(N_h), \quad (8c)$$

(see Figure 7 of Voglis 1994a). Given any pair $(\mathcal{E}, L^2)$, and with the parameters of the linear fits specified, Eqs. 8a,b,c can be inverted to yield the corresponding value of $N_h(\mathcal{E}, L^2)$.

In Fig.2. we plot slices of the numerical histogram of $N(\mathcal{E}, L^2)$ along the energy axis for some fixed values of L^2, for the experiments $n = -3$ (top panel) and $n = 1$ (bottom panel). Because of their different initial density profile, these two experiments are characterized by different initial rates of infall of the matter. In the first experiment all particles have equal infall times. These initial conditions are as close as possible to the conditions for the derivation of Lynden-Bell's statistics. The core formation is thus favoured. In the second experiment the matter collapses gradually and the

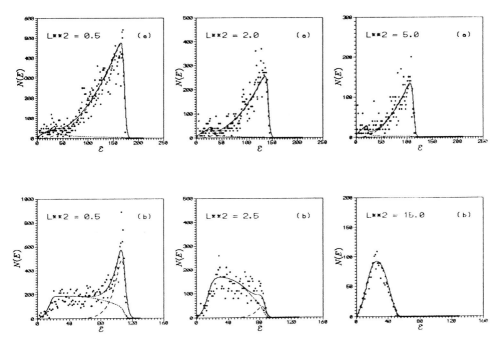

Fig. 2. Slices of $N(\mathcal{E}, L^2)$ for constant angular momentum, for two experiments corresponding to (a) $n = -3$ and (b) $n = 1$. Dots give the N-Body results while solid curves are the model fittings.

halo formation is favoured. The function $N(\mathcal{E}, L^2)$ has different morphology for these two experiments but is nicely fitted by the model if we specify the correct proportion for the core and halo term. Similarly good fits are obtained for the intermediate experiments. The fitting of the distribution function $f(\mathcal{E}, L^2)$, as derived from Eq. 3, is shown in Fig. 3. The model DF falls naturally to zero at $\mathcal{E} = 0$ and $\mathcal{E} = \mathcal{E}_c$, so no cut-off is added by hand as often in simplier (e.g. King) models. A comparison between moments of this model DF and the corresponding numerical data shows an agreement always better than 90%. For example, in figure 4, we plot the numerical vs. predicted spatial density $\rho(r)$ for all the experiments.

The model has three constants in the core term (A, p, β) and six constants in the halo term $(L_o^2, P, \mathcal{E}_0, \nu, \sigma_0, \sigma_1)$. It is emphasized that these are not free constants. Their values, for a particular experiment, are determined by appropriate best fits on the numerical histogram of $N(\mathcal{E}, L^2)$.

It is interesting to compare the above model with simple models used in the literature, such as the Stiavelli-Bertin (1985, SB) model:

$$f(\mathcal{E}, L^2) \propto \mathcal{E}^{3/2} \exp(\beta' \mathcal{E} - |\beta'| \frac{L^2}{2r_a^2}) \qquad (9)$$

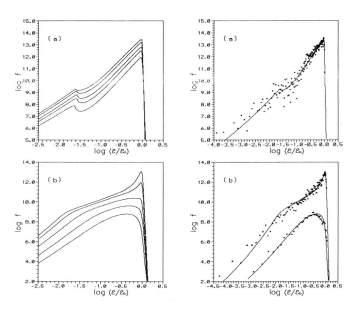

Fig. 3.   Plots of the distribution function $f(\mathcal{E}, L^2)$ for the model (Eqs. 3 to 8): a) several slices of $f(\mathcal{E}, L^2)$ for different constant values of L^2 increasing from top curve to bottom curve (left panel), and a comparison between numerical and model distribution function for constant $L^2 = 0.5$ (right panel), for the $n = -3$ experiment. (b) The same for the $n = 1$ experiment (left panel) and for the constant values $L^2 = 0.7$ and $L^2 = 20.0$ (right panel). Plots (a) and (b) are characteristic to core and halo dominent distribution functions respectively.

According to Merritt et al. (1989), SB models with 'negative temperature' ($\beta' < 0$) better describe the population of stars moving in radial orbits after a cold collapse. Aguilar and Merritt (1990) find that while giving good fits for the density and velocity anisotropy profiles, they produce an excess of particles near $\mathcal{E} = 0$, where the relaxation is not complete (see Fig. 17 of Aguilar & Merritt 1990). This can be understood as follows: assuming $T_r \propto \mathcal{E}^{-3/2}$, in the region of small binding energies, the number density given by Eq.9 is:

$$N(\mathcal{E}, L^2) \propto \exp(\beta'\mathcal{E} - \mid \beta' \mid \frac{L^2}{2r_a^2}) \qquad (10)$$

If $\beta' < 0$, Eq. 10 gives contours of N on the $(\mathcal{E}, L^2)$ plane with negative slope for $0 < \mathcal{E} < \mathcal{E}_c$. On the other hand, the slope of each contour of $N_h(\mathcal{E}, L^2)$ as given by Eq. 6 changes sign at the point $\mathcal{E}_m(N_h)$, $L_m^2(N_h)$ given by Eq. 8a,b. Thus, our model has the flexibility to represent 'negative temperature' models for $\mathcal{E} > \mathcal{E}_m$ and 'positive temperature' models for $\mathcal{E} < \mathcal{E}_m$.

    We have found that the model constants are monotonic functions of the exponent n of the initial density perturbation profile (Efthymiopoulos &

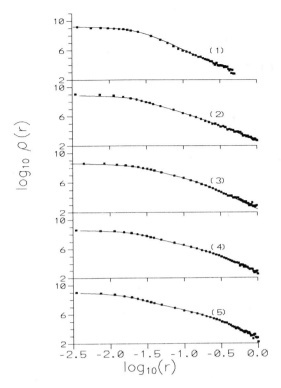

Fig. 4.   Numerical (dot) and model (solid) radial density profiles of the endstates for all the experiments. The numbers 1 to 5 correspond to the sequence of values of $n = -3, -2.5, -2, -1, 1$

Voglis, 1995). This implies that any particular value of n, producing a particular set of initial conditions, can be associated to a unique set of parameter values. Thus, the 'memory' of initial conditions, due to the fact that the system is only partially relaxed, is contained in the parameters of the coarse-grained distribution function of the endstate.

Acknowledgements: We thank Professor G. Contopoulos for presenting this paper, Drs. L. Hernquist, A.J. Allen, P.L. Palmer and J. Papaloizou for providing their codes, and the referee for his helpful comments. C.E. received support by the Greek Foundation of State Scholarships.

References

Aguilar, A., and Merritt, D., 1990, Ap.J., **354**, 33

Allen, A.J., Palmer, P.L., and Papaloizou, J., 1990, MNRAS **242**, 576

Davies, M., Efstathiou, G., Frenk, C.S., and White S.D.M., 1985, Ap.J., **292**, 371

Efthymiopoulos, C., and Voglis, N., 1996 (in preparation)

Hénon, M., 1959, Ann. d'Astrophys., **22**, 126

Hénon, M., 1964, Ann. d'Astrophys., **27**, 83

Hernquist, L., 1987, Ap.J. Suppl., **64**, 715

Lecar, M., and Cohen, L., 1972, in "Gravitational N-Body Problem", M. Lecar (ed.), Reidel, Holland

Lynden-Bell, D., 1967, MNRAS, **136**, 101

May, A., and van Albada, T.S., 1984, MNRAS, **209**, 15

Merritt, D., Tremaine, S., and Johnstone, D., 1989, MNRAS, **236**, 829

Palmer, P.L., and Voglis, N., 1983, MNRAS, **205**, 543

Peebles, P.J.E., 1980, "The Large Scale Structure of the Universe", Princeton University Press.

Shu, F.H., 1969, Ap.J., **158**, 505

Spitzer, L., and Shapiro, S.L., 1972, Ap.J., **173**, 529

Stiavelli, M., and Bertin, G., 1985, MNRAS, **217**, 735

Voglis, N., 1994a, MNRAS, **267**, 379

Voglis, N., 1994b, in "Galactic Dynamics and N-Body Simulations", G. Contopoulos, N.K. Spyrou, and L. Vlahos (Eds), Lecture Notes in Physics, Springer-Verlag

Voglis, N., Hiotelis, N., and Hoflich, P., 1991, A&A **249**, 5

Voglis, N., Hiotelis, N., and Harsoula, M., Astr. Sp. Sci. **226**, 213

Voglis, N., and Efthymiopoulos, C., 1996, preprint

White, S.D.M., 1979, MNRAS, **189**, 831

GLOBAL DYNAMICS OF THE LOGARITHMIC GALACTIC POTENTIAL BY MEANS OF FREQUENCY MAP ANALYSIS

YANNIS PAPAPHILIPPOU AND JACQUES LASKAR
CNRS, Astronomie et Systèmes Dynamiques
Bureau des longitudes
3, rue Mazarine, 75006 Paris, FRANCE

Abstract. By means of Laskar's frequency map analysis, we derive a global vision of the dynamics of the 2 degrees of freedom logarithmic galactic potential, for various values of the perturbation parameter. Some preliminary results on the exploration of the 3 degrees of freedom logarithmic system are also given.

1. Introduction

In the past, the methodology of dynamical systems has been applied with success to systems modelising the motion of stars in galaxies. Nevertheless, these studies present difficulties which are often due to the fact that the majority of galactic systems cannot be treated with perturbation techniques. As a result, the use of the optimal set of action-angle variables of an integrable approximation of the galactic Hamiltonian is rather complex. Although there exist numerical methods which can compute these variables (principally developed in galactic dynamics by Binney and his collaborators [1], [2], [3]), the problem remains unsolved when the perturbation is large. The complications are increased when we pass from two degrees of freedom systems to multi-dimensional ones. In that case, it is quite difficult to visualise the global dynamics of a system in simple graphs.

In this work, a different approach is made by using the frequency map analysis method of Laskar [4], [5], [6], [7], [8], which relies on the computation of a system's more natural feature, its main frequencies of motion. The method is applied here to a widely used galactic model, the logarithmic potential [9].

2. The Method

The first step of the frequency map analysis method is to derive a quasi-periodic approximation

$$f'(t) = \sum_{k=1}^{N} a_k e^{i\omega_k t} \tag{1}$$

of a complex function $f(t)$ which is determined by usual numerical integration. This process relies heavily on the NAFF algorithm developed by Laskar in planetary dynamics [10].

253

J. C. Muzzio et al. (eds.), Chaos in Gravitational N-Body Systems, 253–258.
© *1996 Kluwer Academic Publishers. Printed in the Netherlands.*

Once the algorithm is applied to the numerical solutions of a non-degenerate Hamiltonian system the most interesting aspect of the method can take place, that is the construction of the frequency map [5],[6],[7]. Indeed, out of the quasiperiodic approximations, we just retain the frequency vector $(\nu_1, \nu_2, \ldots, \nu_n)$ which, up to numerical accuracy [8], parametrises the tori of the integrable part of the system. Then, by keeping all the angle-like variables constant, we are able to construct the frequency map :

$$F_T : \mathcal{R}^{n-1} \longrightarrow \mathcal{R}^{n-1}$$
$$(I)_{n-1} \longrightarrow (\nu)_{n-1} \tag{2}$$

which maps the action-like variables $(I)_{n-1} = (I_1, I_2, \ldots, I_{n-1})$ to the rotation vector $(\nu)_{n-1} = (\nu_1/\nu_n, \nu_2/\nu_n, \ldots, \nu_{n-1}/\nu_n)$. The dynamics of the system is then analysed by studying the regularity of this map.

3. The Logarithmic Galactic Potential

This potential is of special importance for galactic dynamics : it represents a rough model of an elliptical galaxy with a bulge of radius R_c and a constant velocity curve at large radii. The Hamiltonian of the system is :

$$H = \frac{1}{2}(X^2 + Y^2 + Z^2) + \ln(x^2 + \frac{y^2}{q_1{}^2} + \frac{z^2}{q_2{}^2} + R_c{}^2) \tag{3}$$

where (x, y, z) are the usual Cartesian coordinates and (X, Y, Z) their conjugate momenta. The softening parameter R_c is kept equal to 0.1 throughout this work. The axial ratios of the equipotential ellipsoids, represented by q_1 and q_2, can be considered as the perturbation parameters.

For specific values of the variables and the parameters the system is integrable :

a) When the motion is restricted on one of the invariant planes $(x, X), (y, Y)$ or (z, Z) the system has only one degree of freedom.

b) When the motion is restricted on one of the 4-dimensional surfaces generated by two of the previous planes the system has 2 degrees of freedom. If, now, q_1 and/or q_2 take specific values (for $q_1 = 1$, $q_2 = 1$ or $q_1 = q_2$) the system becomes integrable as the corresponding component of the angular momentum is a second integral of motion.

c) When $q_1 = q_2 = 1$, two components of the angular momentum in addition to the energy are integrals (spherical system).

4. Two Degrees of Freedom System

Before proceeding to the study of the complete system, we can gain much knowledge from the 2 degrees of freedom system [11], whose Hamiltonian is

$$H = \frac{1}{2}(X^2 + Y^2) + \ln(x^2 + \frac{y^2}{q^2} + R_c{}^2). \tag{4}$$

The parameter q can vary from 1 to q_{cr} (0.6964 in our case), which depends on the energy value h [11]. For $q < q_{cr}$ the density can take negative values and the model looses its physical sense. For values of q close to 1 (integrable case) the phase space of the system is covered with two main types of orbits :

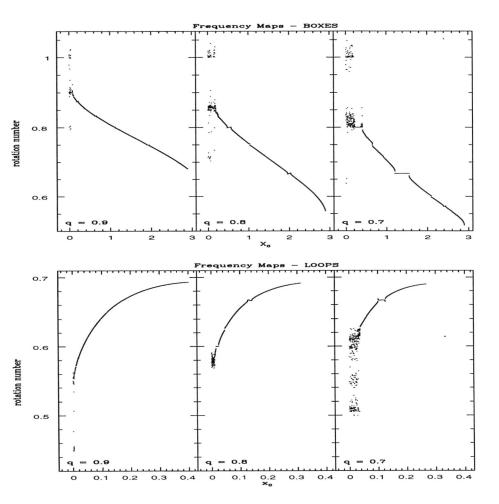

Figure 1. Frequency Maps for box (*a*) and loop orbits (*b*) of the 2D system for different values of the perturbation parameter *q* (galactic cases).

a) boxes which can be thought as perturbations of the rectilinear orbits generated by the 1-degree of freedom system on the invariant planes (x, X) or (y, Y);

b) loops which can be thought as perturbations of the orbits generated by the axissymetric system $(q = 1)$.

Thus, we construct two frequency maps, one for each type of orbit, for various values of the perturbation parameter q and for a fixed energy value $h = 0.4059$ (Fig. 1). These graphs present many interesting features :

First of all, the chaotic regions of the system, are represented by the scattered points and their width can be measured precisely in terms of frequencies or in terms of action-like variables. Moreover, all the important elliptic islands (horizontal plateaux) and hyperbolic points (vertical gaps) can be detected with their actual strength. It is straightforward to recognise to which resonance they correspond. As we follow the evolution of these resonances by changing the perturbation parameter we can easily visualise their overlapping and the diffusion of the corresponding orbits. We should stress that we can also determine the position of almost all the elliptic periodic orbits, even the ones which correspond to very small islands. Thus, the global dynamical structure of the system is represented in these simple graphs.

5. Three Degrees of Freedom System

We give here some preliminary results on the application of the method to the complete system. In that case we cannot explore the phase space by following only one line of initial conditions, as the dimension of the system is magnified, with respect to the 2 degrees of freedom system. Therefore, we integrate 10000 initial conditions on the (X, Y) plane (boxes) and calculate the three frequencies (ν_x, ν_y, ν_z). The global dynamics of the system can be depicted graphically in the complete frequency map, formed by the plane (a_1, a_2), where (a_1, a_2) are the rotation numbers ν_x/ν_y and ν_x/ν_z, respectively (Fig. 2).

The first thing to mention is the existence of 5 main resonant lines ($a_1 = 0$, $a_2 = 0$, $a_1 = 1$, $a_2 = 1$ and $a_1 = a_2$) corresponding to integrable cases to which we referred in section 3. The diagonal $a_1 = a_2$ and the $a_1 = 1$ and $a_2 = 1$ lines correspond to the 2D system previously studied. As the system resides from the two degrees of freedom case, by changing the perturbation parameters, its dynamics becomes less regular. All the vertical and horizontal resonant lines spring from the 2D cases (compare directly with the corresponding maps of Fig. 1 and 2). Lines of the form $\alpha a_1 + \beta a_2 + \gamma = 0$ ($\alpha, \beta \in Z^\star$ and $\gamma \in Z$) are due to the coupling. The intersections of resonant lines are periodic orbits (for example the most visible one in the centre of Fig. 3c corresponds to the 5:4:6 resonance). As expected, the chaotic zones are enlarged with respect to the 2D system and exist even for values of the perturbation parameters close to the integrable cases. When the perturbation is magnified, resonant lines overlap producing a rapid diffusion of orbits.

6. Conclusion

In this work, we presented the first results obtained by applying the frequency map analysis method of Laskar to the logarithmic galactic system. We have shown in the 2D case that all the dynamics of the system is depicted in 1-dimensional frequency curves, gaining much more information than by a simple Poincaré surface of section. All the resonances of some importance are identified as well as the regular and chaotic regions. These last are related to the overlapping of resonances, as the perturbation parameter changes. Finally, we can easily determine the corresponding periodic orbits which appear as the backbone of galactic structure.

On the other hand the preliminary study of the 3D case seems very promising, as the complete frequency maps are a very strong tool for visualising high order systems. Indeed, these graphs correspond directly to the Arnold web of the system.

As we can observe, the addition of another degree of freedom enlarged the chaotic regions of the system. Furthermore, the interaction of all these resonant lines gives rise to rapid diffusion. One of the central questions of galactic dynamics is to what extent we can construct stationary models of galaxies by using triaxial potentials. We expect that for realistic galactic potentials, the existence of chaotic regions should be a necessity and

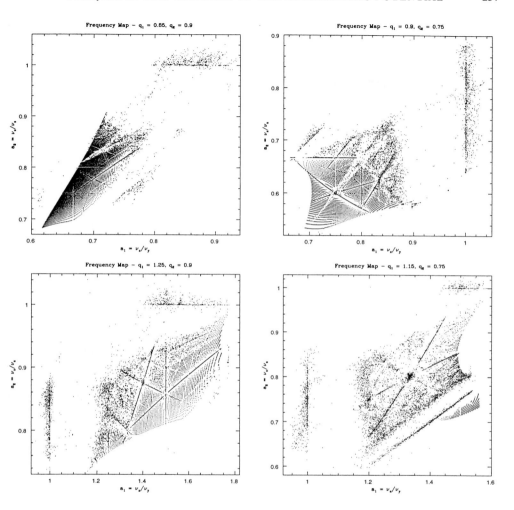

Figure 2.   Complete Frequency Maps for the box orbits of the 3D system for different values of the perturbation parameters q_1, q_2.

should support the self-consistency of the model [12]. The previous remark in addition to the diffusion of galactic orbits are two intriguing questions of galactic dynamics, which we expect to study in the near future.

Acknowledgements

This work was partially supported by a MRES grant, by the CEE contract CHRX-CT94-0460 and by the CNRS.

References

1. McGill, C. and Binney, J. (1990), Torus construction in general gravitational potentials, *MNRAS*, **244**, 634
2. Binney, J. and Kummar, S. (1993), Angle variables for numerically fitted orbital tori, *MNRAS*, **261**, 584
3. Kaasalainen, M. and Binney, J. (1994), Torus construction for potentials supporting different orbit families *MNRAS*, **268**, 1033
4. Laskar, J. (1990), The chaotic motion of the solar system. A numerical estimate of the size of the chaotic zones, *Icarus*, **88**, 266-291
5. Laskar, J., Froeschlé, C. and Celletti, A. (1992), The measure of chaos by the numerical analysis of the fundamental frequencies. Application to the standard map, *Physica D* **56**, 253-269
6. Laskar, J. (1993), Frequency analysis for multi-dimensional systems. Global dynamics and diffusion *Physica D* **67**, 257-281
7. Laskar, J. (1995), Frequency map analysis of a Hamiltonian system, in AIP Conference Proceedings 344, *Non-Linear Dynamics in Particle Accelerators : Theory and Experiments*, Arcidosso Italy 1994, eds. S.Chattopadhyay, M.Cornacchia and C.Pellegrini, AIP Press
8. Laskar, J. (1996), Application of frequency map analysis , in NATO-ASI *Hamiltonian Systems with Three or More Degrees of Freedom*, S'Agaro 19-30 June 1995, eds. C.Simo and A.Delshams, Kluwer, (in press)
9. Binney, J. and Tremaine, S. (1987), *Galactic Dynamics*, Princeton Univ. Press, Princeton N.J.
10. Laskar, J. (1988), Secular evolution of the solar system over 10 million years, *Astron. Astroph.* **198**, 341-362
11. Papaphilippou, Y. and Laskar, J. (1996), Frequency map analysis and global dynamics in a galactic potential with two degrees of freedom, *Astron. Astroph.* (in press)
12. Merritt, D. and Fridman, T. (1996), Triaxial galaxies with cusps, *Astroph.J.*, **456** (in press)

NUMERICAL STABILITY OF OSIPKOV-MERRITT MODELS

ANDRES MEZA* and NELSON ZAMORANO**
Universidad de Chile, Facultad de Ciencias Físicas y Matemáticas,
Departamento de Física, Casilla 487-3, Santiago, Chile

Abstract. Using a series of N-body simulations, we have studied the onset of instabilities in three different models generated with the Osipkov-Merritt distribution function. Contrasting these numerical results with the criterion recently proposed by Hjorth, we have found significative differences. Our results strongly suggest a revision of this criterion.

1. Introduction

Collisionless stellar systems can be described by a phase-space distribution function f which satisfies the coupled Vlasov and Poisson equations

$$\frac{\partial f}{\partial t} + \mathbf{v} \cdot \frac{\partial f}{\partial \mathbf{x}} - \frac{\partial \Phi}{\partial \mathbf{x}} \cdot \frac{\partial f}{\partial \mathbf{v}} = 0, \tag{1}$$

$$\nabla^2 \Phi = 4\pi G \rho \equiv 4\pi G \int f \, d^3\mathbf{v}, \tag{2}$$

where $\Phi(\mathbf{x}, t)$ is the gravitational potential.

According to Jeans' theorem any function of the integrals of motion is a steady state solution of the Vlasov equation (see e. g. Binney and Tremaine 1987). For example, spherically symmetric systems are described by $f = f(E, L)$, where E is the total energy per unit mass and L is the magnitude of the angular momentum per unit mass. However, a general proof for the stability of these solutions is yet to be found.

Recently, a new definition for a macroscopic steady state of a collisionless self-gravitating system in terms of H-functions has been suggested by Hjorth (1994). As a consequence, a necessary and sufficient condition for the stability of any single variable distribution function, $f = f(Q)$, was found: $df/dQ \leq 0$. In particular, this criterion can be applied to models with Osipkov-Merritt distribution functions, taking $Q \equiv E + L^2/2r_a^2$ (Osipkov 1979; Merritt 1985).

The stability criterion proposed by Hjorth (1994) has been applied to discriminate between stable and unstable models. The N-body simulations include the Hernquist (1990) and one of the Dehnen (1993) models (see also Tremaine et al. 1994). For these simulations we have used an N-body code based on the self-consistent field method of Hernquist and Ostriker (1992).

---

* e-mail: ameza@cipres.cec.uchile.cl
** e-mail: nzamora@tamarugo.cec.uchile.cl

J. C. Muzzio et al. (eds.), Chaos in Gravitational N-Body Systems, 259–264.
© 1996 *Kluwer Academic Publishers. Printed in the Netherlands.*

We also have done a set of N-body simulations for the Plummer (1911) model to compare our numerical results with the previous one obtained by Dejonghe and Merritt (1988). In this case we have employed a slightly different N-body code based on the self-consistent field method as given by Clutton-Brock (1973).

2. Numerical Methods

2.1. N-BODY CODE

The N-body simulation method we use is based on the self-consistent approach described by Hernquist and Ostriker (1992). The Poisson equation (2) is solved by expanding the density and potential in a set of basis functions

$$\rho(r,\theta,\phi) = \sum_{nlm} A_{nlm}\rho_{nl}(r)Y_{lm}(\theta,\phi), \tag{3}$$

$$\Phi(r,\theta,\phi) = \sum_{nlm} A_{nlm}\Phi_{nl}(r)Y_{lm}(\theta,\phi). \tag{4}$$

The zeroth order term must be chosen according to the density profile of the model under study and its selection fixes completely the other terms in the basis (see Hernquist and Ostriker 1992; Clutton-Brock 1973).

In their derivation, Hernquist and Ostriker (1992) took the zeroth order basis function to be the Hernquist model (Hernquist 1990), namely

$$\rho_{00} \equiv \frac{1}{2\pi}\frac{1}{r}\frac{1}{(1+r)^3}, \tag{5}$$

$$\Phi_{00} \equiv -\frac{1}{1+r}, \tag{6}$$

where the density and potential are expressed in dimensionless units.

A different set of basis functions is obtained taking as the lower terms the Plummer model (Plummer 1911; Clutton-Brock 1973), defined by

$$\rho_{00} \equiv \frac{3}{4\pi}\frac{1}{(1+r^2)^{5/2}}, \tag{7}$$

$$\Phi_{00} \equiv -\frac{1}{(1+r^2)^{1/2}}. \tag{8}$$

For these two basis sets one of us (AM) has written a code which implements the formulae given by Hernquist and Ostriker (1992). In both codes, the particle positions and velocities are updated using a second order integrator with a fixed time step Δt, given by

$$\mathbf{x}_{i+1} = \mathbf{x}_i + \Delta t \, \mathbf{v}_i + \frac{1}{2} \Delta t^2 \, \mathbf{a}_i, \tag{9}$$

$$\mathbf{v}_{i+1} = \mathbf{v}_i + \frac{1}{2} \Delta t \, (\mathbf{a}_i + \mathbf{a}_{i+1}), \tag{10}$$

where the subscript identifies the time step. We have done several tests to study the numerical properties of this integrator and found that their properties are similar to the standard time centered leap-frog (see Hut, Makino and McMillan 1995).

2.2. Initial Conditions

Dehnen (1993) (see also Tremaine et al. 1994) has obtained a family of models with Osipkov-Merritt distribution functions, starting from following parametric expression for the density

$$\rho(r) = \frac{3 - \gamma}{4\pi} \frac{1}{r^\gamma (r + 1)^{4-\gamma}}, \quad 0 \leq \gamma < 3. \tag{11}$$

The parameter γ determines the slope of the central density cusp. At large radii, all the γ-models have $\rho \propto r^{-4}$. The models of Jaffe (1983) and Hernquist (1990) are recovered for $\gamma = 2$ and $\gamma = 1$, respectively.

A particularly simple model is obtained setting $\gamma = 0$, in this case the density turns to

$$\rho(r) = \frac{3}{4\pi} \frac{1}{(1 + r)^4}. \tag{12}$$

This density profile is the only one among the γ-models that has a (non-isothermal) core. Moreover, this density can be expanded using only two terms of the basis functions given by Hernquist and Ostriker (1992).

To test our code we have done a series of simulations for the Plummer (1911) model, which was already studied by Dejonghe and Merritt (1988) using a multipolar N-body code.

Osipkov-Merritt distribution functions were computed for each of these three density models. Each model is truncated at radius enclosing 99.9% of the total mass. For all the simulations we have employed $N = 5 \times 10^4$ equal mass particles. The potential expansion is truncated at order $n = 4$ and $l = 2$ for the radial and angular functions, respectively. All the runs were evolved using a time step $\Delta t = 0.05$, with a total elapsed time $T_{\text{end}} = 400$.

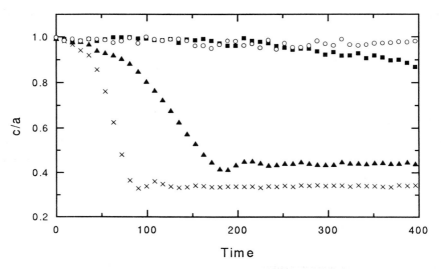

Fig. 1.   Evolution of the axis ratio for the Plummer model. *Circles*: $r_a = 1.2$; *squares*: $r_a = 1.1$; *triangles*: $r_a = 0.9$; *crosses*: $r_a = 0.8$.

3.   Results and Conclusions

For each one of the models we have done several runs using different values for r_a, the anisotropy radius. To provide an accurate description for the spatial distribution of particles, we use an iterative algorithm to obtain the axis ratios of a fitting ellipsoid at a specified radius r_{med}, using the eigenvalues of the modified tensor of inertia

$$I_{ij} = \sum \frac{x_i x_j}{a^2}, \quad a^2 = x^2 + y^2/q_1^2 + z^2/q_2^2. \tag{13}$$

Figure 1 shows the evolution of the axis ratio of the fitting ellipsoid at radius $r_{med} = 2.5$ for the Plummer model. The models with $r_a = \{0.8, 0.9\}$ are strongly unstable and develop a bar. The stability boundary appears to lie near $r_a = 1.1$ and the models with $r_a \gtrsim 1.1$ are stable. Our results agree with those of Dejonghe and Merritt (1988).

The results for the Hernquist model appear at Figure 2, which is a plot of the axis ratio of the fitting ellipsoid at radius $r_{med} = 5$, the radius which encloses $\sim 70\%$ of the total mass. The models with $r_a < 1.0$ are strongly unstable. The stability boundary appears to be close to $r_a = 1.0$ and probably this model is stable.

Figure 3 displays the evolution of the axis ratio for the Dehnen $\gamma = 0$ model. The models with $r_a = \{0.45, 0.75\}$ develop a bar. The model with $r_a = 1.5$ is stable. We conclude that models with $r_a \gtrsim 1.5$ are stable.

A summary of our results are shown in column (3) of Table I. The values obtained using Hjorth's criterion are shown in column (2). In all the cases,

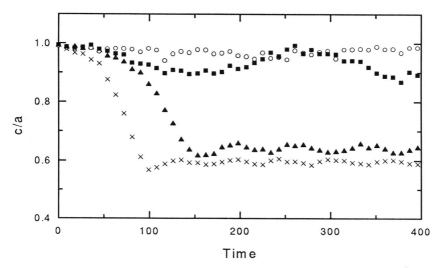

Fig. 2.   Evolution of the axis ratio for the Hernquist model. *Circles*: $r_a = 1.0$; *squares*: $r_a = 0.9$; *triangles*: $r_a = 0.5$; *crosses*: $r_a = 0.3$.

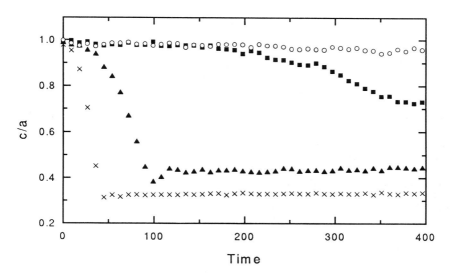

Fig. 3.   Evolution of the axis ratio for the Dehnen $\gamma = 0$ model. *Circles*: $r_a = 1.5$; *squares*: $r_a = 1.4$; *triangles*: $r_a = 0.75$; *crosses*: $r_a = 0.45$.

the numerical values for the stability boundary r_a are different from the one predicted.

TABLE I

Comparison of analytical predictions and numerical results for the stability boundary r_a.

Model	r_a^{theo}	r_a^{num}
Plummer	0.90	1.1
Hernquist	0.52	1.0
Dehnen $\gamma = 0$	0.75	1.5

Acknowledgements

We are grateful to David Merritt for constructive criticism of an early version of this work. AM was supported by a grant from CONICYT, FONDECYT 2950005 and Depto. Postgrado y Postítulo-Universidad de Chile: Beca PG12295. NZ was partially supported by DTI E-3646-9312 and FONDECYT 1950271.

References

Binney, J. J. and Tremaine, S. D.: 1987, *Galactic Dynamics*, Princeton University Press: Princeton

Clutton-Brock, M.: 1973, *Astrophysics and Space Science* **23**, 55

Dehnen, W.: 1993, *Monthly Notices of the RAS* **265**, 250

Dejonghe, H. and Merritt, D.: 1988, *Astrophysical Journal* **328**, 93

Jaffe, W.: 1983, *Monthly Notices of the RAS* **202**, 995

Hernquist, L.: 1990, *Astrophysical Journal* **356**, 359

Hernquist, L. and Ostriker, J. P.: 1992, *Astrophysical Journal* **386**, 375

Hjorth, J.: 1994, *Astrophysical Journal* **424**, 106

Hut, P., Makino, J., and McMillan, S.: 1995, *Astrophysical Journal, Letters to the Editor* **443**, 93

Merritt, D.: 1985, *Astronomical Journal* **90**, 1027

Osipkov, L. P.: 1979, *Pis'ma Astr. Zh.* **5**, 77

Plummer, H. C.: 1911, *Monthly Notices of the RAS* **71**, 460

Sygnet, J. F., Des Forets, G., Lachieze-Rey, M. and Pellat, R.: 1984, *Astrophysical Journal* **276**, 737

Tremaine, S. D., et al.: 1994, *Astronomical Journal* **107**, 634

STRUCTURE OF GALACTIC SATELLITES

LILIA P. BASSINO *, JUAN C. MUZZIO * and FELIPE C. WACHLIN *
Facultad de Ciencias Astronómicas y Geofísicas, Paseo del Bosque s/n,
1900–La Plata, Argentina

Abstract. We use numerical simulations to study the interaction of a galactic satellite, represented by a King model, and a galaxy, represented by the potential of a singular isothermal sphere. For circular orbits, we show that tidal stresses affect the shapes of galactic satellites, leading to prolate figures with their longest axis in the direction to the galactic centre.

Key words: Galactic satellite motion – tidal stresses – satellite deformation – numerical experiments

1. Introduction

It is well known that globular clusters are not spherical, but show different degrees of ellipticity. These deviations from sphericity in such stellar systems (oblate, prolate or more general triaxial figures) may be related to internal rotation, tidal stresses, velocity anisotropy, and so on (see Muzzio, 1995 and references therein).

Heggie and Ramamani (1995) have recently studied, using numerical methods, the interaction between tidally truncated clusters, simulated with King's models where the energy is replaced by Jacobi's integral, and a galaxy which is represented by two different potentials: a point–mass and a disk model. They concluded that the clusters are triaxial, with their longest axes in the direction to the galactic center; moreover, in the case of the point–mass model the axial ratios suggest prolate figures. They also point out that the central regions of the clusters do not differ much from the central regions of the corresponding King's models.

We want to extend the work of Heggie and Ramamani by means of numerical simulations, and to study the deviations from sphericity of the satellites due to their interaction with the galaxy. In particular, we want to find out whether the tidal forces do contribute to the observed shapes of the satellites or if their effect can be neglected, as some authors have done when studying globular clusters' flattening (e.g., Geyer, Hopp and Nelles, 1983; Han and Ryden, 1994). In the present investigation we will only consider circular orbits and King's models; future work will include elliptical orbits and velocity anisotropy (Michie models).

* Consejo Nacional de Investigaciones Científicas y Técnicas, Argentina

J. C. Muzzio et al. (eds.), Chaos in Gravitational N-Body Systems, 265–269.

2. The Models

In our models the galaxy was represented by the fixed potential of a singular isothermal sphere, with a mass of 8.9×10^{11} solar masses, truncated at a radius of 44 kpc, according to the values obtained by Innanen et al. (1983) from observed data of globular clusters. The satellite was modeled as a King sphere made up of 50,000 particles. We adopted for the satellite the observed parameters given in the list of Webbink (1985) for the globular cluster NGC 4833: a mass of 216,600 solar masses, a tidal radius of 33.8 pc and a concentration $C = log(r_t/r_c) = 1.24$. From the equation that gives the tidal radius of a system that orbits within an isothermal sphere we obtained an orbital radius (called p) for the satellite of 2.7 kpc.

The values of the mass and the tidal radius of the satellite were the same for all the experiments. We carried out other simulations for a more concentrated satellite (C=1.86) and a less concentrated one (C=0.59), as well as for two shorter orbital radii (2.1 and 1.5 kpc).

As in our previous paper (Bassino et al. 1994), we followed the evolution of the models with the N–body code of Luis Aguilar, which uses a multipolar expansion to compute the potential of the satellite, and which was modified by us to include the effect of dynamical friction (via Chandrasekhar's equation). All the experiments were run on a Hewlett–Packard 9000–735 Apollo workstation and the conservation of energy (after accounting for the dynamical friction loss) was always better than 10^{-4}.

Adopting a unit length equal to the core radius of the satellite, we took the softening parameters as 0.01 for the interaction among the particles and as 0.10 for that between each particle and the central one (which has a mass 10 times higher than the rest of them).

3. Results

3.1. SEMI–AXIS RATIOS OF THE MOMENT OF INERTIA TENSOR

We present the results of our simulations in Table I. After allowing the satellite to evolve for ten orbital periods around the galaxy, we computed the semi–axes of the moment of inertia tensor of the distribution of bound particles; that is, only particles with negative energy were considered. One must be aware that our estimator favours the outer layers, because the moments of inertia go with the square of the distance to the center.

In the table, roman numerals were assigned to the models according to their orbital radius, followed by the letters "a","b" or "c" according to their concentration. Then, to study the ellipticity of the systems, we considered the semi–axis ratios a/c and b/c, where the longest axis is in the direction to the galactic center. At first sight they seem to be triaxial, as a/c is larger than b/c which, in turn, is larger than one; but the b/c ratios are all very close to one, which suggests that the figures tend to be prolate. The sixth column gives the fraction of particles lost during the 10 periods; the loss is higher as we get nearer to the galactic centre and, for the shorter orbital radius, the model with the lower concentration (model IIIa) disintegrated completly after 8 periods. The seventh column shows the percentage of particles lost between the ninth and tenth periods, to provide some idea of how stable the satellites are: the satellites are losing less particles in that interval than the average value over the ten periods, except for model IIa which was still losing particles at a higher rate than the rest; we then continued its evolution for 5 periods more, and the proportion of lost particles during these 5 periods was 2.4 %, indicating a trend towards a stable state too. We also computed the axis ratios at the end of the 8th and 9th period to obtain a root mean square error for both axis ratios of about 0.01.

TABLE I

Semi–axis ratios of the moment of inertia tensor and ellipticities

Model	p [kpc]	C	a/c	b/c	% lost part.	% lost part. (9-10 P)	ϵ [45% r_t]
Ia	2.7	0.59	1.12	1.06	4	0.3	0.03 ± 0.05
Ib	2.7	1.24	1.10	1.00	6	0.3	0.06 ± 0.04
Ic	2.7	1.86	1.11	1.01	4	0.5	0.10 ± 0.06
IIa	2.1	0.59	1.24	1.03	16	3.0	0.04 ± 0.03
IIc	2.1	1.86	1.25	1.03	31	1.6	0.16 ± 0.03
IIIa	1.5	0.59			disruption		
IIIb	1.5	1.24	1.12	1.04	22	0.8	0.12 ± 0.03

3.2. ELLIPTICITIES

In order to study the ellipticities ($\epsilon = 1 - b/a$) from a different point of view, we generated images projecting the satellites onto the orbital plane (after 10 periods of evolution), again only including the bound particles. We then fitted to these images elliptical isodensity contours by means of the task "ellipse" of the IRAF software for image processing. In this task, which belongs to the package designed for the Hubble Space Telescope data, the image is measured using an iterative method (Jedrzejewski, 1987) that fits elliptical isophotes to galaxies. The results are shown in the last column of Table I; the errors are rather large, but they are probably overestimated as they are obtained from the residual intensity scatter and the internal error of the harmonic fit, and we must take into account that the algorithm was

TABLE II

Ellipticity variation with radius
(Model Ib, after 10 P)

r [% r_t]	r [r_c]	ϵ
20	3	0.03 ± 0.02
30	5	0.04 ± 0.03
45	8	0.06 ± 0.04
67	11	0.15 ± 0.04

devised for real galaxies while we are applying it to theoretically modelled satellites.

We can see two trends from Table I: for a given orbital radius, the ellipticity is higher as the satellites are more concentrated, and for a given concentration the ellipticity is higher as we get nearer to the galactic centre.

3.3. ELLIPTICITY VARIATION WITH SATELLITE'S RADIUS

We also analyzed the variation of the ellipticity with respect to the distance to the satellite centre. Table II corresponds to model Ib, the one with orbital radius equal to 2.7 kpc and the intermediate concentration; we computed the ellipticities at four different distances from the satellite's centre, given as a percentage of the tidal radius in the first column and in units of the core radius in the second one. We can see a clear trend from the table: the ellipticities are higher as we consider larger distances from the satellite centre. All the models showed a similar behaviour.

3.4. ELLIPTICITY VARIATION WITH PROJECTIONS ONTO DIFFERENT PLANES

Up to now we have considered projections onto the orbital plane, but it is interesting to study the projections onto the other planes. We chose the model IIc (the one with intermediate orbital radius and higher concentration), let it evolve for 20 periods and then projected it onto the orbital plane (x,y), the plane (x,z) which is normal to the orbital plane and normal to the direction to the galactic center, and the plane (y,z) which is normal to the orbital plane and contains the direction to the galactic centre. Table III shows the ellipticities for two different distances from the center of the satellite, corresponding to 30 and 45 core radii, respectively. We confirm the trend to get values nearer to zero for the projections onto the (x,z) plane than for those onto the other planes; this agrees with the results from the

TABLE III

Ellipticity variation with projection on
to different planes (Model IIc, after 20 P)

Projection	$\epsilon\ [45\%\ r_t]$	$\epsilon\ [67\%\ r_t]$
(x,y)	0.12 ± 0.03	0.18 ± 0.05
(x,z)	0.03 ± 0.03	0.07 ± 0.08
(y,z)	0.15 ± 0.03	0.22 ± 0.05

tensor of inertia calculations, suggesting that the shape of the satellites is essentially prolate.

4. Conclusions

The semi–axis ratios of the tensor of inertia and the ellipticities computed from images of projections of the satellite (both including only bound particles) led to the conclusion that the satellites are prolate figures with the longest axis in the direction to the galactic centre.

The general trends shown by the ellipticities are that they increase with decreasing orbital radius, with increasing concentration and with increasing distance from the center of the satellite.

Thus, by means of simple models (circular orbits and King spheres), we have shown that tidal stresses do affect the shapes of galactic satellites. Even in the case that the outer layers of the clusters are too faint and cannot be observed, one must keep in mind that they should be affecting the dynamics of the inner layers, that we do observe, and their effects cannot be ruled out.

References

Bassino, L. P., Muzzio, J. C. and Rabolli, M.: 1994, *Astrophys. J.* **431**, 634

Geyer, E. H., Hopp, U. and Nelles, B.: 1983, *Astron. Astrophys.* **125**, 359

Han, C. and Ryden, B. S.: 1994, *Astrophys. J.* **433**, 80

Heggie, D. C. and Ramamani, N.: 1995, *Mon. Not. Roy. Ast. Soc.* , in press

Innanen, K. A., Harris, W. E. and Webbink, R. F.: 1983, *Astron. J.* **88**, 338

Jedrzejewski, R. I.: 1987, *Mon. Not. Roy. Ast. Soc.* **226**, 747

Muzzio, J. C.: 1995, 'Sinking, tidally stripped, galactic satellites' in S. Ferraz–Mello and J. C. Muzzio, ed(s)., *Chaos in Gravitational N–Body Systems*, Kluwer Acad. Publ.:Dordrecht, this volume

Webbink, R. F.: 1985, 'Structure parameters of galactic globular clusters' in J. Goodman and P. Hut, ed(s)., *IAU Symp. No.113, Dynamics of Star Clusters*, D. Reidel Publ. Comp.:Dordrecht, 541

DYNAMICAL FRICTION EFFECTS
ON SINKING SATELLITES

SOFÍA A. CORA, JUAN C. MUZZIO and M. MARCELA VERGNE
Facultad de Ciencias Astronómicas y Geofísicas de la Universidad Nacional de La Plata,
and Programa de Fotometría y Estructura Galáctica del Consejo Nacional de
Investigaciones Científicas y Técnicas

Abstract. We investigated the orbital decay, caused by dynamical friction, of a rigid extended satellite which moves within a larger spherical stellar system (a galaxy). In some of our experiments the satellite and the particles that make up the galaxy interact in a fully self consistent way, while in others the the self consistency of the particles that make up the galaxy is neglected. In our simulations the galaxies were represented by Plummer spheres ($\rho \propto r^{-5}$) with isotropic velocity distribution. For satellites we used similar models moving along elongated orbits, like those of galactic satellites. Satellites of 1, 4 and 9% of the galaxy mass were considered. We found very good agreement between our numerical results and theoretical predictions obtained from a straightforward application of Chandrasekhar's dynamical friction equation. We obtained approximate values of the Coulomb logarithm from the best possible fit between numerical and theoretical results. Finally, we investigated the evolution of the structure of the galaxy as a result of the orbital decay of the satellite.

Key words: dynamical friction – galaxies – galactic satellites – orbital decay

1. Introduction

When a body moves through a medium of (usually) smaller particles, it suffers a deceleration due to dynamical friction. Chandrasekhar (1943) derived an equation to compute such deceleration considering a point mass moving through an infinite uniform medium of identical particles which do not interact with each other. The orbital decay of galactic satellites (i.e., globular clusters or dwarf galaxies) is governed by dynamical friction and it is present in several cases of astrophysical interest such as the spiralling of globular clusters toward the galactic center, the fate of the Magellanic Clouds and galactic cannibalism (see, e.g., Binney and Tremaine, 1987). As numerical simulations became a regular tool of stellar dynamics, several investigations were undertaken to check the validity of Chandrasekhar's analytical results. Unfortunately, some contradictory results have arisen from different studies. White (1978), using a self-consistent code, and Lin and Tremaine (1983), who developed a non-self-consistent code to investigate the orbital decay of a galactic satellite, concluded that Chandrasekhar's equation was adequate to describe this phenomenon in most circumstances. White (1983) used self-consistent and non-self-consistent codes to investigate the same problem, and concluded that dynamical friction was due to a *global* response of the galaxy, and not to *local* effects, such as those considered by Chandrasekhar

J. C. Muzzio et al. (eds.), Chaos in Gravitational N-Body Systems, 271–277.

(1943). His results disagree with those obtained by Bontekoe and van Albada (1987), who concluded that dynamical friction is a local effect, well described by Chandrasekhar's dynamical friction formula. Finally, Zaritsky and White (1988) found that this disagreement was due to subtle aspects of the numerical codes; they concluded that the local description is usually adequate. Later on, this result was challenged by Hernquist and Weinberg (1989).

In the present investigation, we explore the effect of the dynamical friction on a galaxy-satellite system, considering the satellite as a single rigid body. For this purpose we use self-consistent and non-self-consistent simulations. We also analyze the characteristics of the galaxy response as a result of the orbital decay process.

2. Numerical Experiments

We performed several N-body simulations to study the effect of dynamical friction on a rigid satellite passing through a self-gravitating galaxy. The equations of motion were integrated using the program N-BODY2 (Aarseth, 1985).

2.1. THE GALAXY

We modelled four spherical galaxies, one made up of 10,000 particles and the others of 5,000 particles (generated with different seed numbers). The galaxies were initially represented by stellar polytropes of index n=5 (Plummer law), with isotropic velocity distribution (Binney & Tremaine, 1987). The density distribution for that model is given by

$$\rho(r) = \frac{3\alpha}{4\pi G(1 + r^2/\varepsilon^2)^{5/2}} \tag{1}$$

being α a constant related to the central density and ε is the scale length parameter of the galaxy. We truncated the galaxies at a radius that comprises 99.99 per cent of their initial total mass M.

The system of units adopted is such that $G = M = 1$. The internal energy, E, is equal to -0.5 in order to have a crossing time equal to unity. In these conditions, $\alpha =39.1414$ and $\varepsilon =0.2945$. The equations used to obtain these values do not take into account the limiting radius introduced and the fact that the particles interact through a softened potential, so that the generated galaxies may be not exact equilibrium configurations. Due to the small softening parameter and large limiting radius adopted, the departures should be very small, indeed, but for extra safety, the galaxies were allowed to evolve in isolation for a few crossing times before using them in our experiments.

2.2. The Satellite

The satellite was modelled as an extended object with a density distribution given by the Plummer law. We considered experiments with satellites with masses m_s =0.01, 0.04 and 0.09, and corresponding softening parameters ε_s =0.03, 0.06 and 0.09.

The initial position of the satellite coincides with its apocentric distance at the start of the simulation. Because of the dependence of the dynamical friction on the mass of the satellite, the more massive satellites have a larger initial distance from the center of the galaxy than the less massive one. Their initial velocities were chosen in such a way that they have orbits similar to those of the constituent particles of the galaxy.

3. Theoretical Model

We obtained an analytical estimate of the deceleration of the satellite introducing the velocity distribution of the Plummer sphere in the formula of Chandrasekhar. The fact that the galaxy is not an infinite homogeneous medium was accounted for through the maximum distance cutoff which, in our case, was adopted as the scale length of the galaxy (ε); besides, White (1976) showed how to allow for the extension of the satellite (Chandrasekhar had considered a point mass only) and, taking both effects into account, we ended up with the following expresion for the Coulomb logarithm:

$$ln(\Lambda) = \frac{1}{2} \left[ln\left(1 + \frac{\varepsilon^2}{\varepsilon_s^2}\right) - \frac{(\varepsilon^2/\varepsilon_s^2)}{1 + (\varepsilon^2/\varepsilon_s^2)} \right] \qquad (2)$$

The formula so obtained was used to compute *theoretical* orbits for the different initial conditions of our experiments. The value of $\varepsilon/\varepsilon_s$ ($\equiv \Lambda'$) depends on the softening parameter of the satellite (10, 5 and 3.333 for $m_s = 0.01$, 0.04 and 0.09, respectively), but these are initial test values only, because they are later on changed in order to obtain *theoretical* orbits similar to the orbits obtained from the numerical experiments. When choosing a suitable parameter to compare the analytical results with those from the numerical simulations, we found that the precise location of the center of the galaxy caused some trouble. Although, as we considered elongated orbits, we could not use just the distance of the satellite to the center of the galaxy, we noticed that even the angular momentum of the satellite was not a good choice for a comparison parameter because it is much influenced by the definition of the galactic center. In the end, we found that the relative energy of the satellite with respect to the galaxy was a much better comparison parameter, because it is independent of the definition of the galactic center

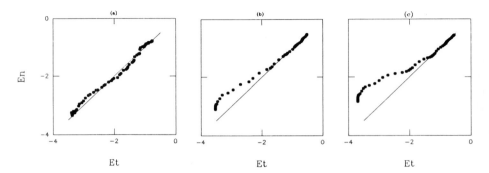

Fig. 1.   Numerical vs. theoretical energies of the satellite (m_s =0.01 (a), 0.04 (b) and 0.09 (c)) with respect to a galaxy of 5,000 particles. These plots show a reasonably good fit at all times for the least massive satellite, and a good fit at the first time steps for the more massive ones, when they are far from the center of the galaxy and the structural changes induced in it are negligibly small.

and only depends on the relative velocity of the satellite with respect to that center. A caveat is needed here: the potential energy was computed from all the particle-satellite interactions for the numerical experiments, but only considering the satellite as a point mass within the field of the galaxy for the analytical solutions; the two energies we compare are thus not exactly the same but, as will be noticed from the figures we will show later, they correlate exceedingly well. Besides, the relative energy computed from the numerical simulations can be affected by changes in the galactic mass distribution, and not only by the location of the satellite; as we will see, those changes are negligible for the smallest satellite, so that they do not affect the comparison and, alternatively, very significant for the more massive satellites turning the comparison irrelevant, no matter how it is actually performed. For each model, we compared plots of the *numerical* energy of the satellite versus the *theoretical* one at every time. Choosing different Λ' values for the theoretical models, we obtained the observed values (Λ'_o) as those that provide the best fits. The criteria for the best fit were different for the models with different satellite masses (Figs. 1a., 1b. and 1c.), because the more massive satellites induce structural changes in the galaxy (it absorbs the orbital energy released by the satellite and reacts expanding itself). This expansion becomes evident when plotting the evolution of the radii that contain different percentages of the mass of the galaxy (Fig. 2).

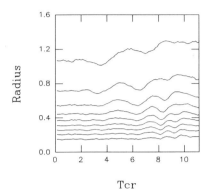

Fig. 2. Evolution of the 10, 20, ..., and 90 per cent mass shells of the galaxy with a satellite of m_s =0.09. The inner shells are altered only slightly, but the outer ones clearly expand their radii.

4. Results

We obtained twenty eight theoretical models with the values of Λ'_o that provide the best fit to the corresponding numerical models. They can be separated into four groups depending on the number of particles used for the galaxy in the numerical experiments and on the mass of the satellite. The groups *I*, *III* and *IV* have 5,000 particles and m_s =0.01, 0.04 and 0.09 respectively, and the group *II* has 10,000 particles and m_s =0.01.

We computed for each group the mean value of Λ'_o, $<\Lambda'_o>$ with their corresponding standard deviation around the mean, and the sample standard deviation, σ_{n-1}. Thus we have:

$$< \Lambda'_o >= 9.67 \pm 0.85 \quad and \quad \sigma_{n-1} = 2.96 \quad for \quad group \ I$$

$$< \Lambda'_o >= 10.00 \pm 0.25 \quad and \quad \sigma_{n-1} = 0.43 \quad for \quad group \ II$$

$$< \Lambda'_o >= 5.78 \pm 0.15 \quad and \quad \sigma_{n-1} = 0.44 \quad for \quad group \ III$$

$$< \Lambda'_o >= 3.26 \pm 0.07 \quad and \quad \sigma_{n-1} = 0.13 \quad for \quad group \ IV$$

The values given above indicate that the mean values of Λ'_o are very similar to the corresponding theoretical values (Λ'_t= 10, 5, 3.333 for m_s= 0.01, 0.04, 0.09 repectively). It must be recalled that in our simulations we considered a rigid satellite, not allowing for the erosion and evolution of it; the inclusion of these effects might introduce further differences between the theoretical and observed values of Λ' (Prugniel and Combes, 1992). The agreement between analytical and numerical results is very good, indeed, despite the fact that we are considering an inhomogeneous finite galaxy rather than Chandrasekhar's homogeneous infinite medium, and one may wonder whether it is fortuitous, as suggested by White (1983). We think that it is not, and that the reason for the good agreement lies in the fact that the main contribution to the friction comes from relatively nearby particles, as stressed by the need of a maximum distance cutoff. In that way, only the local parameters (density, velocity dispersion) matter, and it is not much relevant whether the system within which the satellite moves is inohomogeneous or finite.

We also performed some numerical experiments using a non-self-consistent code in order to check the effect of the self-consistency of the galaxy on the adopted values for Λ'_o. We obtained twenty seven non-self consistent models with the same initial conditions as those of the groups of models I, II, III and IV. From them, we obtained:

$$< \Lambda'_{\rm non} >= 9.44 \pm 0.91 \quad and \quad \sigma_{\rm n-1} = 3.15 \quad for \quad group \; I_{\rm non}$$

$$< \Lambda'_{\rm non} >= 9.25 \pm 0.50 \quad and \quad \sigma_{\rm n-1} = 0.87 \quad for \quad group \; II_{\rm non}$$

$$< \Lambda'_{\rm non} >= 5.75 \pm 0.23 \quad and \quad \sigma_{\rm n-1} = 0.70 \quad for \quad group \; III_{\rm non}$$

$$< \Lambda'_{\rm non} >= 3.44 \pm 0.11 \quad and \quad \sigma_{\rm n-1} = 0.19 \quad for \quad group \; IV_{\rm non}$$

The mean values of Λ'_o and $\Lambda'_{\rm non}$ do not show any statistically significant differences; therefore, we can conclude that the self-gravity is irrelevant for the dynamical friction process.

4.1. LOCAL OR GLOBAL RESPONSE?

The fact that the orbital decay can be well described by Chandrasehar's formula, implies that the decay process through dynamical friction can be characterized as a local process. To verify this, we analyzed both the set of self-consistent and non-self-consistent models. The latter can be expected to have density distortions, due to the effect of the satellite, less pronounced than the former. If this were so, we could conclude that the global response would be unimportant, since in both cases the satellite decays in a similar way. Many attemps were done to check this effect. First, the distorsions in the mass distribution of the galaxy caused by the satellite were measured by a one-dimensional Fourier analysis of the azimutal distribution of the particles projected on the orbital plane of the satellite. The coefficients obtained indicate that there is no significant difference between the cuadrupolar responses of the self-consistent and non-self-consistent galaxies. The same result was obtained when comparing the results for galaxies with and without satellites.

In view of the failure of the foregoing analysis to allow any conclusion, we run another set of simulations considering two satellites diametrically opposed. In this case the galaxy response should be altered with respect to the model with only one satellite, and since the decay rate of the satellites were very similar in these two cases, we tentatively conclude that the global response of the galaxy is not important.

5. Conclusion

We performed a large number of self and non-self-consistent simulations of the dynamical friction effects on a satellite moving within a galaxy. Our main conclusions can be summarized as follows:

(i) The structure of the galaxy is considerably altered by the energy and angular momentum released by the more massive satellites ($m_s = 0.04$ and 0.09). The structural changes are negligibly small for the least massive satellite ($m_s = 0.01$). For them, the Chandrasekhar's formula gives very good results.

(ii) The self gravity and the global response of the galaxy play an unimportant role in the dynamical friction process.

Acknowledgements

We are very grateful to S.D. Abal de Rocha, M.C. Fanjul de Correbo, R.E. Martínez and H.R. Viturro for their technical assistance. Special thanks go to an anonymous referee for his very detailed comments that helped to improve the present paper. This work was supported by grants from the Fundación Antorchas and from the CONICET.

References

Aarseth, S. J.: 1985, '' in J. U. Brackbill and B. I. Cohen, ed(s)., *Multiple Time Scales*, Academic Press: New York, 377
Binney, J. and Tremaine, S.: 1987, *Galactic Dynamics*, Princeton Univ. Press: Princeton
Bontekoe, Tj. R. and van Albada, T. S.: 1987, *M.N.R.A.S.* **224**, 349
Chandrasekhar, S.: 1943, *ApJ* **97**, 255
Hernquist, L. and Weinberg, M. D.: 1989, *M.N.R.A.S.* **238**, 407
Lin, D. N. C. and Tremaine, S.: 1983, *ApJ* **264**, 364
Prugniel, Ph. and Combes, F.: 1992, *A&A* **259**, 25
White, S. D. M.: 1976, *M.N.R.A.S.* **174**, 467
White, S. D. M.: 1978, *M.N.R.A.S.* **184**, 185
White, S. D. M.: 1983, *ApJ* **274**, 53
Zaritsky, D. y White, S. D. M.: 1988, *M.N.R.A.S.* **235**, 289

GRAVITATIONAL SCATTERING EXPERIMENTS IN INFINITE HOMOGENEOUS N-BODY SYSTEMS

TOSHIYUKI FUKUSHIGE
Department of Earth Science and Astronomy,
College of Arts and Sciences, University of Tokyo,
3-8-1 Komaba, Meguro-ku, Tokyo 153, Japan

Abstract.
I performed a series of numerical experiments of the gravitational scattering in infinite homogeneous N-body systems. The infinite homogeneous system is expressed using periodic boundary condition. In the homogeneous systems we can obtain results more straightforword than in spherical systems. Firstly, I investigated upper cutoff of impact parameter in the Coulomb logarithm. There has been two arguments that the upper cutoff should be system size (R) and it should be mean particle distance ($RN^{1/3}$). I obtained the result that the upper cutoff is roughly equal to the system size R. Secondly, I investigated exponential divergence of initial small difference. I confirmed that this exponential divergence saturates at a scale of $\sim RN^{-1/2}$.

1. Upper Cutoff of Coulomb Logarithm

In order to estimate the total effect of two-body encounters the magnitude of the Coulomb logarithm has to be estimated. The Coulomb logarithm has a parameter, which is the upper cutoff, b_{max}, of the impact parameter of the encounter.

The correct determination of the upper cutoff, b_{max}, is important because of the following two reasons. Firstly, b_{max} gives correct evaluation of strength of effects associated with two-body encounter, such as two-body relaxation and dynamical friction. Secondly, b_{max} gives the relative importance of near and distant encounters in the two-body encounters.

Until recent there has been two arguments on this upper cutoff b_{max}. Cohen, Spitzer and Routly (1950), and Farouki and Salpeter (1982,1994) argued that it is roughly equal to system size ($\sim R$). They preferred this choice even though "many-body effects" influence scattering events at impact parameter over interparticle distance.

On the other hand, Chandrasekhar(1941), Kandrup(1980), and Smith(1992) argued that the upper cutoff b_{max} should be roughly equal to an interparticle distance ($\sim RN^{-\frac{1}{3}}$). They argued it at least in the sense that there is no ambiguity concerning the appropriateness of the "independent two-body encounters" approximation.

In order to determine the upper cutoff, I performed numerical experiments of the gravitational scattering in the infinite homogeneous N-body system. My evaluation of the Coulomb Logarithm is more straightforward than that by previous researchers (Farouki and Salpeter 1984, 1994; Smith

J. C. Muzzio et al. (eds.), Chaos in Gravitational N-Body Systems, 279–284.
© *1996 Kluwer Academic Publishers. Printed in the Netherlands.*

1992) in the following two points. Firstly, I evaluated it by calculating change in velocity. In the homogeneous system the change in velocity only due to two-body encounter can be obtained. The previous researchers evaluated it using time scale of mass segregation and cumulative squared changes in orbital energies. Secondly, I artificially changed the upper cutoff of the gravitational interaction using a truncated Coulomb potential and Ewald summation (Ewald 1921), and calcualted the change in velocity. The previous researchers indirectly evaluated it by changing the number of particles.

I calculated the changes in velocities of test particles moving through infinite homogeneous distribution of field particles. I found that the upper cutoff, b_{max}, is roughly equal to the system size, R. In section 1.1 I desribe the model of my numerical experiments and in section 1.2 I show the results of experiments.

1.1. MODEL

I calculated the orbits of the test particles moving through infinite homogeneous distribution of the field particles. The field particles are randomly distributed in a cube, and fixed. In order to express the infinite distribution I used periodic boundary condition. I used the system of units in which $L = M = G = 1$, where L is the length of a side of the cube, M is total mass of the system and G is the gravitational constant. I used truncated Coulomb potential in order to change artificially the maximum impact parameter, b_{max}. This potential is given by

$$\phi(r) = -\frac{m[1 - \mathrm{erf}(r/\eta)]}{r} \tag{1}$$

where m is the mass of the field particle, $\mathrm{erf}(r)$ is the error function and η is a cutoff parameter. I also used the Ewald method (Ewald 1921) in order to calculate the Coulomb potential under periodic boundary condition. I integrated the orbits of the test particles using Runge-Kutta method with automatically adjustment of time step (Press et al 1986). In order to accelerate force calculations, I used a special purpose computer of many-body simulation: MD-GRAPE(Fukushige et al. 1995).

The number of the test particles N_t is 1000. The number of the field particles N_f are 1024 and 8192. I set the mass of the field particles to be equal each other. I softened potential at small r in order to fix the minimum impact parameter. In my softening law, the distance r and parameter η in equation (1) are replaced with $\sqrt{r^2 + \varepsilon^2}$ and $\eta + \varepsilon$, respectively, where ε is the softening parameter and was set to be $1/128$. I performed experiments up to 10 time unit and sampled change in velocity each 0.5 time unit. I set the speed of all test particles to be 1.0.

1.2. RESULT

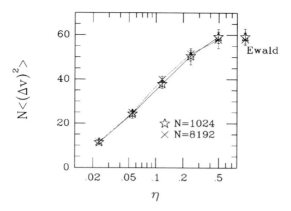

Fig. 1. The average of squared change in velocity $\langle(\Delta\mathbf{v})^2\rangle$ as a function of the cutoff parameter η. The crosses and stars are for $N_f = 1024$ and 8192, respectively. The change in obtained with the Ewald method is also plotted.

I artificially changed the maximum impact parameter of interaction with changing the parameter η. Figure 1 shows $\langle(\Delta\mathbf{v})^2\rangle$ scaled by $1/N$ as a function of η. The stars and crosses are for $N_f = 1024$ and 8192, respectively. I also plot $\langle(\Delta\mathbf{v})^2\rangle$ calculated with the Ewald method.

The value $\langle(\Delta\mathbf{v})^2\rangle$ increases logarithmically as η increases up to $\eta = 0.5$. The value $\langle(\Delta\mathbf{v})^2\rangle$ obtained with the Ewald method corresponds to the case without cutoff, and is nearly equal to the that for $\eta = 0.5$. This means the encounter with an impact parameter over 0.5 has no contribution to $\langle(\Delta\mathbf{v})^2\rangle$. The value $\eta = 0.5$ corresponds to the system size R in periodic boundary system. Under periodic boundary condition the distribution of particles whose impact parameter is over 0.5 is replica of the distribution in the original cube, and there is no density strucuture whose wave length is over 0.5. In Figure 1 there are no saturation of $\langle(\Delta\mathbf{v})^2\rangle$ at a scale of the interparticle distance $RN^{-\frac{1}{3}}$ (~ 0.1 for $N = 1024$, ~ 0.05 for $N = 8192$). In conclusion, the upper cutoff $b_{\max}$ is roughly equal to the system size R.

2. Exponential Divergence of Initial Small Difference

In the field of stellar dynamics, exponential divergence of a small difference of initial condition has been well known. Since the first discovery by Miller (1964) many researchers studied this subject(Lecar 1968; Sakagami and Gouda 1990; Suto 1991; Kandrup and Smith 1991; Kandrup, Smith and Willmes 1992; Quinlan and Tremaine 1992; Huang, Dubinski and Carlberg 1993). A clear theoretical understanding of this exponential growth was given recently by Goodman, Heggie and Hut (1993). They analyzed the effects of successive encounters and showed that these tend to magnify errors on a time scale which is comparable with the crossing time of system.

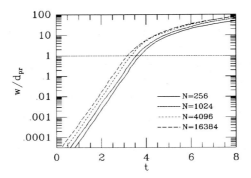

Fig. 2.   Increase of initial smell difference between test particles for N_f = 256, 1024, 4096, and 16384. Each curve represents the median value of 1000 experiments.

Besides stellar dynamics, another important application of the exponential divergence is the gravitational lensing. Since photons also move according to gravitational potential, distance between two photons also increases exponentially when it is small. This effects may affect observations of high redshift objects (Fukushige and Makino 1994; Funato et al. 1994), and anisotropy of cosmic background radiation (Fukushige et al. 1994,1995).

I performed numerical experiments to investigate further two aspects on the exponential divergence. I calculated the orbits of test particles moving through an infinite homogeneous distribution of the field particles.

Firstly, I investigated when the exponential divergence stops. This saturation means that the exponential divergence is not related to any new relaxation process. Goodman et al. (1993) argued that the exponential divergence is true when the separation of the two stars is small. Kandrup et al.(1994) also argued that the expeontial divergence decelerates when it becomes comparable to the size of the system. I numerically comfirmed that the saturation of the divergence occurs when it become comparable to the projected mean saparation d_{pr}, which given by

$$d_{\mathrm{pr}} = RN^{-\frac{1}{2}}. \tag{2}$$

Secondly, I investigated the effect of softening on the exponential divergence. There are two reasons why the introduction of softening is interesting. One is that softening is often used in N-body simulations to avoid numerical difficulties caused by close encounters. The other is that that introduction of the softening gives the relative importance of close and distant encounters in the exponential divergence. Kandrup et al. (1992), Goodman et al. (1993), and Hung et al (1993) discussed the effect of softening in spherical N-body systems. I found that the exponential divergence is surpressed if the softening parameter is larger than mean projected separation d_{pr} .

The above two results comfirmed that the main contribution of the divergence comes from encounters at an impact parameter of order d_{pr} (Goodman et al. 1993). The exponential divergence is due to coherent scattering by the same scatterer. When the difference increases up to a size comparable to d_{pr}, the scattering becomes incoherent and the exponential divergence stops. If the softening parameter is larger than a size comparable to d_{pr}, encounters with an inpact parameter of d_{pr}, which is the main contribution of the exponential divergence, are softened and the exponential divergence is supressed.

In section 2.1 I desribe the model of my numerical experiments and in section 2.2 I show the results of experiments.

2.1. MODEL

The model for this section is almost as same as that discussed in section 1.1. In the following I describe only different points from section 1.1. I calculated the orbits of 1000 pairs of two test particles. I softened the gravitational potential so that the mass distribution of field particle can be expressed by King model ($W_0 = 5$). I used the half mass radius of mass distribution as a softening parameter ε. At initial I gave small difference in velocities of pair of test particles. I sampled the distance between test particles, w, and used the median of 1000 pairs as a representative value.

2.2. RESULT

Figure 2 shows the increase of small difference between test particles. The difference w is scaled by projected mean particle distance d_{pr}. The number of field particles N_{f} are 256, 1024, 4096, and 16384. We see in figure 2 that the difference increases exponentially, and the exponential divergence stops when the difference becomes $w \sim d_{\mathrm{pr}}$.

The e-folding time, t_{e}, is evaluated as $t_{\mathrm{e}}/t_{\mathrm{dy}} \sim 0.27$ where t_{dy} is dynamical time. The e-folding time is defined by $\ln \alpha$ where α is a growth factor and the dynamical time is defined by $1/\sqrt{G\rho}$, where ρ is density of the field particles. The e-folding time is almost independent of the number of particle N.

Figure 3 shows the e-folding time t_{e} scaled by t_{dy} as a function of a softening parameter ε scaled by d_{pr}. The number of field particles N_{f} are 1024 and 8192. We see in figure 3 that the exponential divergence are surpressed if the softening parameter is larger than the projected mean separation d_{pr}, not mean separation($\sim RN^{-1/3}$). I also plot the fitting curve expressed by $t_{\mathrm{e}}/t_d = 0.28 + 0.57(\varepsilon/d_{\mathrm{pr}})^{1.5}$.

3. Summary

I performed numerical experiments of the gravitational scattering in infinite homogeneous N-body systems. Firstly, I investigated the upper cutoff b_{max}

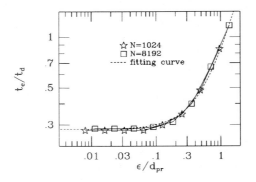

Fig. 3. The e-folding time scaled by dynamical time plotted against $\varepsilon/d_{\mathrm{pr}}$ for $N_{\mathrm{f}} = 1024$ and 8192. The dotted curve indicates the fitting formula given in section 2.2.2.

of the Coulomb logarithm. I found that the upper cutoff b_{max} is roughly equal to the system size R. Secondly, I investigated the exponential divergence of initial small difference. I found that the e-folding time is well fitted with $t_e/t_d = 0.28 + 0.57(\varepsilon/d_{\mathrm{pr}})^{1.5}$ and the divergence stops at a scale comparable to d_{pr}.

I am grateful to Junichiro Makino for helpful discussion, and to Makoto Taiji, who developed MD-GRAPE system.

References

Chandrasekhar, S.: 1942, *Principles of Stellar Dynamics*, Dover:New York
Cohen, R. S., Spitzer, L., and Routly, P. McR., 1950, Phys. Rev., 80, 230.
Ewald, P. P., 1921, Ann. Phy., 64, 253.
Farouki, R., and Salpeter, E. E., 1982, ApJ, 253, 512.
Farouki, R., and Salpeter, E. E., 1994, ApJ, 427, 676.
Fukushige, T., and Makino, J., 1994, ApJL, 436, L111.
Fukushige, T., Makino, J., and Ebisuzaki, T., 1994, ApJL, 436, L107
Fukushige, T., Makino, J., Nishimura, O., and Ebisuzaki, T., 1995a, PASJ, 47, 493.
Fukushige, T., Taiji, M., Makino, J., Ebisuzaki, T., and Sugimoto, D., 1995b, submitted to ApJ.
Funato, Y., Makino, J., and Ebisuzaki, T., 1994, ApJL, 424, L17.
Goodman, J., Heggie, D. C., Hut, P., 1993, ApJ, 415, 715.
Huang S., and Dubinski J., Carlberg R. G. 1993, ApJ, 404, 73.
Kandrup, H. E., 1980, Phys. Rep., 63, 1.
Kandrup, H. E., and Smith, H., 1991, ApJ, 374, 255.
Kandrup, H. E., Smith, H. Jr., and Willmes, D. E., 1992, ApJ, 399, 627.
Kandrup, H. E., Mahon, M. E., and Smith, H. Jr., 1994, ApJ, 428, 458.
Lecar M., 1968, Bull. Astron. 3, 91.
Miller R. H., 1964, ApJ 140, 250.
Press W. H., Flannery B. P., Teukolsky S. A., Vetterling W. T.: 1986, *Numerical Recipes*, Cambridge University Press: London/New York, ch 16.2
Quinlan G. D., and Tremaine S., 1992, MNRAS, 259, 505.
Sakagami M., and Gouda, N. 1991, MNRAS, 249, 241.
Smith, H., 1992, ApJ, 398, 519.
Suto Y., 1991, PASJ, 43, L9.

DOMINANT ROLES OF BINARY AND TRIPLE COLLISIONS IN THE FREE-FALL THREE-BODY DISRUPTION

H. UMEHARA and K. TANIKAWA

Department of Astronomical Science, Graduate School of Advanced Studies
National Astronomical Observatory, Mitaka, Tokyo, 181 Japan

Abstract. In the three-body problem with zero initial velocities and equal masses, the structures of the initial-value space are studied. Most of escapes are found to be directly related to triple and particular binary collisions. Escapes of exchange type are found. The escape probability at the first collapse is computed with the aid of scaling-laws among the regions leading to escape.

Key words: Three-body problem – Collision orbit – Chaos

1. Introduction

Triple collision is crucial to the evolution of the three-body systems, and it is considered as the origin of chaos in the field of mathematical theory of the three-body problem. Various authors studied orbits which experience triple collisions and close triple encounters (see Waldvogel (1973), McGehee (1974), Devaney (1980), Simó (1980), and Moeckel (1983)). One of their interesting results is that the closer an orbit approaches a triple collision, the larger velocity one particle may get. This indicates a close relation between triple collision and escape (disruptions of the system with negative energy).

Triple collisions may occur in the free-fall problem, that is, in the systems with zero initial velocities, because of zero angular momentum. In the free-fall problem, the global initial-value-dependence of escape has been intensively studied numerically by the Russian school (see Agekyan and Anosova (1968), Anosova and Zavalov (1989), Anosova (1991)). They observed escape phenomena in relation to behavior of close triple encounter.

Tanikawa et al. (1995) suggested that binary collisions as well as triple collisions are directly related to escapes, and carried out a search for the distribution of binary and triple collision orbits. Umehara et al. (1995) found the relation between collisions and escapes numerically. They show distribution of escape and binary collision orbits around some triple collisions in the initial-value spaces. Zare and Szebehely (1995) obtained a similar result independently. Broucke (1995) investigated the case with unequal masses, and refered to importance of triple collision.

The present paper explores the case where all particles have equal masses. Three particles m_1, m_2 and m_3 are located at $P(x, y)$, $A(-0.5, 0)$ and

J. C. Muzzio et al. (eds.), Chaos in Gravitational N-Body Systems, 285–290.

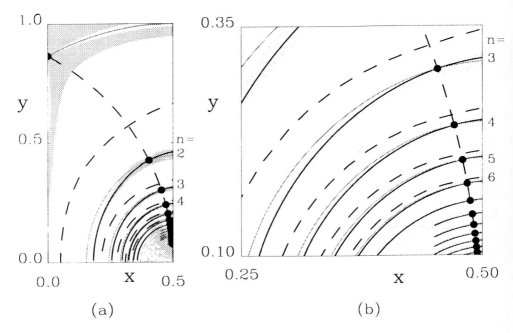

Fig. 1.   Triple collision points, binary collision curves and escape regions in a part of the initial-value space. Curves represent binary collisions. A solid bold curve shows collision between m_2 and m_3 (*perfect slingshot type*). A dashed bold and a solid fine cuves are collisions between m_3 and m_1, m_1 and m_2, respectively. A filled circle • shows the triple collision point. A gray region represents the escape region after the first triple encounter. (a) General features in the initial-value space. (b) Magnification of (a).

$B(0.5, 0)$, respectively in the (x, y) plane where

$$(x, y) \in D, \ D = \{(x, y) | x \geq 0, y \geq 0, (x + 0.5)^2 + y^2 \leq 1\}. \tag{1}$$

The region D is the initial-value space of the free-fall three-body problem with equal masses since triangles formed by three points exhaust all possible initial configurations.

We use the numerical method with Aarseth-Zare (1974) regularization and Bulirsch-Stoer (1966) integrator, and adopt the escape criterion of Yoshida (1972). The method of search for collision orbits is due to Tanikawa et al. (1995).

2.   Slingshot and Exchange

We call the initial values leading to binary and triple collisions, respectively, the binary and triple collision points. If the binary collision points form a curve, we call it the binary collision curve. Connected sets of initial values

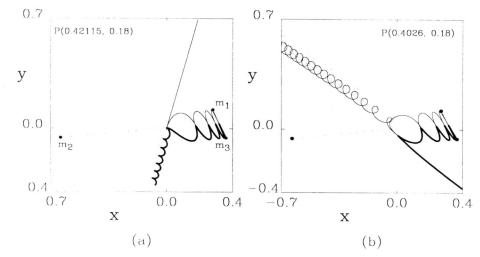

Fig. 2.   Binary collision orbits in the respective escape regions. Solid fine, dotted and solid bold curves are trajectories of m_1, m_2 and m_3, respectively. (a) The orbit in the band-like region D_5 (perfect slingshot). (b) The orbit in the tongue-like region E_5 (exchange).

leading to escapes are called the escape regions.

General features of collisions and escapes are shown in Fig.1(a). For convenience of understanding, we also show the initial-value-dependence beyond D. Anosova and Zavalov (1989) observed a sequence of sub-regions $D_n \subset D$ for $n = 1, 2, \cdots, 5$ that lead to escape after the first triple encounter at the time of the n-th binary encounter between m_1 and m_3. In Fig.1(a) band-like sub-regions D_n for n larger than five can be seen.

We found new regions E_n in which m_3 escapes (see Fig.1(b) which is the magnification of the lower-right part of Fig.1(a)). These are between D_{n-1} and D_n for $n = 5, 6, \cdots$. These regions have tongue-like shapes. We call these *tongue-like regions*.

The relation between escape and triple collision is the following. At least one triple collision point sits in each band-like escape region D_n. Umehara et al. (1995) found it for $n = 1, 2$. In the present paper we confirmed it for $n = 3, 4, \cdots, 20$.

On the other hand, no triple collision points searched by Tanikawa et al. (1995) exist in any tongue-like region E_n. D_n and E_n touch each other for $n = 6, 7, \cdots$. Their boundary is determined as follows: in E_n the distance between m_2 and m_3 never become the minimum of the mutual distances; whereas it becomes the minimum in the part of D_n neghboring E_n.

Before considering the relation between binary collision and escape, we refer to a remarkable behavior of binary collision. The trajectories of m_1, m_2 and m_3 which experience binary collision between m_2 and m_3 in the band-like region are shown in Fig.2(a). The particle m_1 passes through between m_2 and m_3 which are approaching each other. While m_2 and m_3 collide and recede from each other, m_1 is decelerated suddenly and returns, and passes through the gravity center of the binary again. Getting enough energy from the binary, m_1 escapes maintaining isosceles configurations of three particles. All binary collisions between m_2 and m_3 exhibit the same configurations as the above. We call these collision curves *perfect slingshot type*.

The distribution of escape regions is also remarkable. The centers of all band-like regions D_n for $n = 1, 2, \cdots, 6$ are not along the circles as suggested by Anosova and Zavalov (1989), but along the binary collision curves of the *perfect slingshot type*. The triple encounter in D_n is of slingshot type.

Finally, let us consider the orbits in tongue-like regions E_n. The triple encounter in E_n is of exchange type. The trajectories of three particles of a collision orbit in the tongue-like region E_5 are shown in Fig.2(b). Just before the n-th approach of the binary m_1 and m_3, the third particle m_2 approaches m_1. The trajectory of m_1 is deflected by the existence of m_2 and collides with m_3. Then m_3 is reflected to the opposite side of m_2 with respect to m_1 and escapes. The particles m_1 and m_2 approach each other again and become a binary at last.

3. Escape probability

Umehara et al. (1995) found the similarity of structures between D_1 and D_2, where structure means binary collision curves and escape sub-regions around the respective triple collision points in the initial-value space. Also, they suggested that probability of escape can be computed if there is any scaling law among escaping sub-regions. In the present paper we confirmed the similarity of D_n, and that of E_n, for $n = 1, 2, \cdots, 6$. According to Tanikawa et al. (1995), there is an infinite sequence of the binary collision curves of *perfect slingshot type* in the initial-value space D, and it converges at the lower-right corner of D. This assures the existence of an infinite sequence of D_n and E_n. We attempt to formulate the areas of escape regions as functions of n.

The n-dependence of the areas is shown in Fig.3. The relation becomes linear very quickly in the log-log plot. The total area T_n of escape and the area S_n of escape due to slingshot as functions of n are given by

$$T_n = 0.0408 \cdot n^{-2.02} \ (n \in [8, 18]), \quad S_n = 0.0290 \cdot n^{-2.21} \ (n \in [2, 18]). \quad (2)$$

The size of grid elements of initial-value space is made smaller as n increases: 10^{-3} for $n = 1, 2, \cdots, 8$; 0.5×10^{-3} for $n = 9, 10, \cdots, 14$; and 0.2×10^{-3} for $n = 14, 15 \cdots, 19$. The area of sub-regions for larger n can be estimated by extrapolation of the above formulae.

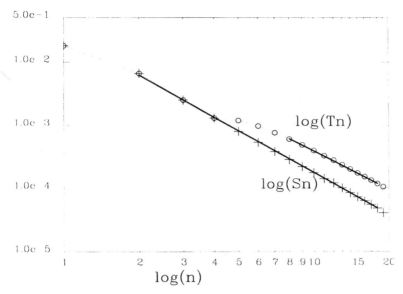

Fig. 3.   Areas of escape regions after the first triple encounters. The abscissa is n. The ordinate is the area of escape sub-regions as functions of n.

The escape probability is defined as the ratio of the area of escape region to the area of initial-value space $D = \pi/6 - \sqrt{3}/8$. The probability P_T of escape, the probability P_S of escape due to slingshot and the probability P_E of escape due to exchange are

$$P_T = \sum_{n=1}^{\infty} T_n/D = 0.1197 \quad , \qquad P_S = \sum_{n=1}^{\infty} S_n/D = 0.1049,$$

$$P_E = P_T - P_S = 0.0148. \tag{3}$$

The relative probability of escape due to slingshot is

$$P_S/P_T = 0.8764. \tag{4}$$

This large ratio confirms quantitatively that slingshot plays the dominant role in escape phanomena.

Finally, we discuss the small angular momentum case. The n-dependence of the relative probability of exchange escape to slingshot escape is evaluated as

$$(T_n - S_n)/T_n = 1 - 0.711 \cdot n^{-0.19}. \tag{5}$$

In the free-fall system with larger n, regions of initial values leading to escape after the exchange is larger than that after the slingshot. However, according to Anosova (1992) who added small angular momentum to the equal-mass system, motion starting from the lower-right part of D is bounded and does not escape. On the other hand, the behavior in the upper part of D is expected not to change with addition of small angular momentum. So we can say that slingshot is more important in the escape phenomena than exchange in small nonzero angular momentum cases.

Acknowledgements

We express our thanks to Dr. J.P.Anosova for discussions. We also appreciate Dr. S.J.Aarseth for allowing to use his program.

References

Aarseth,S.J. and Zare,K.: 1974, 'A regularization of the three-body problem', *Celestial Mechanics* **10**, 185-205.

Agekyan,T.A and Anosova,J.P.: 1968, 'A study of the dynamics of triple systems by means of statistical sampli , *Soviet Physics - Astronomy* **11**, 1006-1014.

Anosova,J.P. and Za v,N.N.: 1989, 'States of strong gravitational interaction in the general three problem', *Soviet Astron.* **33**, 79-83.

Anosova,J.P.: 19 'Strong triple interactions in the general three-body problem', *Celestial Mechanics ar Dynamical Astronomy* **51**, 1-15.

Anosova,J.P. and Orlov,V.V.: 1992, 'The types of motion in hierarchical and non-hierarchical triple systems: numerical experiments', *Astron. Astrophys.* **260**, 473-484.

Broucke,R.A.: 1995, 'On the role of the moment of inertia in three-body scattering' in Roy,A. and Steves,B., ed(s)., *From Newton to Chaos*, Plenum Press: New York, 327-341.

Bulirsch,R. and Stoer,J.: 1966, 'Numerical treatment of ordinary differential equations by extrapolation methods', *Numer. Math.* **8**, 1-13.

Devaney,R.L.: 1980, 'Triple collision in the planar isosceles three body problem', Inven. Math. **60**, 249-267.

McGehee,R.: 1974, 'Triple collision in the collinear three-body problem', *Inven. Math.* **27**, 191-227.

Moeckel,R.: 1983, 'Orbits near triple collision in the three-body problem', *Indiana Univ. Math. Journ.* **32**, 221-240.

Simó,C.: 1980, 'Masses for which triple collision is regularizable', *Celestial Mechanics* **21**, 25-36.

Tanikawa,K., Umehara,H. and Abe,H.: 1995, 'A search for collision orbits in the free-fall three-body problem I. numerical procedure', *Celestial Mechanics and Dynamical Astronomy* **62**, 335-362.

Umehara,H., Tanikawa,K and Aizawa,Y.: 1995, 'Triple collision and escape in the three-body problem' in Aizawa,Y., Saito,S. and Shiraiwa,K., ed(s)., *Dynamical Systems and Chaos Vollume 2: physics*, World Scientific: Tokyo, 404-407.

Waldvogel,J.: 1973, 'Collision singularities in gravitational problems' in Tapley,B.D. and Szebehely,V., ed(s)., *Recent Advances in Dynamical Astronomy*, D.Reidel Publishing Company: Dordrecht-Holland, 21-33.

Yoshida,J.: 1972, 'Improved criteria for hyperbolic-elliptic motion in the general three-body problem', *Publ. Astron. Soc. Japan* **24**, 391-408.

Zare,K. and Szebehely,V.: 1 , 'Order out of Chaos in the three-body problem: Regions of escape' in Roy,A. and Steves,B., ed(s)., *From Newton to Chaos*, Plenum Press: New York, 299-313.

DYNAMICAL CONSEQUENCES OF URANUS' GREAT COLLISION

MIRTA GABRIELA PARISI and ADRIÁN BRUNINI

Facultad de Ciencias Astronómicas y Geofísicas. Universidad Nacional de La Plata,
Paseo del bosque S/N, La Plata (1900). ARGENTINA

and

PROFOEG, CONICET.

April 18, 1996

Abstract.
  It is generally assumed that the Jovian planets share common features as a consequence of similar formation and evolutionary processes. Nevertheless, Uranus is the only one without observed outer satellites and with a large spin axis inclination. This obliquity is usually attributed to a great tangential collision with another protoplanet during the stage of planetary formation. If the Great Collision took place, it imparted enough orbital impulse to unbind the orbits of preexisting outer satellites.
  In this work we settle dynamical constraints to Uranus' Great Collision in connection with the observational evidence.

Key words: Cosmogony – Planetary Satellites – Uranus – Collisions

1. INTRODUCTION

The Jovian planets have a rich system of regular satellites. Jupiter, Saturn and Neptune besides, have satellites orbiting in elongated and inclined orbits, far away from their parent planets (far beyond 60 planetary radii); they are called "irregular satellites". The orbital characteristics of the outer satellites of Jupiter and the outermost satellite of Saturn and Neptune, suggest a capture origin (Pollack et al. 1991, Saha & Tremaine 1993, Goldreich et al. 1989).

There is no apparent reason explainig why Uranus had not been able to capture irregular satellites (Pollack et al. 1991). Nevertheless, there are not known up to date outer satellites of this planet.

Kinoshita and Nakai (1991), have presented a scenario to explain the absence of Uranus outer satellites. They have found that the coupling between the solar perturbation and the oblateness perturbation on outer uranian satellites, orbiting near the orbital plane of Uranus, causes a jump in their orbital excentricity, high enough as to extinct them as outer satellites. Nevertheless, their model is not able to extinct outer satellites outside Uranus' orbital plane. Thus the problem of the absence of Uranus' outer satellites is still open.

J. C. Muzzio et al. (eds.), Chaos in Gravitational N-Body Systems, 291–296.

Other peculiar characteristic of Uranus is the large obliquity of its rotation axis. For the rest of the Jovian planets this inclination is less than, or equal to 30°, but for Uranus this angle is of 98°.

This large obliquity might be due to a tangential collision with another large proto-planet at the stage of planetary formation (Safronov 1972), when presumably Uranus accreted 90-95% of its present mass (Koricansky et al. 1990, Slattery et al. 1992).

If the Great Collision took place, it not only tilted Uranus' spin axis but also imparted enough orbital impulse to unbind the orbits of preexisting distant satellites (Brunini 1995).

In this work, we are interested in settling dynamical constraints to Uranus' Great Collision in connection with the observational evidence.

2. ANGULAR MOMENTUM TRANSFER AND ENERGY CONSIDERATIONS

We are interested in setting a lower limit in the impactor mass. In addition, by angular momentum conservation considerations, a relation between the mass of the impactor m_i and its incident velocity v_i, is found (assuming that the spin axis inclination of Uranus was caused by an off-center collision).

Hydrodynamical calculations have shown that only a small fraction of Uranus' envelope was dispersed by the collision (Koricansky et al. 1990). Within this scenario, the dynamical consequences of the collision may be estimated assuming an inelastic impact (Brunini 1995).

At this stage of Uranus' evolution the envelope was extended and of very low density. In this situation, a collision with the core itself was necessary in order to impart the required additional mass and angular momentum.

We have calculated the angular momentum transfer at collision modelling Uranus as a homogeneous spherical body in rigid rotation. We have considered the total angular momentum of the system to be conserved at collision. Also, we have neglected any angular momentum loss during the evolution of the planet until its present day structure is reached.

Within the previously outlined scenario, we have obtained the following equation for the impactor incident velocity v_i as a function of its mass m_i:

$$v_i = \frac{3m_U}{5m_i R_c} F_m R_U^2 [\Omega^2 + \frac{\Omega_0^2}{(F_m)^2(1 + \frac{m_i}{3m_U})^4} - \frac{2\Omega\Omega_0 \cos\alpha}{(F_m)(1 + \frac{m_i}{3m_U})^2}]^{1/2}, \quad (1)$$

where

$$F_m = 1 + m_i/m_U,$$

and m_U is the pre-collision mass of the planet, R_c is the core radius of the planet at the moment of collision, R_U and Ω are the present radius and angular rotation velocity respectively (the last one corresponding to the present period $T = 2\pi/\Omega$ of 17.3 hs.), Ω_0 is the angular rotation velocity that Uranus would have today if the collision had never existed and α is the angle between the vectors Ω and Ω_0. It should be noted that Ω_0 and α are unknown quantities. For Ω_0 we have chosen two constraint values corresponding to initial periods $T_0 = 2\pi/\Omega_0$ of 9 and 20 hs. The constraints values for α are $\sim 70^o$ and $\sim 130^o$. The values of v_i computed from Eq.(1) are shown in Figure 1 for $T_0 = 20hs.$ and $\alpha = 70$ and 130°.

It was also possible to obtain an upper limit for v_i under simplifying energy considerations, by requiring that the difference of mechanical energy before and after the collision should be less than, or equal to the binding energy of the core. For this purpose, we have neglected any rotation energy dissipation which could have taken place during the later evolution of the planet, until its present day structure is reached. We have also neglected the structural properties of the planet. The velocity limit obtained in this way is displayed in Figure 1, for $T_0 = 20hs.$ From this figure it is possible the determination of a minimum allowed value for m_i, given by the intersection of the function obtained though energy conservation and the curve obtained under angular momentum transfer for $\alpha = 70^o$; this minimum possible mass is $\sim 0.2 - 0.3m_\oplus$.

Nevertheless, the existence of an impactor on an initially unbound heliocentric orbit is hardly expected from a cosmogonic point of view. Because of this reason, we favor the situation, where the impactor was bound to the Solar System. In this case the maximum possible relative velocity corresponds to a parabolic orbit and the impactor incident velocity, including the acceleration caused by the planet, is ~ 31.54 Km/s. This value is also shown in Figure 1 and the corresponding minimum allowed impactor mass is $\sim 1 - 1.1m_\oplus$. It is worth noting, however, that the most favourable case is an impactor on an orbit similar to Uranus' one. In this case the incident velocity, enhanced by the gravitational acceleration, would be of ~ 24.2 Km/s. This velocity corresponds to an impactor of ~ 1.3 to 2.0 $m_\oplus$.

3. CONSEQUENCES ON PREEXISTING OUTER SATELLITES

In a reference frame centered in Uranus' center of mass just before the impact, the conservation of momentum requires that:

$$(m_i + m_U)\Delta V = m_i v_i. \qquad (2)$$

Where ΔV is the orbital velocity change that suffers Uranus and $m_i + m_U$ is the present mass of the planet. Any preexisting Uranus' satellite suffers the same orbital velocity change. If for a satellite in circular orbit with velocity

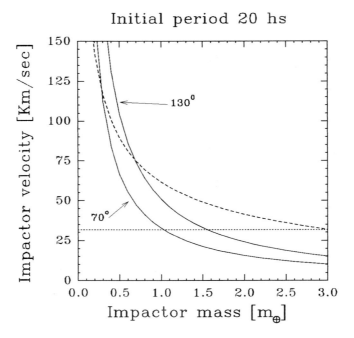

Fig. 1.   Impactor incident velocity v_i as a function of its mass m_i for an initial period $T_0 = $ 20hs. and $\alpha = 70°, 130°$, obtained though angular momentum transfer (full lines). Upper constraint of v_i vs. m_i obtained with energy considerations (dash line) for $T_0 = $ 20hs. Maximum allowed value for v_i for an impactor bound to the Solar System (dot line).

v_{orb} this change is $\Delta V \geq (\sqrt{2F_m} + 1)v_{orb}$, the satellite acquires a velocity greater than, or equal to, the escape velocity and it is lost from the uranian system. If $\Delta V < (\sqrt{2F_m} - 1)v_{orb}$, the orbit of the satellite remains bound to the system.

In this model, we state that a satellite in circular orbit with semiaxis r_m and orbital velocity v_C, acquires through the impulse imparted in the collision, the necessary energy to reach the escape velocity v_e.

Through the energy integrals of the satellite orbit before and after collision, it is possible to obtain the minimun orbital semiaxis r_m, from which any circular orbit with semiaxis $r \geq r_m$ would be unbound, as a funtion of m_i and ΔV:

$$r_m = \frac{Gm_U}{(\Delta V)^2} \frac{(1 + 2\frac{m_i}{m_U})^2}{(\cos\psi + \sqrt{\cos^2\psi + 1 + 2\frac{m_i}{m_U}})^2},   (3)$$

where we define ψ as the angle between the vectors $\mathbf{v}_C$ and $\Delta\mathbf{V}$ and G is the constant of gravitation.

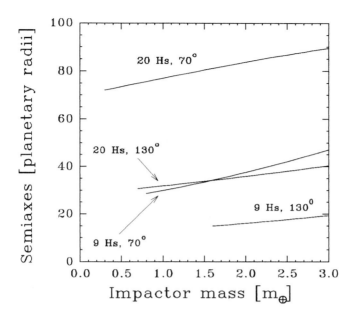

Fig. 2.   Each curve represents the most probable value of the minimum orbital semiaxis r_m (for each α and T_0), as a function of the impactor mass. The upper constraint corresponds to $T_0 = 20$ hs. and $\alpha = 70°$, where every preexisting satellite of the planet with semiaxis between $\sim 70\text{-}90$ planetary radii would have been unbound at collision. The lower constraint corresponds to $T_0 = 9$ hs. and $\alpha = 130°$, where the orbits which would not have been unbound have semiaxes less than $\sim 15\text{-}20$ planetary radii.

Being the direction of $\mathbf{\Delta V}$ unknown, it was necessary to average r_m over all the possible incident directions of the impactor, obtaining the most probable value for r_m as a function of $\mathbf{\Delta V}$ and m_i. Through Eqs.(1) and (2), it was possible to obtain the results as a function of m_i only, for each α and T_0. The corresponding curves are shown in Figure 2, using the constraints values for α and T_0 given in the previous section. The most important meaning of these results is that every hypothetical preexisting satellite of Uranus orbiting beyond $\sim 70\text{-}90$ planetary radii would have escaped from the system, if the collision had taken place.

    It is worth noting that the irregular satellites of the outer planets have large orbital eccentricities and so, it would be important to take it into account. However, as the energy in the two body problem is a function of the semiaxis only, the most probable value of r_m would be independent of the orbital eccentricities of the satellites.

4. CONCLUSIONS

An impact could have occurred at any time during Uranus' evolution, although the requirement that the impact mass be significant suggest that the possible impact could have occurred late in the history of the solar nebulae (Koricansky et. al 1990). Capture theories favor capture at the final stages of the accretion process. The main result of this paper is that any massless particle orbiting beyond $\sim$ 70-90 planetary radii would have been swept out from the system by the Great Collision.

In addition, it was possible to obtain a minimum allowed value for the mass of the impactor, if it was bound to the Solar System of $\sim 1 - 1.1 m_\oplus$. Nevertheless, in the most probable case of a planetoid in heliocentric near-circular orbit the mass is $\sim 1.3 - 2.0 m_\oplus$.

Our two body assumptions is limited in some aspects. It has been long known that retrograde satellites are more stables than direct ones (Henón 1970, Huang & Innanen 1983, Brunini 1996), but our theoretical approach is unable to differentiate between the two cases. Most accurate estimates should be maden through a complete Three-Body Monte Carlo simulation, studying the long term stability of the outer satellites after collision (Huang & Innanen 1983, Brunini 1996), instead of assuming an ad-hoc condition for the satellite loss, but this is out of the scope of the present work.

References

A. Brunini, (1995) *Planet. Space Sci.* (in press).
A. Brunini, (1996) *in this proceedings.*
P. Goldreich, N. Murray, P. Y. Longaretti & D. Bandfield, (1989) *Science* **245**, 500.
M. Hénon, (1970) *Astron. Astrophys.* **9**, 24.
T. -H. Huang & K. A. Innanen, (1983) *Astron. J.* **88**, 1537.
H. Kinoshita & H. Nakai, (1991) *Cel. Mech. Dyn. Astron.* **52**, 293.
D. G. Koricansky, P. Bodenheimer, P. Cassen & J. Pollak, (1990) *Icarus* **84**, 528.
J. B. Pollack, J. I. Lunine & W. C. Tittemore, (1991) In: "Uranus Colloquium (Eds. J. T. Bergtralh, E. D. Miner y M. S. Mattews) Univ. of Arizona Press, 409.
V. S. Safronov, (1972) "Evolution of the Protoplanetary Cloud and Formation of The Earth and the Planets", NASA TTF-677
P. Saha & S. Tremaine, (1993) *Icarus* **106**, 549.
W. L. Slattery, W. Benz & A. G. W. Cameron, (1992) *Icarus* **99**, 167

RUNAWAY GROWTH OF PLANETESIMALS

E. KOKUBO

Department of Earth Science and Astronomy, University of Tokyo
Komaba, Meguro-ku, Tokyo 153, Japan

and

S. IDA

Department of Earth and Planetary Science, Tokyo Institute of Technology
Ookayama, Meguro-ku, Tokyo 152, Japan

Abstract. We present the results of three-dimensional N-body simulations of planetary accretion. We confirm that runaway growth is a dominant mode of planetary accretion in a three-dimensional system where orbits of planetesimals have inclinations. In runaway growth, larger planetesimals grow more rapidly than smaller ones and the mass ratio between them increases monotonically. On the other hand, in a two-dimensional system where all orbits are coplanar, we find that runaway growth does not occur. These results are explained by the simple analytical argument.

Key words: Planetary Formation – Runaway Growth – Planetesimals

1. Introduction and Summary

In the standard scenario of planetary formation, planets are formed through the accretion of "planetesimals". By solving the coagulation equation of planetesimals, Wetherill and Stewart (1989) showed that larger planetesimals grow more rapidly than smaller ones, resulting "runaway growth" of larger planetesimals. However, we have not had a conclusive evidence that runaway growth occurs in a realistic planetesimal system, since Wetherill and Stewart (1989) assumed the infinite homogeneous swarm of planetesimals. Some pioneers (e.g., Lecar and Aarseth 1986) performed two-dimensional (2-D) N-body simulations of planetary accretion. Unfortunately, as will be shown later, 2-D accretion is qualitatively different from three-dimensional (3-D) one. The only 3-D simulation was done with $N = 100$ by Aarseth *et al.* (1993). They showed only the onset of runaway growth since the number of planetesimals was small.

Here we present the results of 3-D N-body simulations of planetary accretion starting with $N = 2000$. We confirm that runaway growth is a dominant mode of planetary accretion in a 3-D system where orbits of planetesimals have inclinations. On the other hand, in a 2-D system where all orbits are coplanar, we find that accretion proceeds in rather an orderly way where all planetesimals grow equally. The difference between 3-D and 2-D accretions is well explained by the simple two-body formulae.

J. C. Muzzio et al. (eds.), Chaos in Gravitational N-Body Systems, 297–302.

2. Method of Calculation

Planetesimals are distributed randomly in a ring-like region at $a = 1$ AU, where a is the semimajor axis. Their initial eccentricities and inclinations are given by Gaussian distribution with dispersions $\langle e^2 \rangle^{1/2} = 2\langle i^2 \rangle^{1/2} \simeq 10^{-3}$ (Ida and Makino 1992). We set the surface mass density of solid substance of the solar nebula as $\Sigma_s = 10$ gcm^{-2}, which is consistent with the standard model (Hayashi 1981). Orbits of planetesimals are integrated by the forth-order Hermite integrator (Makino and Aarseth 1992). All gravitational interactions between planetesimals are calculated using the special-purpose computer "HARP" (Makino et al. 1993). We assume the perfect accretion under which two planetesimals always accrete when their spheres overlap. We do not take gas drag due to the solar nebula into account.

In order to accelerate the accretion process, we use larger radii of planetesimals than realistic ones obtained by assuming the material density $\rho = 2$ gcm^{-3} (Kokubo and Ida 1995). Note that this acceleration of accretion does not change the qualitative feature of accretion in contrast to reduction to 2-D accretion. However, the growth time scale should be rescaled according to the scaling law of radii; in our simulation multiply time by about the enlargement factor.

3. Results and Discussion

The results of the planetesimal system that initially consists of 2000 equal-mass (10^{24} g) bodies are shown in Figs. 1-2. We used the 5-fold radii of planetesimals. The necessary condition for runaway growth, $v_{rel} < v_{esc}$ (gravitational focusing is effective), and the sufficient condition that larger planetesimals have smaller eccentricities (dynamical friction is effective) are kept through the simulation, where v_{rel} is the relative velocity of colliding planetesimals and v_{esc} is the escape velocity from a planetesimal (Ohtsuki et al. 1993). In Fig. 1, a body grows rapidly around $a = 1$AU. The body becomes about 200 times of its initial mass in 20000 years, while the mean mass reaches about twice the initial mass (Fig. 2). It is clearly shown in Fig. 2 that the ratio of the maximum mass to the mean mass grows with time, in other words, runaway occurs. This is the first apparent confirmation by 3-D N-body simulation.

We also performed the 2-D simulation to compare with the result of the 3-D simulation. The initial condition of the planetesimal system is the same as that of the 3-D simulation except that the inclinations of planetesimals are zero and a large planetesimal whose mass is 30 times as large as that of planetesimals initially is introduced to see the growth mode clearly. We used the realistic radii of planetesimals since the 2-D growth rate is high enough.

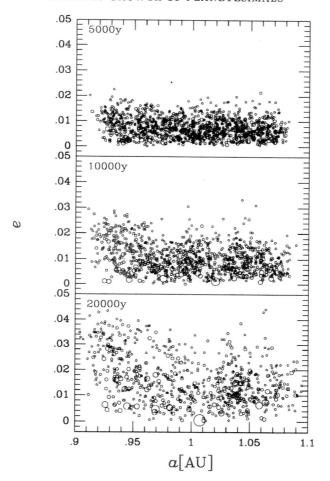

Fig. 1.   Time evolution of a planetesimal system on the $a-e$ plane. The circles represent planetesimals and their radii are proportional to the radii of planetesimals. The numbers of planetesimals are 1147 ($t = 5000$ y), 926 ($t = 10000$ y), and 715 ($t = 20000$ y). We omit the figure on $a-i$ plane, where i is the inclination, since it is similar to that on the $a-e$ plane with $i \simeq 0.5e$.

We found that in a 2-D system runaway growth does not occur (Fig. 3). As shown in Fig. 4, the ratio of the maximum mass to the mean mass decreases with time This shows that in a 2-D system accretion proceeds in an orderly way at least under the perfect accretion.

The above results are explained with the simple two-body formulae. Let us consider that two test bodies with mass M_1 and M_2 grow by capturing field bodies with mass m ($M_1 > M_2 > m$). Runaway growth occurs when

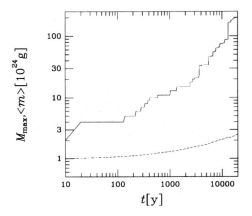

Fig. 2.   The maximum mass of the planetesimals M_{max} and their mean mass except the maximum one $\langle m \rangle$ are plotted as a function of time. The solid line shows the maximum mass and the dashed line shows the mean mass.

$$\frac{d}{dt}\left(\frac{M_1}{M_2}\right) = \frac{M_1}{M_2}\left(\frac{1}{M_1}\frac{dM_1}{dt} - \frac{1}{M_2}\frac{dM_2}{dt}\right) > 0. \tag{1}$$

The key that determines the growth mode is the mass-dependence of the relative growth rate $(1/M)dM/dt$. The growth rate of a test body with mass M in a 3-D system is given by (e.g., Lissauer and Stewart 1993)

$$\frac{dM}{dt} \simeq \frac{\Sigma_{\mathrm{s}}}{h}\pi r^2\left(1 + \frac{v_{\mathrm{esc}}^2}{v_{\mathrm{rel}}^2}\right)v_{\mathrm{rel}}, \tag{2}$$

where h is the scale height of field bodies and r is the radius of the test body. The relative velocity between the test body and field bodies is given by $v_{\mathrm{rel}} \simeq \sqrt{v_M^2 + v_m^2}$, where v_M is the velocity dispersion of the test body and v_m is that of field bodies. We consider the case where gravitational focusing is effective ($v_{\mathrm{rel}} < v_{\mathrm{esc}}$) and dynamical friction is effective ($v_M < v_m$). In this case, 1 in parentheses of Eq. (2) can be neglected and the relative velocity is given by $v_{\mathrm{rel}} \simeq v_m$. When the mass of the test body is not so large as to raise the velocity dispersion of field bodies, the velocity dispersion does not depend on the mass of the test body (Ida and Makino 1993). In this case, we obtain the relative growth rate as

$$\frac{1}{M}\frac{dM}{dt} \propto M^{\frac{1}{3}}, \tag{3}$$

where we used $h \propto v_m$ and $r, v_{\mathrm{esc}} \propto M^{1/3}$. This mass-dependence clearly satisfies inequality (1) and thus runaway growth occurs in a 3-D system. On the other hand, the growth rate in a 2-D system is given by

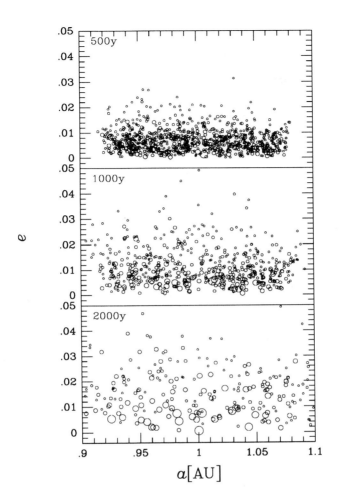

Fig. 3.   The same as Fig. 1 but for the 2-D system. The numbers of planetesimals are 897 $(t = 500$ y$)$, 540 $(t = 1000$ y$)$, and 242 $(t = 2000$ y$)$.

$$\frac{dM}{dt} \simeq \Sigma_{\rm s} 2r \left(1 + \frac{v_{\rm esc}^2}{v_{\rm rel}^2} \right)^{\frac{1}{2}} v_{\rm rel}, \qquad (4)$$

which in a similar way to a 3-D system is reduced to the relative growth rate

$$\frac{1}{M} \frac{dM}{dt} \propto M^{-\frac{1}{3}} \qquad (5)$$

This does not satisfy inequality (1), which means that runaway growth does not occur in a 2-D system under the perfect accretion. These analytical results agree well with the results of the N-body simulations.

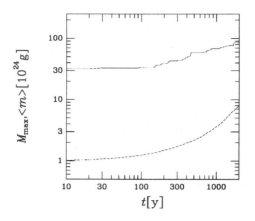

$$M_{max}, <m>\,[10^{24}\,g]$$

$$t[y]$$

Fig. 4.   The same as Fig. 2 but for the 2-D system.

Acknowledgements

We would like to express our sincere gratitude to Junichiro Makino for useful comments. We also thank Makoto Taiji and Izumi Hachisu for development of "HARP".

References

Aarseth, S. J., Lin, D. N. C., and Palmer, P. L.: 1993, *Astrophys. J.* **403**, 351
Hayashi, C.: 1981, *Prog. Theor. Phys. Suppl.* **70**, 35
Ida, S. and Makino, J.: Icarus, *1992* **96**, 107
Ida, S. and Makino, J.: 1993, *Icarus* **106**, 210
Kokubo, E. and Ida, S.: 1995, *Icarus* , in press
Lecar, M. and Aarseth, S. J.: 1986, *Astrophys. J.* **305**, 564
Lissauer, J. J. and Stewart, G. R.: 1993, *Protostars and Planets III*, Univ. of Arizona
    Press: Tucson, 1061
Makino, J. and Aarseth, S. J.: 1992, *Publ. Astron. Soc. Jpn.* **44**, 141
Makino, J., Kokubo, E., and Taiji, M.: 1993, *Publ. Astron. Soc. Jpn.* **45**, 349
Ohtsuki, K., Ida, S., Nakagawa, Y., and Nakazawa, K.: 1993, *Protostars and Planets III*,
    Univ. of Arizona Press: Tucson, 1089
Wetherill, G. W. and Stewart, G. R.: 1989, *Icarus* **77**, 330

SECONDARY RESONANCES INSIDE
THE 2/1 JOVIAN COMMENSURABILITY

An application of the frequency map analysis

DAVID NESVORNÝ *

Instituto Astronômico e Geofísico,
Universidade de São Paulo, CEP 04301, São Paulo, Brasil

The Hecuba gap centered at the 2/1 mean motion commensurability with Jupiter is still an unexplained feature of the asteroid belt. The low-eccentricity region is populated by secondary resonances there (Lemaître and Henrard 1990), which overlap and form strongly chaotic zone (Giffen 1973). An asteroid can start with low eccentricity and be transferred, in several million years, to high eccentricity through the bridge between secondary and secular resonances at large libration amplitudes and inclinations. This was observed by Wisdom (1987) and studied by Henrard et al. (1995). It results in an unstable orbit crossing trajectories of the inner planets and explains how the low-eccentricity region of the 2/1 commensurability was emptied.

On the other hand, the region at eccentricity $0.2 - 0.4$, low inclination and small amplitude of libration is very stable. The high-order secondary resonances do not overlap there and produce only narrow chaotic layers along their own separatrices. This was observed by Nesvorný and Ferraz-Mello (1996) in the three-body elliptic model. Although the additional perturbation of Saturn turns this region chaotic (Ferraz-Mello 1994), the diffusion is very slow and orbits are usually stable on several 10^7 yr interval. However, recent integrations of mapping model over the age of the solar system (Ferraz-Mello 1995) and direct 100 Myr integrations of the exact dynamical model (Ferraz-Mello, personal communication) showed that the diffusion effect is significant on longer time intervals in the four-body model and that the majority of initial conditions leads to the final high-eccentricity orbit.

But still, there is a question of validity of these extremely long numerical integrations. The effect of error propagation on mentioned timescales is unknown and must be carefully considered.

One of the alternative methods for systematic study of the chaotic diffusion is Laskar's frequency map analysis (Laskar 1990, Laskar et al. 1992, Laskar 1993). It was applied to the 2/1 Jovian commensurability by Nesvorný and Ferraz-Mello (1996) and in the following we present an excerpt of this article.

* e-mail address: david@vax.iagusp.usp.br

J. C. Muzzio et al. (eds.), Chaos in Gravitational N-Body Systems, 303–305.
© 1996 *Kluwer Academic Publishers. Printed in the Netherlands.*

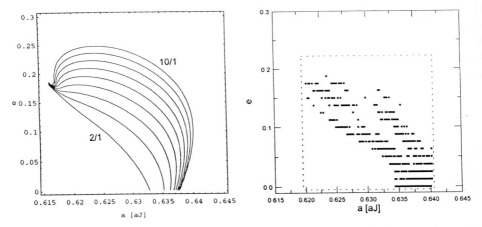

Fig. 1. The 'zero' approximation of secondary resonances positions (left) and corresponding chaotic region in the elliptic model with Jupiter's eccentricity 0.048 (right). The lowest
and highest secondary resonances are $f_\sigma/f_{\bar\omega} = 2/1$ and $10/1$. The squares are the initial
condition for which $f_{\bar\omega}$ was time-dependent. Only the initial conditions in dotted rectangle were studied (all initial angles were zero). The semimajor axis is given in multiples of
Jupiter's semimajor axis a_J.

For various initial conditions, we integrated the exact equations of motion
using the symmetric multistep method of Quinlan and Tremaine (1990), fil-
tered out the high-frequency oscillations (Quinn et $al.$ 1991) and constructed
the frequency map. Its components are the frequencies of asteroid resonant
angle f_σ, longitude of perihelion $f_{\bar\omega}$ and longitude of node f_Ω.

Without going into details (see Nesvorný and Ferraz-Mello 1996) we
present here the following figures. Fig. 1 (left) is position of the secondary
resonances computed by semi-analytical method as in the planar circular
three-body model, which is a 'zero' approximation of their actual positions
(Lemaître and Henrard 1990, Moons and Morbidelli 1993).

These resonances between f_σ and $f_{\bar\omega}$ generate chaos in low eccentricity
in the elliptic model, what can be detected as time variation of the frequency map. The result is in Fig. 1 (right), there the squares mark the initial
conditions with time-dependent $f_{\bar\omega}$, that means the chaotic behaviour.

The secular resonances above 6/1 do not provoke a large scale chaos and
there are only narrow chaotic strips around their separatrices there. But if
one increases Jupiter's eccentricity to 0.061 in the three-body model, the
region up to 9/1 becomes chaotic, because the secondary resonances enlarge
and overlap.

The destruction of the continuous frequency map in the whole phase
space of the 2/1 Jovian commensurability was observed when an additional

effect of Saturn was included in the model. We studied the chaotic diffusion in this restricted four-body model by means of frequency map analysis and concluded that at least some of the asteroids, which would be originally located in moderate eccentricities, had to be transferred, during the evolution of the solar system, to high-eccentricity orbits by the perturbations of outer planets.

References

Ferraz-Mello, S.: 1994, 'Dynamics of the asteroidal 2/1 resonance', *Astron.J.* **108**, pp. 2330–2337

Ferraz-Mello, S.: 1995, IAU Symposium 172, in press

Giffen, R.: 1973, 'A study of commensurable motion in the asteroid belt', *Astron.Astrophys.* **23**, pp. 387–403

Henrard, J., Watanabe, N. and Moons, M.: 1995, 'A bridge between secondary and secular resonances inside the Hecuba gap', *Icarus* **115**, pp. 336–346

Laskar, J.: 1990, 'The chaotic motion of the solar system: A numerical estimate of the size of the chaotic zones', *Icarus* **88**, pp. 266–291

Laskar, J.: 1993, 'Frequency analysis for multi-dimensional systems. Global dynamics and diffusion', *Physica D* **67**, pp. 257–281

Laskar, J., Froeschlé, C. and Celletti, A.: 1992, 'The measure of chaos by the numerical analysis of the fundamental frequencies. Application to the standard mapping', *Physica D* **56**, pp. 253–269

Lemaître, A. and Henrard, J.: 1990, 'On the origin of the chaotic behaviour in the 2/1 Kirkwood gap', *Icarus* **83**, pp. 391–409

Moons, M. and Morbidelli, A.: 1993, 'The main mean motion commensurabilities in the planar circular and elliptic problem', *Celest.Mech. & Dynam.Astron.* **57**, pp. 99–108

Nesvorný, D. and Ferraz-Mello, S.: 1996, 'Chaotic diffusion in the 2/1 asteroidal resonance. An application of the frequency map analysis', submitted to *Astron.Astrophys.*

Quinlan, G.D. and Tremaine, S.: 1990, 'Symmetric multistep methods for the numerical integration of planetary orbits', *Astron. J.* **100**, pp. 1694–1700

Quinn, T.R., Tremaine, S. and Duncan, M.: 1991, 'A three million year integration of the Earth's orbit', *Astron.J.* **101**, pp. 2287–2305

Wisdom, J.: 1987, 'Urey Prize Lecture: Chaotic dynamics in the solar system', *Icarus* **72**, pp. 241–275

APPLICATIONS OF FOURIER AND WAVELET ANALYSES TO THE RESONANT ASTEROIDAL MOTION

T. A. MICHTCHENKO

Instituto Astronômico e Geofísico, Universidade de São Paulo, Caixa Postal 9638,
CEP 01065, São Paulo, Brazil

The Fourier transform is a fundamental technique in oscillatory signal processing and it has been widely and successfully used for many years. When applied to the study of the resonant asteroidal motion Fourier analysis techniques allow to:

- measure precisely the fundamental frequencies and the corresponding amplitudes of oscillatory components of the orbital elements (Ferraz-Mello 1981, Laskar 1990) and, consequently, to study the existence of both secondary and secular resonances;
- study the long periodic effects on the motion of the asteroid separately, by applying a digital low-pass filter (Rabiner et al. 1975, Carpino et al. 1987);
- distinguish the regions of regular and chaotic motion in phase space. In general, Fourier analysis reveals itself to be a good alternative to the Lyapunov exponent calculations, the main advantage being the considerably shorter time intervals needed to be analyzed (Powell & Percival 1979, Michtchenko & Ferraz-Mello 1995).

However, in respect to the quantitative analysis of chaotic motion, Fourier analysis is of limited use since the output of the Fourier transform does not describe the time dependence of the frequency components. A full description of chaotic effects requires the description of non-stationary signals both in time and frequency domains. Recently, Laskar (1990, 1993a,b) proposed a time-dependent frequency analysis, which he applied to investigate the planetary motion in the Solar System. Laskar's method, based on a fixed-size window filter of the data prior to the Fourier transformation, is very effective in the cases when the independent frequencies vary sufficiently slowly, so that their good measure can be obtained from Fourier analysis.

When independent frequencies of the dynamical system vary greatly and rapidly (for example, in the case of the overlap of various secondary resonances), the application of wavelet analysis has been suggested (Grossmann et al. 1987, Bendjoya & Slezak 1993, Michtchenko & Nesvorný 1996). The advantage of wavelet analysis is its ability to detect transient fluctuations, which is achieved by using a special wavelet function set, which automatically adjusts its size and position with respect to frequency and time. The wavelet transform decomposes the time series by using a basis of wavelets which are localized both in time and frequency. Wavelet analysis reveals good

J. C. Muzzio et al. (eds.), Chaos in Gravitational N-Body Systems, 307–310.
© 1996 *Kluwer Academic Publishers. Printed in the Netherlands.*

performance when applied to both weakly and strongly chaotic motions, which makes it advantageous compared to the methods based on the fixed-size window filter.

In this work, the technical details of the calculation of the wavelet transform (WT) are omitted, since they have already been described in (Michtchenko & Nesvorný 1996). In summary, the WT of a real signal $x(t)$ is obtained in the grid of time b and frequency f using abbreviated Morlet analysing wavelet (Grossmann et al. 1987), so as:

$$C(f,b) = \sqrt{\frac{2}{\pi}}\, f \int_{b-\Delta t}^{b+\Delta t} x(t)\, e^{-f^2 \frac{(t-b)^2}{2}}\, e^{-2\pi i f (t-b)}\, d t, \tag{1}$$

where Δt is the width of the window filter. The range of values of f contained in the signal $x(t)$ is obtained from the FT power spectrum of $x(t)$. The wavelet transform $C(f,b)$ is a complex function of two parameters: time b and frequency f. Here the time dependence of the frequency is represented by plotting in the (b,f)-plane the values of frequency corresponding to the maximum of $|C(f,b)|$ obtained at fixed points of time b.

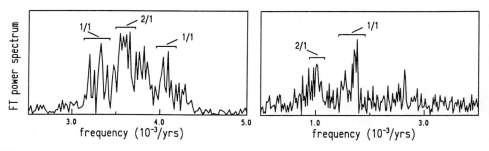

Fig. 1.   Fourier power spectra of the semi-major axis (left) and of the longitude of perihelion (right)

To illustrate the WT performance, we have analysed one asteroidal orbit obtained by integrating numerically the equations of motion for the 3:2 mean motion resonance with Jupiter, with inclusion of the action of Saturn. The phase space of the 3:2 resonance presents several regions of chaotic motion, which have their origin in the action of the secondary resonances of the type $\frac{f_\sigma}{2(f_\varpi - f_\Omega)}$, where the frequency of libration f_σ is the dominant oscillation frequency of the semi-major axis and f_ϖ and f_Ω are the frequencies of the longitude of perihelion and of the longitude of the ascending node, respectively (Michtchenko & Ferraz-Mello 1996).

In order to simulate strongly chaotic motion, the initial conditions of the fictitious asteroid were chosen in the region of the overlap of the secondary resonances ($a_0 = 0.763\, a_J, e_0 = 0.10, I_0 = 15°, \sigma_0 = \Delta\varpi_0 = \Delta\Omega_0 = 0°$).

The asteroid studied shows a chaotic behavior and is ejected from the 3:2 resonance after ~ 11 Myr. For the sake of the illustration of the WT performance, the results obtained over the arbitrarily chosen time segment of 50,000 years are shown here. The Fourier power spectra of the semi-major axis and of the longitude of perihelion (Fig. 1) display broad-band components which characterize the chaotic motion. The intervals of variation of the spectral components are marked by the corresponding ratio of the secondary resonance: $\frac{1}{1}$ and $\frac{2}{1}$. The $\frac{1}{1}$ secondary resonance is characterized by two libration frequencies: $\sim 1/250yrs$ and $\sim 1/300yrs$, the latter being equal to $2(f_\varpi - f_\Omega)$.

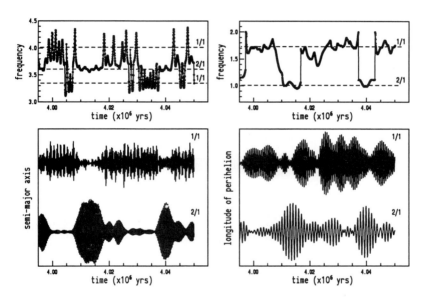

Fig. 2.   Time evolution of the frequency of libration (top left) and the frequency of the longitude of perihelion (top right). Frequency units are $10^{-3}/yrs$. Two oscillatory components of the semi-major axis (bottom left) and of the longitude of perihelion (bottom right) corresponding to the $\frac{1}{1}$ and $\frac{2}{1}$ secondary resonances

The evolution in time of f_σ obtained by means of the WT is plotted in Fig. 2 (top left). The loci of the secondary resonances traced by horizontal lines are indicated by the corresponding resonance ratio. Using a band-pass filtering procedure, we decompose the $a-$oscillation into two components which correspond to the $\frac{1}{1}$ and $\frac{2}{1}$ secondary resonances marked in Fig. 1. Figure 2 (bottom left) depicts these two oscillatory components of the semi-major axis; the resonance ratio corresponding to each curve is indicated in the Figure. The transient regime of the semi-major axis oscillation when a jump from one resonance to another occurs can be clearly observed.

Figure 2 (right), which is analogous to Fig. 2 (left), shows the time evolution of f_ϖ (top right) and two oscillatory ϖ-components corresponding to the $\frac{1}{1}$ and $\frac{2}{1}$ secondary resonances. Note the large range of values detected in the frequency f_ϖ and also the jumps from one secondary resonance to another. This behavior of the semi-major axis and of the longitude of perihelion is related to the overlap of $\frac{1}{1}$ and $\frac{2}{1}$ secondary resonances when the orbit jumps from one resonant region to another.

Finally, the evolution of the ratio $\frac{f_\sigma}{2(f_\varpi - f_\Omega)}$ over $11,000,000$ years is shown in Fig. 3. The rapid transitions between various secondary resonances drive the orbit along the overlapping region to the 3:2 resonance border where the asteroid is then ejected from resonant zone (see Fig. 6 from Ferraz-Mello et al. 1996).

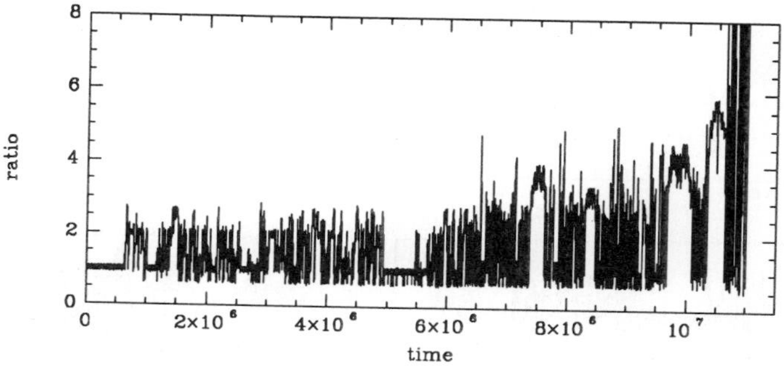

Fig. 3.   Time evolution of the ratio $\frac{f_\sigma}{2(f_\varpi - f_\Omega)}$

References

Bendjoya, Ph., Slezak, E., 1993, Celest. Mech. Dynam. Astron. 56, 231

Carpino, M., Milani, A., Nobili, A.M. 1987, A&A 181, 182

Ferraz-Mello, S. 1981, Astron.J. 86, 619

Ferraz-Mello, S., Klafke, J., Michtchenko, T., Nesvorný, D. 1996, Celest. Mech. Dynam. Astron. (submitted)

Grossmann,A., Kronland-Martinet,R., Morlet,J., 1987. In: Combes,J.M., Grossmann,A. and Tchamitchian,Ph. (eds.), Wavelets Time-Frequency Methods in Phase Space, Springer-Verlag

Laskar, J. 1990, Icarus 88, 266

Laskar,J., 1993a, Celest. Mech. Dynam. Astron. 56, 191

Laskar,J., 1993b, Phisica D 67, 257

Michtchenko, T., Ferraz-Mello, S., 1995, A&A, 303, 945

Michtchenko, T., Ferraz-Mello, S., 1996, A&A (in press)

Michtchenko, T., Nesvorný, D., 1996, A&A (in press)

Powell G.E., Percival, I.C. 1979, J.Phys.A: Math. Gen. 12, 2053

Rabiner, L.R., Gold, B. 1975, 'Theory and Application of Digital Signal Processing', Prentice Hall, Englewood Cliffs, New York, USA, 187

SPECTRA OF WINDING NUMBERS
OF CHAOTIC ASTEROIDAL MOTION

IVAN I. SHEVCHENKO

Institute of Theoretical Astronomy, Russian Academy of Sciences
Nab. Kutuzova 10, St.Petersburg 191187, Russia
E-mail: iis@ita.spb.su

Abstract. Spectra of winding numbers, visualizing the resonant structure of chaotic motion, are constructed for asteroidal orbits in the 3/1 mean motion commensurability with Jupiter; namely, for intermittent trajectories displaying multiple eccentricity bursts. Computation of orbits is performed in the framework of the planar-elliptic Sun–Jupiter–asteroid problem. A major feature of the asteroidal spectra is a peak, emerging due to trajectories' sticking to the border of the chaotic layer. It is shown that it is naturally approximated by the spectrum of the separatrix mapping with definite values of parameters.

Chaotic asteroidal trajectories often exhibit transitions between various resonant states. A particular resonant state is associated with eccentricity bursts (see e.g. Wisdom, 1985). The motivation of the following study is to adequately display the resonant structure of asteroidal motion, especially when the eccentricity is low, and to explain it when possible.

The analysis is performed for the case of the 3/1 commensurability of mean motions of an asteroid and Jupiter, in the planar-elliptic restricted three-body problem. Orbits exhibiting the mode of jumps to very high eccentricities $e \approx 1$ (see Ferraz-Mello, 1993) are not considered. The Hamiltonian of the problem represents, after averaging on the orbital time scale, the time-independent Hamiltonian of a system with two degrees of freedom (Wisdom, 1985):

$$
H = -\frac{3\mu_1^2}{2\Phi_{res}^4}\tilde{\Phi}^2 + 2\mu F\rho + \mu e_J G\sqrt{2\rho}\cos\omega -
$$
$$
- 2\mu C\rho\cos(\varphi - 2\omega) - \mu e_J D\sqrt{2\rho}\cos(\varphi - \omega) - \mu e_J^2 E\cos\varphi, \qquad (1)
$$

where $\tilde{\Phi} = \Phi - \Phi_{res}$, $\Phi = \sqrt{\mu_1 a}$, $\Phi_{res} = (\mu_1^2/3)^{1/3}$, $\rho = \sqrt{\mu_1 a}(1 - \sqrt{1 - e^2})$, $\varphi = l - 3l_J$; l and l_J are mean longitudes of an asteroid and Jupiter; $\omega = -\tilde{\omega}$, where $\tilde{\omega}$ is the longitude of perihelion of the asteroid; a and e are its semimajor axis and eccentricity. Canonical conjugates to $\tilde{\Phi}$ and ρ are φ and ω respectively. The coefficients C, D, E, F, G are numerical constants (cf. Wisdom, 1985); $\mu_1 = 1 - \mu$, where $\mu = 1/1047.355$ is the mass of Jupiter; its semimajor axis $a_J = 1$, mean longitude $l_J = t$. Jupiter's eccentricity e_J is a free parameter.

Let us make a canonical transformation, with valence -1, to new variables $\Sigma = \tilde{\Phi} + \rho$, $K = -2\tilde{\Phi} - \rho$, $\sigma = -2\omega + \varphi$, $\kappa = -\omega + \varphi$. Then, in order to disjoin

311

J. C. Muzzio et al. (eds.), Chaos in Gravitational N-Body Systems, 311–314.
© *1996 Kluwer Academic Publishers. Printed in the Netherlands.*

two degrees of freedom, set for the 'fast' one $\Sigma = \Sigma_0 = const.$, $\sigma = \Omega t + \sigma_0$. Then change to $\tilde{K} = K + \Sigma_0 - \frac{2\mu F}{\mathcal{G}}$, where $\mathcal{G} = 3\mu_1^2/\Phi_{res}^4 = 12.99$. Omitting constant terms, finally

$$
\begin{aligned}
H = {} & \frac{\mathcal{G}\tilde{K}^2}{2} + \mu e_J D\sqrt{2(s_0 + \tilde{K})}\cos\kappa - \mu e_J G\sqrt{2(s_0 + \tilde{K})}\cos(\kappa - \sigma) + \\
& + 2\mu C(s_0 + \tilde{K})\cos\sigma + \mu e_J^2 E\cos(2\kappa - \sigma),
\end{aligned} \tag{2}
$$

where $s_0 = \Sigma_0 + \frac{2\mu F}{\mathcal{G}}$.

Hamiltonian (2) is that of a pendulum with weakly (in case of co-directional rotation of angles κ and σ, and in case of motion near the separatrix) varying separatrix energy, and with periodic disturbing terms. Motion of a pendulum with periodic perturbation in the vicinity of the separatrix is described by the separatrix mapping (Chirikov, 1979):

$$
\begin{aligned}
y_{n+1} &= y_n + \sin x_n, \\
x_{n+1} &= x_n - \lambda \ln|y_{n+1}| + c,
\end{aligned} \tag{3}
$$

where y denotes relative pendulum's energy normalized by the characteristic amplitude of its variation, and x is the phase angle; λ and c are constants. Eq. (3) maps the state of the system every time the pendulum's angle goes through zero. Motion is chaotic inside the borders $|y| \approx \lambda$ (Chirikov, 1990).

By means of introduction of a surface of section, a Hamiltonian system with two degrees of freedom can be reduced to a two-dimensional area-preserving mapping. Hamiltonian (2) is qualitatively analogous to that from which the separatrix mapping was derived by Chirikov (1979). The near-separatrix motion with Hamiltonian (2) can be shown to be reducible to the separatrix mapping, when this motion takes place in the vicinity of the chaotic layer's border in the regime of co-directional rotation of angles κ and σ. Then, κ would correspond to the pendulum's angle, and σ to the phase angle of perturbation. Only the fact of this correspondence is employed in what follows. The approximate analytical reduction to the separatrix mapping will be given elsewhere.

As soon as appropriate angles are chosen, spectra of winding numbers can be built. Construction of these spectra allows one to visualize the resonant structure of motion. A modified spectrum (MSWN) is defined here as follows. For each interval between eccentricity bursts (i.e. between crossings of the separatrix), the duration computed in circulations of angle κ is denoted as N, and that computed in circulations of angle σ, as N_σ. The winding number R for each interburst interval is defined as the ratio N_σ/N. The

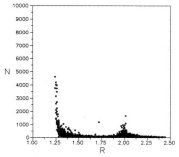 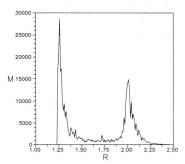

Fig. 1.   (a) The MSWN for an asteroidal intermittent trajectory computed by Wisdom's mapping; $n_{it} = 10^9$, $N > 50$, number of points (interburst intervals) $n_p = 2360$. (b) An ordinary SWN converted from the MSWN.

dependence of N on R represents the MSWN for interburst intervals, i.e. for the circulation side of the separatrix. Bursts themselves take place on the other (libration) side of the separatrix.

Consider an intermittent orbit found by Wisdom (1983). It has the following starting values for Wisdom's (1983) mapping: $l_0 = \pi$, $\tilde{\omega}_0 = 0$, $a_0 = 0.4806$, $e_0 = 0.05$; $e_J = 0.048$, $\tilde{\omega}_J = 0$. The MSWN in Fig. 1a is computed for $n_{it} = 10^9$ iterations of Wisdom's mapping (i.e. Jupiter periods). It has two peaks, a major one at $R \approx 5/4$ and one more at $R \approx 2$. The critical value of R, corresponding to sticking to the layer's border, is that which is minimal: $R_{cr} \approx 5/4$.

In Fig. 1b, an ordinary SWN, converted from the MSWN in Fig. 1a, is shown. The quantity M (vertical axis) denotes the number of κ circulations with the winding number between R, $R+0.01$, summed up for all interburst intervals with $N > 50$. The peaks of the MSWN are easily located in the ordinary SWN.

The MSWN can be as well built for the separatrix mapping. The winding number is defined accordingly as $R = N_x/N$, where N_x is the duration of an interval between crossings of the separatrix measured as a total sum of variation in the phase x, and N is this duration measured in the mapping's iterations. The vertical scale of an MSWN for the separatrix mapping is mostly controlled by the parameter λ, since the mean value of N is $\langle N \rangle \approx 3\lambda$ (Chirikov, 1990). Along the horisontal axis, location of a peak corresponding to the chaotic layer's border is given by a formula, which is easily deduced from Eq. (3). There one should set $y = y_{cr} = \lambda$, for the border. Hence $R_{cr} \approx \frac{1}{2\pi}(c - \lambda \ln \lambda)$. Therefore, as soon as λ is fixed, the parameter c controls the shift of peaks along the horisontal axis.

The separatrix mapping's MSWNs are constructed here to model that of the asteroidal motion in the vicinity of the chaotic layer's border, to which

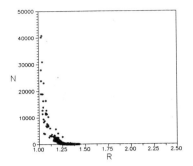 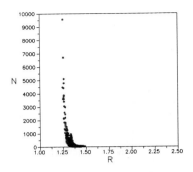

Fig. 2.   The MSWN for the separatrix mapping with $\lambda = 1.34$; $n_{it} = 10^6$, $N > 50$.
(a) $c = 8.3$; number of points (separatrix crossings) $n_p = 1318$. (b) $c = 8.6$; $n_p = 1930$;
three points, with $R = 1.24$ and $10000 < N < 21000$, are beyond figure's frame.

the peak at $R = 5/4$ in Fig. 1a corresponds. The peak at $R = 2$ is not
reproduced in the modelling, since this minor peak corresponds to sticking
to an island.

For the asteroidal orbit considered above, one circulation of the κ angle
takes on the average 1000 iterations of Wisdom's mapping. On this reason,
the number of iterations n_{it} for the construction of the separatrix mapping's
MSWNs was taken 1000 times less than in the asteroidal case, i.e. equal to
10^6. In Fig. 2, the MSWNs for the separatrix mapping (3) are shown for
$\lambda = 1.34$, $c = 8.3$ and 8.6. In case of $c = 8.6$ the peak is located at $R = 5/4$
(Fig. 2b). The choice of a slightly different value of c results in its location
not at $R = 5/4$, but at $R \approx 1$ (Fig. 2a).

Comparison of Figs. 1a and 2b shows that the major peak, emerging in the
asteroidal MSWN due to trajectories' sticking to the border of the chaotic
layer, is naturally approximated by the spectrum of the separatrix mapping
with definite values of parameters λ and c. By means of this approximation,
values of both parameters can be estimated.

It is a pleasure to thank Hans Scholl for helpful discussions. This work
was supported in part by the Russian Foundation of Fundamental Research
(Grant 95-02-05301-a), and by the American Astronomical Society (Grant
717000).

References

Chirikov, B. V.: 1979, *Phys. Reports* **52**, 263.
Chirikov, B. V.: 1990, *Patterns in Chaos*, INP Preprint 90-109, Novosibirsk.
Ferraz-Mello, S., 1993, in A. Milani et al. (eds.), *ACM 1993*, Kluwer, Dordrecht, p. 175.
Wisdom, J.: 1983, *Icarus* **56**, 51.
Wisdom, J.: 1985, *Icarus* **63**, 272.